FIRE MOUNTAINS OF THE WEST:
The Cascade and
Mono Lake Volcanoes

D0029815

Stephen L. Harris

MOUNTAIN PRESS PUBLISHING COMPANY
1988

Library of Congress Cataloging in Publication Data

Harris, Stephen L.
  Fire mountains of the west.

  Rev. ed. of: Fire and ice. Seattle Wash. :The Mountaineers, 1976.
  Bibliography: p.
  Includes index.
  1. Volcanoes — Cascade Range.   2. Cascade Range.
I. Harris, Stephen L., 1937-   . Fire and ice.
II. Title
QE524.H18   1988   551.2'1'09795   88-13526
ISBN 0-87842-220-X

PRINTED IN THE U.S.A.

MOUNTAIN PRESS PUBLISHING COMPANY
P. O. Box 2399
Missoula, Montana 59806

To Doug

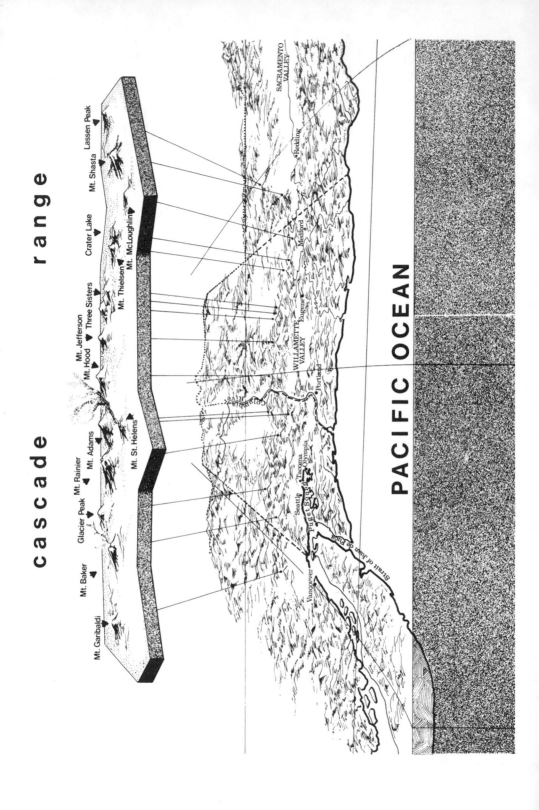

cascade range

Mt. Garibaldi
Mt. Baker
Glacier Peak
Mt. Rainier
Mt. Adams
Mt. St. Helens
Mt. Hood
Mt. Jefferson
Three Sisters
Mt. Thielsen
Mt. McLoughlin
Crater Lake
Mt. Shasta
Lassen Peak

Vancouver
Strait of Juan de Fuca
Puget Sound
Seattle
Tacoma
Olympia
Columbia River
Portland
WILLAMETTE VALLEY
Eugene
Medford
Redding
SACRAMENTO VALLEY

PACIFIC OCEAN

Preface

When St. Helens exploded with deadly effect in 1980, millions of Americans were reminded that some of the world's most potentially dangerous volcanoes lie along the U.S. Pacific Coast. St. Helens is expected to remain intermittently active for years or decades to come. It may produce more eruptions that will equal or surpass that of May 18, 1980, its unpredictability matched only by its reputation among geologists as one of the most violently explosive volcanoes in the "Ring of Fire" that encircles the Pacific Ocean.

This new book, a complete revision of *Fire and Ice*, includes a full account of St. Helens' latest activity, and a thoroughly up-to-date picture of her fellow fire-mountains in the Cascade Range and other parts of the West. Since *Fire and Ice* first appeared in 1976, there has been a virtual explosion of information about our Western volcanoes and the hazards they pose. This book makes this wealth of material available to the general reader.

The only single volume to present a comprehensive survey of all the West's volcanic centers that geologists now view as potential danger spots, *Fire Mountains of the West* is non-technical and intended for the general reader who wants to learn more about the geologic forces that shaped our Western mountainscape. Providing a carefully-researched biography of each of the major Cascade volcanoes, as well as those in California's Mono-Mammoth Lake area, the book incorporates virtually all published—and much hitherto unpublished—data on the most recently active volcanoes in the 48 adjacent states. Concentrating on the individual mountains—from Long Valley and Mammoth Mountain in California to Garibaldi in British Columbia—the book also includes discussions of specific hazards involved when the volcanoes next erupt.

Although not conceived as a field guide or reference book on the weather or plant and animal. life in the Cascade and other volcanic peaks, *Fire Mountains of the West* adds to almost every chapter directions on how one can enjoy a particular mountain by car, by foot, or, in some cases, by climbing to its summit.

In acknowledging my general debt to the many Earth scientists and historians whose work is cited in the bibliography, I would like to give my special thanks to Dwight R. Crandell, now partly retired from the U.S. Geological Survey's Federal Center in Denver, Colorado, who edited the original edition for technical accuracy and contributed many valuable suggestions while the book was being written. I am also grateful to many other U.S.G.S. geologists, particularly those at Menlo Park, California, and the David A. Johnston Cascades Volcano Observatory in Vancouver, Washington. Donald Peterson, former head of the Observatory, graciously provided helpful information, as did Norman MacLeod, present director of the Observatory. MacLeod's recent work on the Newberry Volcano in Oregon is gratefully acknowledged.

Charles R. Bacon, whose work has greatly enhanced our understanding of the processes that formed Crater Lake, generously shared the results of his labors in the field. Wes Hildreth, also of the U.S.G.S., Menlo Park, repeatedly provided unpublished data on Adams' previously unknown eruptive history. Wendell A. Duffield, Coordinator of the U.S.G.S. Geothermal Research Program—which is searching for sources of heat energy in the Cascades—provided helpful material, as did Edward Sammel, who successfully conducted drilling for subsurface thermal power at Newberry. Patrick Muffler and Michael Clynne shared their findings at Lassen Volcanic National Park. Julie Donnelly-Nolan gave valuable information about the Medicine Lake Volcano, while James G. Moore supplied insight into the mechanism of the 1980 St. Helens' avalanche and eruption. Robert L. Christiansen, who was the geologist in charge during St. Helens' climactic activity, offered helpful facts about Shasta and Holocene activity at Lassen Peak. Dean Eppler, who has just completed a doctorate on Lassen's historic eruptions, kindly shared his recent studies.

In addition, I would like to thank C. Dan Miller for his earlier review of Shasta's geology and for his current work on volcanic hazards in the Mono Lake-Long Valley region. David Hill's review of the chapter dealing with the Mono area geology was also welcome.

A number of university researchers and other geologists also contributed to this edition. James Begét, now at the University of Alaska, Fairbanks, deserves tremendous credit for achieving a new understanding of Glacier Peak's recent eruptive history and for his work on Jefferson's last tephra eruptions. Edward Taylor, of Oregon State University, Corvallis, communicated some essential dating of the South Sister's latest eruptions. William Scott, U.S.G.S. Federal Center in Denver, provided a detailed account of the South Sister's last eruptive cycle.

Catherine Hickson, of Canada's Department of Energy, Mines and Resources, supplied helpful data on the Garibaldi Belt of volcanoes, while John Souther kindly reviewed the chapter on Garibaldi.

Alexander R. McBirney, of the University of Oregon, Eugene, reviewed the section on the geological evolution of the Cascade Range and made very useful suggestions.

Carolyn L. Dreidger of the U.S. Geological Survey's Glaciology Project in Tacoma, Washington, provided much indispensable information on Cascade glaciers. Her colleagues, Austin Post and Dee Molenaar, also shared their expertise. Eugene P. Kiver, of Eastern Washington State University, Cheney, also offered helpful suggestions in presenting glacial processes. David Frank, at the Seattle headquarters of the U.S.G.S., gave data on recent changes in Baker's thermal activity and its effect on the crater glacier.

Other contributors to the earlier edition are also gratefully remembered: W.S. Wise of the University of California, Santa Barbara; Kenneth Hopkins of the University of Northern Colorado; the late LeRoy Maynard, formerly of the University of Oregon, for his enthusiastic help in preparing the chapter on Oregon's McLoughlin; Kenneth Sutton, presently at the University of Hawaii, for permission to use his unpublished research and field studies on Jefferson.

Fire Mountains of the West is greatly enhanced by the new maps, diagrams, and drawings of Chris Hunter, a talented California artist and cartographer. Finally, I would like to thank David Alt, Professor of Geology at the University of Montana, for his masterful editing of the manuscript, and his colleagues at Mountain Press Publishing Company for their help in design and production.

Because this book is designed for the general reader, its editors decided to eliminate footnotes crediting individual sources. The author expresses his appreciation to the many Earth scientists who generously shared their research by citing their work in the bibliographies listed at the end. A major research tool, each bibliography combines all known historical and geological references to each major western volcano. These compilations are the most comprehensive and complete reference lists currently in print.

Contents

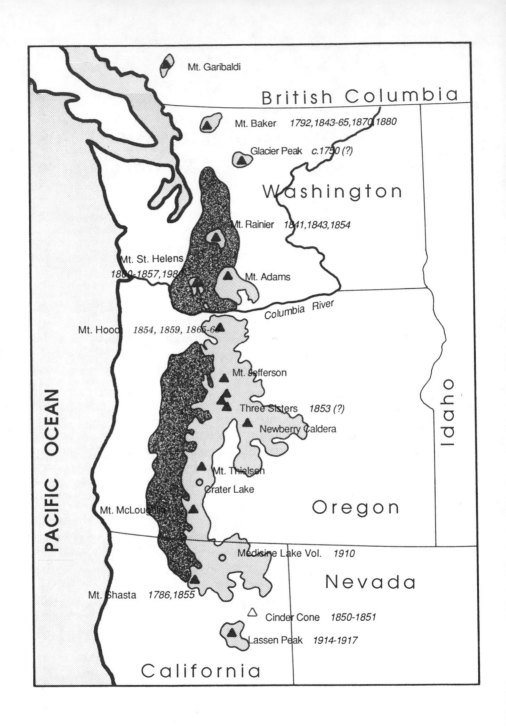

Map of the Cascade Range showing the principal volcanic cones and the dates
of their known historic eruptions.

I
Our Western Volcanoes:
An Introduction

The American West is rich in volcanoes. They range strikingly in size and shape—from small piles of volcanic cinders to towering giants mantled in glacial ice—but they are one of the major natural dangers to human life and property in the United States.

The destructive power of a previously little-known Western volcano was demonstrated on May 18, 1980 when St. Helens exploded with lethal fury, devastating 200 square miles and killing 57 people. Explosions releasing energy equivalent to 27,000 Hiroshima-sized atomic bombs detonated in rapid succession ejected an immense ash cloud 15 miles into the stratosphere. Ash blanketed hundreds of thousands of square miles east of the volcano, transforming day to night in large sections of eastern Washington, northern Idaho, and western Montana. Since then St. Helens' shattered cone has been rocked by six more explosive eruptions, three of which scattered ash over western Washington and Oregon, including the Portland metropolitan area.

Since 1980, St. Helens has remained almost constantly active. Sporadic explosive bursts send plumes of gray ash high into the air, while thick tongues of glowing lava repeatedly ooze into its mile-wide crater. A jagged lava dome, 3,000 feet long and 800 feet high has grown inside the crater during the last eight years, but, like its two predecessors, it could be blasted skyward at any time. This young and unpredictable volcano has had an unusually violent history. In the past it produced much larger outbursts than those of 1980, and may do so again, perhaps in our lifetime.

St. Helens is only one of the many potentially explosive young volcanoes that dot the Pacific Coast from California to British Columbia, Canada. Another "hot spot" that worries geologists is the volcanic field in the Mono Lake-Long Valley region of east-central California. Since 1978 repeated swarms of earthquakes have shaken the area, warning that an eruption may be imminent.

All of the historic eruptions recorded in the 48 adjacent states, however, have occurred in the Cascade Range, of which St. Helens is a part. Extending from northern California to southwestern Canada, the Cascades are distinguished by a chain of high-standing volcanoes. These fire-born, ice-carved giants dominate the range and give it its unique character. Nowhere else in the United States has nature so dramatically linked these two opposing forces—volcanic fire and glacial ice.

With St. Helens' reawakening—and that of similar volcanoes elsewhere, such as Nevado del Ruiz in Columbia—the question is raised: which of *our* volcanoes will erupt next? The question expresses more than mere curiosity. Recent geologic events give the issue a genuine urgency. When St. Helens erupted during the last century, it was joined by seven other Cascade volcanoes. Future eruptions are geologically inevitable and may occur in clusters as they did during St. Helens' previous eruptive cycle (1800-1857) when volcanoes in three states blazed into life. It is possible that the coming decades may witness a volcanic renaissance comparable to that of mid-1800s.

Historic Eruption in the West

Thus far in the 20th century two Cascade volcanoes—St. Helens and Lassen Peak—have erupted, but as many as nine others have also been active during historic time. Between the late 1780s and the mid-1860s—the span of a single lifetime—at least eight vents in three states revealed their fiery nature. In Washington, Baker erupted several times between 1792 and 1880; Glacier Peak, remarkable for the enormous quantities of material it has ejected in the past, was active in the late 18th or early 19th century, while Rainier spewed pumice on at least one occasion between 1820 and 1854. Baker dramatically increased its emission of heat, steam, and sulphur only five years, almost to the day, before St. Helens reawakened. Several times during the 1840s and 1850s, Baker erupted almost simultaneously with St. Helens. It is too early to tell, but the two volcanoes may yet restage their 19th century joint performance.

Oregon's Mt. Hood, the highest peak in the state, also has a record of historic activity. A recent U.S. Geological Survey report indicates that Crater Rock, a craggy lava dome looming high on Hood's south face, was emplaced only about 200 years ago. Although Hood's last reported eruptions occurred in 1866, vents near Crater Rock occasionally produce enough steam and sulphurous fumes to create plumes visible from Portland, 50 miles to the west. The South Sister, a glaciated cone in central Oregon, may have been the source of an eruption witnessed in 1853.

California has five historically active volcanoes, plus several others that erupted shortly before the state was settled. These include Glass Mountain on the Medicine Lake volcano in northeastern California and Cinder Cone in Lassen Volcanic National Park. During 1850-51, the glow from the erupting cone was bright enough to be seen from many miles away. The nearby Chaos Crags were also steaming vigorously in the 1850s. Glacier-capped Shasta, the second highest peak in the Cascade Range, was probably the active volcano sighted by the explorer La Perouse in 1786. Shasta was mildly active again in the mid-1850s, and still produces a hot sulphur spring at the summit.

Until upstaged by St. Helens, Lassen Peak was celebrated as the most recently active volcano, south of Alaska, in the continental United States. Beginning in May 1914, and continuing intermittently until 1921, Lassen ejected steam, ash, and lava; the climax occurred a year after the activity began, when it blew an enormous mushroom cloud an estimated seven miles into the air.

Although the Cascade volcanoes have produced the most recent eruptions in the continental United States, a volcanic field in east-central California also poses a major hazard. The Mono Lake-Long Valley area on the east flank of the Sierra Nevada Range is the site of one of the most catastrophic eruptions in geologic history. During the last 2,000 years the region has produced at least 30 additional eruptions, building a chain of cinder cones and lava domes from Mono Lake, directly east of Yosemite National Park, southward to the Mammoth Lakes in Long Valley. Beginning in 1978, swarms of earthquakes struck the area, culminating in 1980 with a series of heavy shocks registering 6.0 or higher on the Richter Scale. The seismic activity, together with ground swelling and an increase in steam vent discharge, caused the U.S. Geological Survey to designate the area a prime source of future volcanic eruptions.

After reaching a peak during the middle decades of the 19th century, Cascade volcanic activity declined precipitately. The long quiet interval between about 1880 and 1980—broken only briefly by the Medicine Lake volcano and Lassen Peak—gave the Western volcanoes an ill-deserved reputation as a dying breed. Until recently many geologists tended to discount most of the reports of historic eruptions, which were generally viewed at a considerable distance by untrained observers, concluding that the silent snow-mantled peaks were defunct dragons that had breathed their last metamorphic fire. During the last two decades, however, geologists have studied and

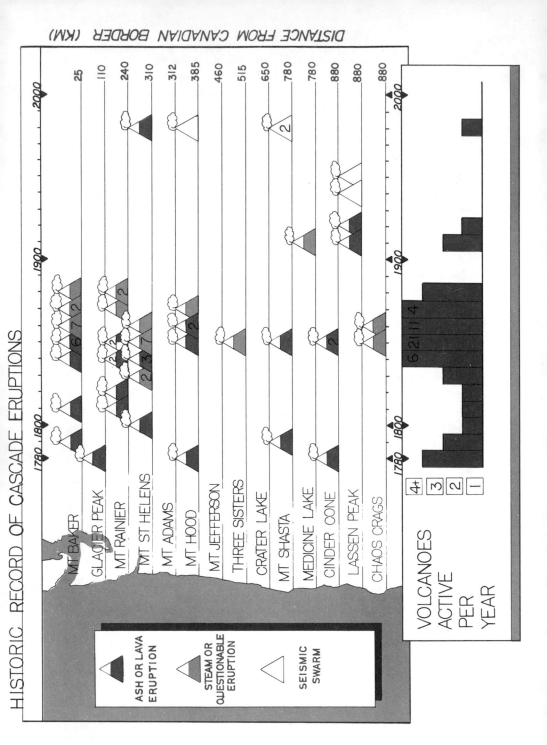

Chart illustrating the temporal clusters of historic Cascade eruptions between about 1780 and 1980.

dated deposits around many of the major volcanic centers—from Baker near the Canadian boundary to the Long Valley caldera in central California—and found that nearly all had erupted repeatedly—in some cases catastrophically—during the recent geologic past. Since the end of the Ice Age, the Cascade volcanoes alone have produced, on the average, at least one major outburst per century.

The Cascade Landscape

When Lewis and Clark journeyed to the Pacific in 1805, they followed the Columbia River. The explorers were greatly hindered in the last stages of their trek by a series of whitewater rapids where the Columbia narrowed to run at dizzying speeds through a steep rocky gorge. Amid the lava cliffs and roaring cataracts, Lewis and Clark encountered the last serious obstacle to the success of their mission.

Because the route Lewis and Clark chose was the only practicable one and was consequently followed by traders and settlers, the mountains that loomed above the Columbia River rapids became known as *the mountains by the cascades*. Although that part of the range which gave it its name is not typical, the designation proved appropriate; few spots in the Cascades are far from the music of running water.

The Cascades offer today's visitors a spectacularly scenic recreational area. The range is unsurpassed for the variety of its topography, its extensive glaciers, and its dense evergreeen forests, which provide an emerald background for the countless lakes, streams, and waterfalls. About 30 to 50 miles wide and 4500 to 5000 feet high at its southern extremity, the range expands in northern Washington to a glaciated wilderness at least 80 miles across, where scores of peaks exceed 8000 feet. Individual volcanic cones tower above the range itself.

The highest points in the Rockies or Sierras tend to be only slightly higher than other elevations around them. The mightiest Cascade summits, however, dominate the landscape for 50 to 100 miles in every direction. Rainier, for example, rises abruptly above the Puget Sound lowland and has a visual height nearly equal to its actual vertical height of 14,410 feet. Spreading over 100 square miles of lesser mountains which themselves rise more than a mile above sea level, Rainier soars high above its nearest competitors. Baker, Adams, Shasta, Jefferson, and Hood likewise visually overwhelm encircling peaks, many of which, if placed elsewhere in the nation, would be regarded as major landmarks.

Historic Eruptions in the Cascade Range

YEAR	VOLCANO	NATURE AND/OR PRODUCTS OF ERUPTIONS	SELECTED REFERENCES
1760-1810	Hood	Old Maid eruptive period: pyroclastic flows; mudflows; Crater Rock dome	Crandell 1977; 1980 Cameron & Pringle, 1987
c. 1750	Glacier Peak	Ash	Begét 1982; Majors 1980
c. 1780	Cinder Cone	Lava flow	Finch 1937
1786	Shasta	Pyroclastic flows from Hotlum cone; hot mudflows	Finch 1930; Miller 1980
1792	Baker	"Flame" and "ramblings"	Cook 1973: 354
c. 1800-1804	St. Helens	Ash, lava flows	Lawrence 1941, 1954; Hoblitt & others 1980; Crandell & Mullineaux, 1981
c. 1810 or 1820	Baker Rainier	Ash (?) Ash (?)	Plummer 1893; Majors 1978 Plummer 1893
1820-1854	Rainier	Andesite pumice	Mullineaux and others 1969
1831	St. Helens	Ashclouds darken skies	Holmes 1955, 1980
1835	St. Helens	Ashclouds darken skies	Holmes 1955, 1980
1841	Rainier	Ash (?)	Plummer 1898;
1842	St. Helens	Ash, growth of Goat Rocks dome (?) ; debris flow or mudflows into Toutle River.	Holmes 1955, 1980; Hoblitt & others 1980.
1843	St. Helens	Ash, Goat Rocks dome (?)	Holmes 1955, 1980
1843	Baker	Ash; debris flow and mudflows down Boulder Glacier into Baker River and Skagit River	Fremont 1845; Gibbs 1871
1843	Rainier	Ash (?)	Fremont 1845
1844	St. Helens	Goat Rocks dome elevation	Burnett 1845: 109-110; 1902: 423-4; Gary 1923: 76-77

YEAR	VOLCANO	NATURE AND/OR PRODUCTS OF ERUPTIONS	SELECTED REFERENCES
1846	St. Helens	Ash	Coombs 1960: 13
1846	Baker	Ash (?)	Majors 1978: 43
1846	Rainier	Ash (?)	Plummer 1893
1847	St. Helens	Ash	Kane 1925: 136; Holmes 1980
1848	St. Helens	Ash ("fire")	Harris 1976: 185
1849	St. Helens	Ash	Gibbs 1855: 469
1850	Baker	Ash	Harris 1976: 241
1850	St. Helens	Ash	Harris 1976: 188
1850	Cinder Cone	Ash	Harkness 1875: 411; Finch 1937
1851	Cinder Cone	Ash	Harkness 1875: 411; Finch 1937
1852	Baker	Ash	Winthrop 1913: 38, 280
1853	Baker	Rockfall; debris flow (?)	Plummer 1893: 3
1853	St. Helens	Lithic Ash	Stevens 1936: 251
1853	St. Helens	Lithic ash	Stevens 1936: 251
1853	Hood	Ash (?)	Miller 1853
1853	South Sister	Ash (?)	Miller 1853
1854	Baker	"Vast rolling masses of dense smoke"	Gibbs, ed. Clark 1955: 496; Davidson 1885: 262
1854	St. Helens	Ash	Plummer 1898: 42; Harris 1976: 186
1854	Rainier	Ash (?)	Gibbs, ed. Clark 1955: 321 Matthes 1914

YEAR	VOLCANO	NATURE AND/OR PRODUCTS OF ERUPTIONS	SELECTED REFERENCES
1854-1857	Chaos Crags	Steam; ash (?)	Williams, 1932
1855	Shasta	Ash (?)	Eichorn 1857: 13
1855	Baker	Small clouds of ash	Davidson 1885; Majors 1978
1856	Baker	"Dense Smoke"	Majors 1978
1857	St. Helens	Ashclouds	Holmes 1955; 1980
1857	Chaos Crags	Steam	Williams 1932: 347
1858	Baker	"Fire and smoke"	Davidson 1885; Majors 1978
1859	Hood	"Column of fire," ashfall on snow; mudflow (?)	Steel 1907; Harris 1976
1859	Baker	"Jets of flame," "dense clouds of smoke"	Coombs 1960
1860	Baker	Ashclouds	Plummer 1898; Majors 1978
1863	Baker	"Flames"	Harris 1976
1864	Baker	Earthquake and rockfall	Majors 1978
1865	Baker	"Dense clouds of smoke"	Harris 1976
1865	Hood	Ash; "jets of flame;" "smoke"	Steel 1906: 23
1867	Baker	"Fire", "dense volumes of smoke"	Majors 1978
1870	Baker	"Great volumes of smoke"	Davidson 1885: 262
1873	Rainier	"Clouds of smoke pouring from highest peak"	Plummer 1898
1879	Rainier	"Brown, billowy clouds"	Coombs 1960
1880	Baker	"Flames streaming up . . . . large volumes of smoke"	Majors 1978: 60-62
1882	Rainier	"Brown, billowy clouds"	Coombs 1960

8

YEAR	VOLCANO	NATURE AND/OR PRODUCTS OF ERUPTIONS	SELECTED REFERENCES
1910	Glass Mountain (Medicine Lake Volcano)	Ash; "flame"	Finch 1928
1914-1917	Lassen Peak	Ash, lava flow, mudflows	Day and Allen 1925; Loomis 1926; Willendrup 1983
1980	St. Helens	Ash; pyroclastic flows; domes; mudflows	Lipman and Mullineaux 1981; Foxworthy and Hill 1982

Note: When sources are contemporary newspaper accounts, the reader is directed to more accessible compilations of journalistic reports, such as Holmes (1955, 1980), Harris (1976) and Majors (1978 and 1980).

9

Because of their altitude, the Cascades form a formidable climatic barrier. Running roughly parallel to the Pacific Coast on a generally north-south axis, the range intercepts moisture-laden winds moving eastward from the Pacific Coast. Rapid cooling causes the marine air to condense and fall as rain or snow. Although precipitation is less in the southern Cascades, it can reach over 150 inches annually on the wettest slopes; winter snowcover in the Pacific Northwest is often heavy at elevations as low as 2000 feet. As a result, the western slope of the range is abundantly watered and thickly forested.

East of the Cascade crest, the scenery and vegetation change abruptly. The eastside forest is drier, free of the lush ferns, briars, and brush that entangle the western timberlands. Instead of the moisture-loving Douglas fir, pine becomes the dominant species. Because the annual rainfall dwindles to as little as eight inches east of the divide, the inland foothills are typically brown in summer, spotted only with dry grass, sagebrush, and occasional scrub growth. Almost nowhere else in the nation do a few miles mark so complete a transformation of the landscape.

To people in western Washington and Oregon, the country "East of the Mountains," as it is popularly called, is a different world. Instead of the western greenery, the traveler encounters an arid plateau. This is the great basalt desert created during an extended period of volcanism that began about 16,000,000 years ago—long before the present Cascades rose. In Miocene and Pliocene time enormous floods of liquid rock poured from cracks in the earth's surface to bury almost 200,000 square miles of eastern Washington, Oregon, and parts of Idaho and California.

Draining the lava highlands and moving irresistibly westward to the Pacific, the Columbia River cuts the only breach through the American Cascades in almost their entire 700-mile length. The Columbia existed before the Cascades began to rise during Pliocene time, about 7,000,000 years ago. Its eroding power kept pace with the growing mountains which eventually reached about 4500 feet above sea level. To the north, uplift was much greater.

The Columbia River Gorge provides a geologic window into the interior of the range. As the river sliced through the layered basalts on its way to the ocean, it exposed folds and arches of these ancient strata, showing how pressures within the earth warped its crust. These upwarped rocks, eroded into peaks and ridges, became the Cascades of southern Washington and northern Oregon.

Near the westward-flowing Columbia grew three of the largest Cascade volcanoes, Adams and St. Helens in Washington and Hood in Oregon. Because they stand like white sentinels above the river, they are locally known as the Guardians of the Columbia. A favorite subject of Indian lore, the guardian peaks compensate for the otherwise undistinguished topography of the Cascades in this area.

History and Exploration

Although earlier explorers had noted them, it was not until the British navigator George Vancouver sailed into Puget Sound in the spring of 1792 that the Cascade peaks began to receive their present names. Baker, northernmost of the U.S. volcanoes, was named for Vancouver's third lieutenant. Rainier honored Admiral Peter Rainier, while St. Helens commemorated a then-famous diplomat. While exploring the Columbia River, Vancouver's first lieutenant, Broughton, sighted a gleaming white spire which he mistakenly took to be at least 25,000 feet high, and named it after Samuel Hood, another high-ranking naval officer. Although he guessed that all these individual peaks were probably part of a continuous chain, Vancouver made no attempt to christen the range.

On their return trip, Lewis and Clark observed still another high snowy pinnacle, standing about 46 miles south of Hood. This Matterhorn-like peak they named after Thomas Jefferson, the sponsor of their expedition.

The remaining high peaks received their names almost casually from early explorers or settlers during the 19th century. Oregon's Three Sisters were first called Faith, Hope, and Charity by wagon train members who used them as landmarks. Shasta, the exact source of which name is unknown, was first sighted by Russian or Spanish scouting parties in northern California. Adams was named purely by accident; and Crater Lake was not even discovered until 1853.

The National Parks

Congress recognized the unique qualities of the Cascade Range by establishing four major national parks within its boundaries. Lassen Volcanic National Park was created in 1916 while the volcano was still in the midst of its latest series of eruptions. The Park encompasses 10,457-foot Lassen Peak and dozens of associated volcanoes, such as Cinder Cone and Chaos Crags. Lassen Park includes extensive thermal areas, marked by sizzling steam jets, boiling sulphur pits, and miniature mud volcanoes with craters full of bubbling, spattering, hot mud. These thermal displays are the largest and most varied in the country outside of Yellowstone National Park.

One hundred ninety miles north of Lassen Peak, in southern Oregon, lie the indigo waters of Crater Lake, cradled in the six-mile-wide basin of a collapsed volcanic giant. This ruined volcano, posthumously named Mazama, contains what is possibly the most magnificently situated body of water on earth. Crater Lake National Park also preserves evidence of a volcanic explosion that may have been greater than any other of post-glacial time. Even that of Krakatau, the volcanic island near Java which, in 1883, produced the most tremendous outburst of historic times, was perhaps less violent. The titanic blasts that led to the destruction of ancient Mazama ejected over 42 cubic miles of material. That cataclysm occurred about 6900 years ago, but even today geologists watch warily lest old Mazama reawaken.

In size, height, and grandeur the climax of the Cascade Range is Rainier. The mountain's enormous bulk is visible from Canada to Oregon to the Pacific. With 45 square miles of unmelting snow and ice covering its lava cone, Rainier supports the largest glacier system in the United States south of Alaska, yet the summit cone is indented by two geologically youthful craters, still hot enough to keep the crater rims free of snow and generate scores of steam vents and fumaroles. One of these craters contains a lake which the steam has melted out *beneath* the summit icecap, 14,000 feet above sea level.

The North Cascades

North of Rainier the Cascades change radically in character. Composed chiefly of extremely old metamorphic and sedimentary rock, the North Cascades have been eroded into a maze of sharp peaks and twisting narrow valleys. Seemingly locked in a contemporary ice age, this remote and stormy terrain is one of the last true wilderness areas in the United States. Its austere and primitive qualities have been safeguarded by the establishment in 1968 of a fourth national park, the North Cascades.

Although this part of the range is essentially nonvolcanic, its highest elevations are the two northernmost volcanoes in the American Cascades, Baker and Glacier Peak. Both unfortunately, lie outside the boundaries of North Cascades National Park. While Glacier Peak is isolated in the center of the range, Baker is readily accesible. Baker's crater, the source of several historic eruptions, still seethes with escaping steam, which sometimes rises in billowing, fleecy clouds.

About 50 miles beyond the Canadian boundary stands the Garibaldi group of volcanoes. Although most maps place Garibaldi in the "Coast Mountains" of southern British Columbia, it actually represents a northern spur of the Cascade Range. The Fraser River valley, which separates the two mountain areas, is an erosional feature, not a structural division.

Cascade Prototypes and Personalities

Of all the high volcanoes that crown the Cascades, no two are precisely alike, either in appearance or geologic history. The Pacific Northwest Indians recognized this individuality and spun some of their most attractive legends about the white giants who overshadowed their daily lives. Several of the peaks figure conspicuously in native stories of the Great Flood during which the mountains played the role of Noah's ark; Baker, Jefferson, and Shasta offered their summits as places of refuge when the Deluge swept away all life below.

In myth, the mountains were also warrior gods, given to catapulting red-hot boulders at each other. Sometimes their aim was accurate enough to decapitate a mountain rival. Adams' flat top is accounted for in this picturesque fashion. The Indians also invested the peaks with romantic interests. They said that Hood and Adams feuded over the love of the youngest and fairest of all mountain deities, St. Helens.

Volcanoes as Benefactors

As farmers and ranchers of the Pacific Northwest are aware, the volcanoes' activity has not been wholly destructive. Volcanic rocks are typically rich in minerals and are quickly weathered into unusually fertile soil. Near Rainier formerly uneven terrain has been leveled and filled by huge mudflow deposits, which make excellent grazing and farmland. Widespread ash blankets from St. Helens, Mazama, and Glacier Peak have further enriched the ground with potassium and other nutrients. In addition, the volcanoes' covering of snow and glacial ice supplies meltwater to fill reservoirs and provide both electric power and water to irrigate parched ranchlands east of the mountains.

Volcanoes may provide a partial solution to the present energy crisis. Enormous quantities of thermal energy, which can be used to generate electric power, are available in the Cascades. The U.S. Geological Survey's Geothermal Research Program is now exploring ways to discover and harness it. Accumulations of magma lying beneath the range heat adjacent crustal rocks. Where fractured and porous rock contains water or steam, a natural medium exists to transfer this energy to the earth's surface. When bore holes are drilled into water-bearing hot rocks, the rising steam or hot water supplies the power to operate turbines and generators. Already the southern Oregon town of Klamath Falls uses volcanic steam to heat public buildings, and Hood may similarly provide a new energy source for Portland. The highest temperatures found thus far—510 degrees F—occur at a depth of only 3075 feet in the Newberry caldera. Despite severe budget cuts in the 1980s, the U.S.G.S. Research Program plans to continue studying the Cascades' thermal energy potential.

Among the benefits of tapping geothermal energy are its relative cheapness and safety. It also creates significantly less environmental pollution than nuclear reactors or utilities burning oil and coal.

Because there are literally thousands of volcanoes in the American West, it is impossible for one book to describe them all. This volume concentrates on the principal volcanoes of the Cascade Range and the Mono Lake-Long Valley area, devoting a chapter to each of the most distinctive landforms.

Before beginning our tour of the Western volcanoes, we first will survey briefly the larger geologic system of which our volcanoes are an integral part—the "Ring of Fire."

II
Our Place in the Ring of Fire:
How We Must Learn to Live with Earthquakes and Volcanoes

St. Helens in 1980, El Chichon, Mexico, in 1982, Nevado del Ruiz, Columbia—killer of 25,000 persons in 1985—what do these volcanoes have in common? All belong to a long chain of highly explosive volcanoes that encircles the Pacific Ocean. Known as the "Ring of Fire," this circum-Pacific belt contains about 75 percent of the world's approximately 520 active volcanoes.

Volcanoes are not scattered haphazardly around the globe, but tend to occur in clusters or in chains. In addition to the "Ring of Fire," which runs like a crimson thread around the Pacific basin, most of the world's volcanoes are in three well-defined zones. One encompasses the volcanoes of the Mediterranean Sea that have been famous since antiquity— Vesuvius, Etna, Stromboli, and Santorini (Thera). Another volcanic belt runs through northeast Africa, while a third includes the explosive vents of the Caribbean, such as Mount Pelée, which killed 30,000 people during an eruption in 1902.

Geologists believe that volcanoes lie along the edges of rocky slabs of the earth's crust called "plates" and that most earthquake and volcanic activity is triggered by plate movement. According to the theory of plate tectonics, the entire outer shell of the earth—the lithosphere—consists of about 12 major slabs that constantly move: tearing apart, colliding, slipping under or overriding each other.

The crustal plates drift over an underlying zone of hot plastic material within the mantle. The largest section of the earth's interior, the mantle, approximately 1,800 miles thick, is composed of dark green and black rock called peridotite. It surrounds the earth's metallic core, which is about the size of the moon.

Continents, composed mostly of granite and granite-related rocks, float like giant rafts on the denser peridotite of the mantle. The plates

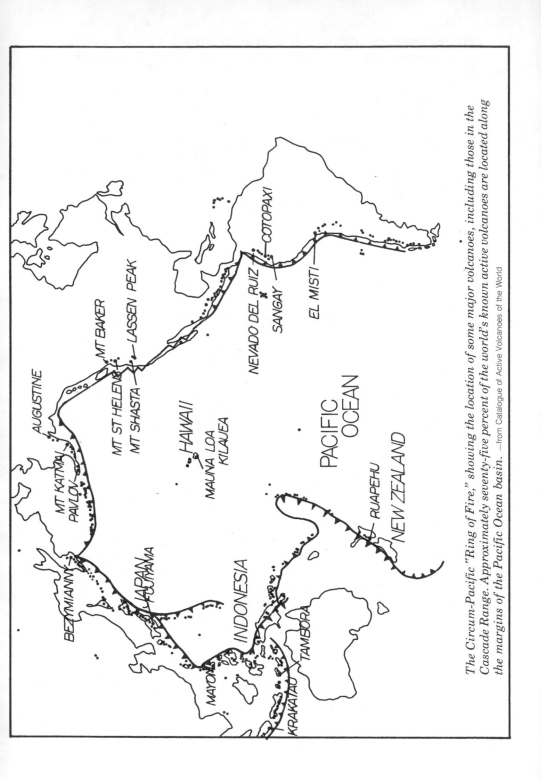

The Circum-Pacific "Ring of Fire," showing the location of some major volcanoes, including those in the Cascade Range. Approximately seventy-five percent of the world's known active volcanoes are located along the margins of the Pacific Ocean basin. —from Catalogue of Active Volcanoes of the World carrying ocean basins on their back form a much thinner oceanic crust composed of basalt, a dark volcanic rock derived from peridotite.

Earth's plates continue to change and move because new oceanic crust is almost constantly being created through volcanic action. The great submarine mountain ranges that wind about the ocean floors like seams on a baseball are linear zones along which new material from the earth's interior erupts onto the surface. Giant fractures along the ocean floor allow liquid basalt to rise to the surface. As new crust forms along these fissures—known as "spreading centers"—older crust is pushed away toward neighboring continents.

The best-known submarine range is the Mid-Atlantic Ridge, part of which rises above the ocean surface as the volcanic island of Iceland. Similar ridges criss-cross the Pacific Ocean floor, marking the sites of underwater eruptions. When the expanding ocean floor encounters a continent, it is thrust under or "subducted" beneath the continental mass. The heavier basalt sinks beneath the continent because the granitic land mass is lighter and rides higher on the underlying mantle.

As the oceanic plate descends beneath the continental margin, it encounters increasingly high temperatures and pressures, which melt the basalt and drive water out of the rocks as red hot steam. Underground pockets of gas-rich molten rock called magma form. During its descent beneath a continent, the oceanic plate commonly fractures adjacent rock, generating earthquakes and opening fractures in the overlying crust. Because it is hotter than the surrounding rock, magma rises into the crust, where it displaces older rock to form a magma chamber, the subterranean reservoir of molten rock that feeds volcanoes. The accumulation of magma along continental margins where they override oceanic plates explains why chains of volcanoes—paralleling the coastline—border almost all sides of the Pacific Ocean basin.

The Pacific plate is broken into several interlocking plates, including two small segments that are currently sliding beneath the northwest coast of the United States. About 250 miles off the North American coast lies the Juan de Fuca Ridge spreading zone that is pushing the Juan de Fuca plate beneath the coast of southern British Columbia, Washington, and Oregon. A smaller spreading center, immediately to the south, is simultaneously thrusting the tiny Gorda plate beneath the edge of Northern California. Moving at the slow rate of one or two inches per year, the subducting plates generate the magma that erupts to build the stately procession of snow-capped volcanoes from California's Lassen Peak to Canada's Garibaldi and its northern neighbors.

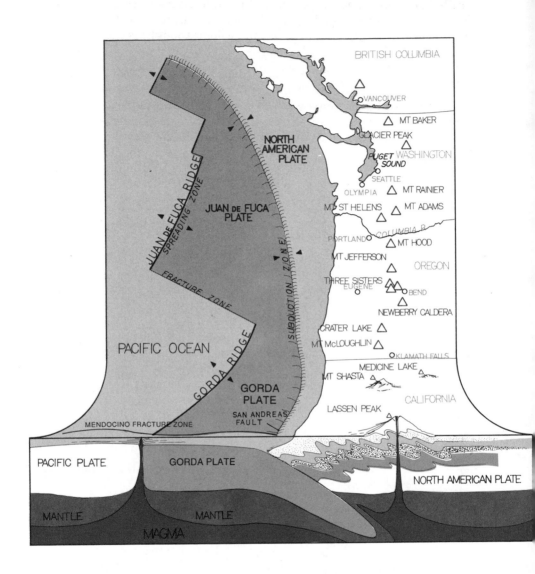

Plate movement along the northwest coast of the United States: the Juan de Fuca and Gorda crustal plates are slowly sliding under the western margin of North America. Subduction generates magma beneath the edge of the continent and probably also creates zones of weakness through which magma rises to the surface, forming the Cascade volcanoes. —After Foxworthy and Hill, 1982.

Subduction has occurred periodically along western North America for many tens of millions of years, but subduction zones have shifted position radically as ancient oceanic plates were gobbled up by the westward-advancing North American continent. One of the most exciting developments in recent geologic thought is the recognition that sea floor beneath North America carried with it all kinds of alien material that was scraped off the descending plate and fused against the westward edge of the continent.

Acting like giant conveyor belts, the eastward-moving oceanic plates brought ancient island chains, submarine volcanoes, and even mini-continents from the Pacific and attached them to western North America. Composed of lighter material than the dense sea floor, these "exotic terranes" did not sink into the mantle with the descending oceanic plate but floated above it to become welded to the continental front. The realization that many of the formations comprising the Pacific Coast states are of "foreign" origin, created somewhere unknown in the mid-Pacific and imported to our shores by moving plates, has revolutionized our understanding of Western geology.

The North Cascades, for example, largely composed of metamorphic rocks long recognized as much older than the volcanic rocks making up most of the Cascade Range, are an ancient mini-continent that rafted ashore about 50,000,000 years ago. This once-separate continent carried its own subduction zone, which continues to function off the Pacific Northwest coast today. Washington state's Olympic Mountains, as well as other segments of the coast ranges, apparently consist of sediments deposited on the sea floor and then scraped off against the continental margin as the oceanic plate was subducted. Some geologists have identified as many as 75 separate exotic terranes in California alone, while other researchers suggest that parts of Idaho may once have been connected to geologically similar fragments that are now part of Alaska. Our familiar Western landscape is thus a mosaic of fragments imported to the West coast during many millions of years.

Although plate movement seems slow in human terms—one to three inches per year in various parts of the world—it is swift by geological standards. At this rate a subduction zone can devour 32 miles of sea floor every million years, about 320 miles in ten million years. Thus, plate relationships can change drastically.

This change has taken place in California, where the tectonic situation is strikingly different from that in the Pacific Northwest. South of the Cascade Range terminus at Lassen Peak, subduction no longer occurs. The Mendocino Fracture Zone, an east-west trending line of faults extending west from the California coast, marks the southern edge of the Gorda plate. South of the Mendocino line, the Pacific plate is not being thrust beneath the continental landmass but is gradually moving alongside it—in a generally northwesterly direction. The boundary dividing this part of the Pacific plate from the North American plate is the notorious San Andreas fault, a giant fracture in the earth's crust that slices through California from the Mexican border to north of Point Arena, where it passes into the sea floor.

Plate movement continues as the rocky slab underlying the Pacific basin grinds inexorably northwestward. Earthquakes occur as sections bordering the fault temporarily lock together, and then break apart. When pressure builds to the breaking point, the crustal rocks slip suddenly along the fault-plane, producing seismic waves that range in intensity from scarcely perceptible quivers to the violent shaking that cripples whole cities. Sudden movement along two different sections of the 600 mile-long San Andreas fault generated two of the greatest earthquakes in United States history. In 1857 the southern segment of the fault broke, devastating a wide region, including the Los Angeles area, where millions of people now live. In 1906 the northern San Andreas shifted, shattering San Francisco and several other northern California towns. Both of these quakes are thought to have had magnitudes of 8.0 or higher on the Richter scale. Numerous other earthquakes along the San Andreas system have taken their toll in life and property. Future jolts of high magnitude are inevitable.

The Mono Lake-Long Valley volcanic area is not related to subduction or the San Andreas fault. Lying along the eastern flank of the Sierra Nevada Range, it occupies a highly unstable region of active mountain-building and frequent earthquake activity. The Sierra consist of huge blocks of crustal granite that have been tilted to the west and lifted to heights of 12,000 to over 14,000 feet above sea level. The east face of the range is a steep fault scarp thousands of feet high that looms above the Long Valley and Owens Valley region. These valleys lie in a large depression formed by the downdropping of a crustal block east of the Sierra uplift. This downdropping may have been caused by the stretching of the continental crust as North America lurches westward over the Pacific plate.

Magma has leaked to the surface at many points along the fractured east margin of the Sierra, but the largest concentration of recent volcanic activity is in the Mono Lake-Long Valley area. A region where severe earthquakes are common-place, this "hot spot" retains a high potential for producing the next major eruption in the continental United States.

Learning to Live with Volcanoes

If we have learned anything from St. Helens or the disaster at Columbia's Nevado del Ruiz, it should be a recognition of the swiftness and awesome power with which a volcano can take human lives and radically alter the landscape. These recent eruptions vividly illustrate that volcanic mountains are not, as some poets would have us believe, changeless or eternal. The lofty glacier-crowned peaks that give much of the West its distinctive character and matchless scenery also have the largely ignored potential to decimate vast areas and endanger some of our towns and cities. Just as people in other parts of our nation must learn to cope with the natural disasters— floods, tornadoes, hurricanes—that threaten their particular region, we on the Pacific Coast need to heighten our awareness of the hazards to life and property that the Cascade and other volcanoes pose.

To help reduce losses from future eruptions, the U.S Geological Survey operates a program to monitor potentially active volcanoes. In 1980 the survey established the David A. Johnston Cascade Volcano Observatory in Vancouver, Washington, named in honor of the brilliant young geologist killed in the May 18 eruption of St. Helens. The Observatory maintains an interdisciplinary staff of geologists, geophysicists, and chemists to monitor the continuing activity of St. Helens and to warn the public of impending eruptions. Using a variety of seismic, ground-deformation, and geochemical techniques, the Observatory scientists were able to predict all of the most significant eruptive events since May, 1980, several hours to three weeks in advance. After the five explosive outbursts in 1980, most of the eruptions have consisted of dome-building effusions of viscous dacite lava, with occasional explosions of gas and tephra.

The U.S. Geological Survey also installed various monitoring devices at other Cascade volcanoes and in the Mono Lake-Long Valley region. The monitoring system includes electronic distance meters and tilt leveling stations to measure changes in ground level, the swelling that typically precedes a volcanic eruption. As of late 1988 tilt measuring networks have been established at Lassen Peak, Shasta, the South Sister, Hood, Rainier, and Baker, with a smaller net at Crater Lake.

The University of Washington maintains an earthquake-monitoring station supervised by Stephen D. Malone and his colleagues. Seismographs at the University record earthquake activity at most of the Washington state volcanoes and Hood. The rest of Oregon's many volcanoes are not under regular surveillance. The U.S. Geological Survey Menlo Park office has a seismic network covering Shasta, Lassen Peak, and the Long Valley area. Any volcano manifesting an increase in earthquakes, indicating the movement of molten rock undergound, or in heat and steam emission immediately becomes the focus of scientific scrutiny.

Our Western volcanoes form some of the nation's finest scenic attractions, drawing millions of visitors annually to the forest, streams, and lakes at their feet. With their seeming solidity and permanence, it is easy to forget that they are part of the "Ring of Fire" that embodies one of nature's most deadly forces. Recognition of the volcanoes' destructive potential and careful monitoring of their restless movements may save thousands of lives in the future.

III
The Fire:
How a Volcano Works

To primitive man, the volcanoes that intermittently blazed along the Pacific Coast were a source of terror and superstition. The Oregon Indians who survived the paroxysms that destroyed the summit of ancient Mazama later spun fascinating legends about the event. To them, volcanoes like Mazama and Shasta were divine and therefore unknowable—supernatural personalities who thundered in wrath and often devastated the land. Science has stripped the volcano of its divinity; no one any longer believes that gods inhabit its molten interior or stand hurling thunderbolts from its flaming crest. Yet the volcano remains one of the most awesome as well as one of the most destructive forces in nature.

Defining a Volcano

A volcano is both an opening in the earth's crust through which gas and hot rock are emitted, and the hill or mountain formed by the accumulation of ejected material. Molten rock undergound is *magma*, which becomes *lava* when it erupts. Geologists usually divide volcanic rock into two broad categories. A molten stream of lava is known as a *lava flow*, which solidifies into *lava rock*. When the lava erupts in fragments, the general term *pyroclastic* is used, from the Greek, meaning "fire-broken." *Tephra* refers specifically to rock fragments that were blown into the air.

Volcanoes built entirely of tephra are called *cinder cones* because the volcanic fragments composing them resemble cinders. Cinder cones contain rocks of many different sizes and shapes. Extremely fine material, lava pulverized to the size of sand grains or smaller, commonly is carried by the wind away from the erupting vent and settles to form a blanket of *volcanic ash*. Fragments between 0.1 inch

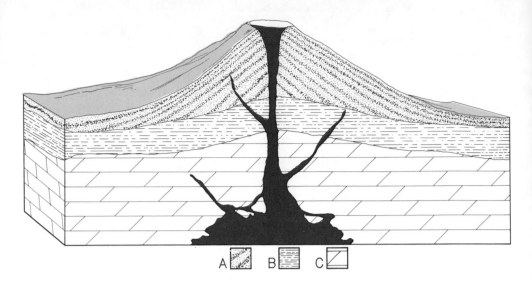

Simplified cross-section of a typical cinder cone. A. Basement rock; B. Pre-cinder cone deposits; C. Pyroclastic material. Solid red denotes magma chamber and feeder pipes.

and about 2.5 inches in diameter are *lapilli*, from the Latin for "little stones." Angular fragments ranging from about 2.5 inches to many feet in diameter are called *blocks*. Rocks of the same dimension are called *bombs* if their streamlining shows that they erupted in a molten condition. When lapilli or bombs are glassy and rich in gas, which escapes leaving behind empty spaces or vesicles in the rock fragments, pumice is formed. Light-weight and porous (vesicular), this frothy material is typically buoyant in water. *Scoria* is somewhat less porous and darker in color.

Although cinder cones are common throughout the Western States, the Three Sisters region in central Oregon has a particularly large number of youthful examples. The most accessible is Lava Butte, a charcoal and orange streaked cone standing at the edge of U.S. Highway 97 about ten miles south of Bend. Named for the long streams of lava that issued from its base, this cinder cone offers a paved road circling to its summit and a Forest Service museum that displays local volcanic rocks. Wizard Island, another cinder cone, rises above the water of Crater Lake.

The summits of both Lava Butte and Wizard Island contain a circular depression called a *crater*. This bowl-like hollow found at the top of most volcanoes encloses the vent through which volcanic rocks erupt. Typically funnel-shaped in cross-section, the crater is the mouth of the volcano's internal pipe that connects the vent to an underground supply of molten rock. The channelway inside the volcano is called the *conduit*.

Volcanic cones built around a single central opening typically have a circular ground plan and a generally conical form. Other eruptive vents may appear at any point along the volcano's slopes. Such *lateral* or *peripheral* eruptions can build a mountain composed of several superimposed cones, such as Shasta or Adams in Washington state.

Shield volcanoes, typically much larger than fragmental cones, are built by a succession of extremely fluid lava flows. All of the volcanoes in the Hawaiian Islands—Mauna Loa, Mauna Kea, Kilauea—are shields. Such volcanoes are rarely explosive, although they often generate enough gas pressure to shoot up brilliant fountains of molten rock. These fiery displays commonly accompany eruptions in Hawaii's Volcanoes National Park.

Because highly fluid lavas typically flow a considerable distance from their source, the shield volcano characteristically builds a broad-based, gently-sloping mound that resembles an ancient warrior's shield laid flat with the curved side upward. During the early

Simplified cross-section of a typical shield volcano. Built of extremely fluid lava, shield volcanoes typically have broad bases and gentle slopes. A. Lava flows; B. Ancestral deposits; C. Non-volcanic rock. Solid red denotes magma chamber and feeder pipes.

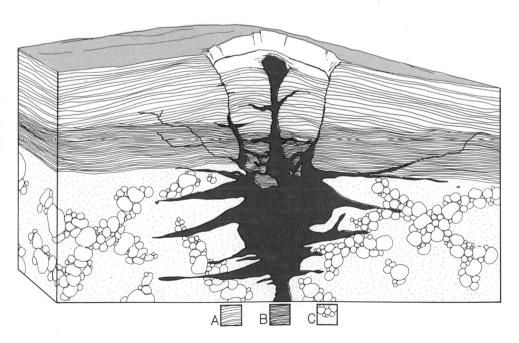

A ▨ B ▨ C ▨

part of Pleistocene time, which began about 2 million years ago, many shield volcanoes formed in the Cascade Range, particularly in northern California and Oregon. Although some retain their original mound-like profiles, many are now so deeply eroded that the masses of solidified lava that plugged their central conduits are exposed.

The third and most impressive volcanic form is the *composite* or *stratovolcano*, so-called because it consists of lava flows interlayered with tephra and other pyroclastic material. All of the loftiest Cascade peaks—Rainier, Shasta, Hood, and Adams—are composite volcanoes. Although most of the large composite cones consist principally of lava flows, they are higher and steeper than any of the shield volcanoes. The original conical forms of most of these mountains have been extensively modified by Pleistocene glaciation. A few built the visible parts of their cones during the 10,000 to 12,000 years since Pleistocene time ended. Of these, Mt. St. Helens, before the 1980 destruction of its summit, offered the most striking example of how the large stratovolcanoes may have looked in their prime.

A fourth kind of volcano found in the Cascades and in the Mono Craters chain is the lava *dome*. Unlike the cinder cone, shield or stratovolcano, it is not made of layers of material added during successive eruptions. Instead, it is a single mass of lava rock extruded during one or more pulses of activity. The lava was too stiff and pasty to flow away from the opening, so it piled up over the vent.

Simplified cross-section of a typical composite cone (stratovolcano). A. Lava flows; B. Pyroclastic and mud flow deposits; C. Basement rock. Solid red denotes magma chamber and feeder pipes.

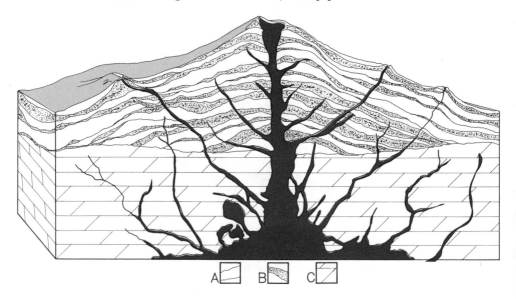

Agglomerate (rubble) over air-fall scoria over layers of dark gray lapilli and ash. —photo courtesy D.W. Hyndman

For the first time in history, scientists are able to observe the growth, as well as the destruction by explosion, of several Cascade domes. During 1980-85, a series of dacite domes oozed from the central conduit of Mt. St. Helens.

When such a viscous mass fills or overflows a previously existing crater, it forms a *plug dome*. Most of these steep-sided mounds rise no higher than about a thousand feet, although Lassen Peak, one of the world's largest lava domes, stands perhaps two or three times as high over the crater it buried. The slopes of Lassen Peak, as on other similar volcanoes, formed through disintegration of the dome surface. Fragments crumbling from the dome's crust eventually created aprons of material, *crumble breccia*, that give the otherwise almost vertical-sided plug the appearance of a smoothly sloping cone. Lassen Peak has several craters at the summit, although many plug domes, such as Black Butte, at the edge of Interstate Highway 5 near the city of Mt. Shasta, have none.

What Makes a Volcano?

Cascade volcanoes are the result of *subduction*, a geologic process in which the Pacific Ocean floor is slowly sinking or being pulled down by its own weight beneath the western edge of North America. As the water-saturated former seafloor sinks beneath the continental crust it encounters increasingly high temperatures. Forty miles down, the temperature probably exceeds 1200 degrees Centigrade, hot enough to melt most rock. But normally the rock remains relatively solid because the great pressures exerted by the surrounding rock prevent liquefaction.

When a crack or fault in the earth's crust allows room for the superheated rock to expand and liquify, it does so immediately, moving upward along narrow fissures. A few miles below the surface it may displace the enclosing rocks to form an underground reservoir of molten rock called a *magma chamber*.

Initial eruptions are typically violent and release large volumes of gas. The most abundant volcanic gas is water vapor in the form of red hot steam derived from water previously held in solution in the magma. Steam is the main driving force in bringing magma to the surface, for when water changes to a gaseous state, its volume increases instantly about 1,000 times, providing the tremendous power needed to create a volcanic eruption. Besides steam, volcanic gases commonly include carbon dioxide, carbon monoxide, hydrogen, sulphur dioxide, sulphur, chlorine, and nitrogen. Ash and other rock fragments carried up in the rising column of gas give the volcano its familiar plume of "smoke." Electrical discharge or the glow reflected from incandescent rock onto water vapor give the appearance of fire, although no actual combustion occurs during an eruption. When the gas pressure is less intense, the rising magma is not blown to bits but instead flows from the vent as streams of lava.

Cascade Lavas

Lava flows sometimes display a smooth ropy surface, which hardens into billowy, wave-like undulations. Called *pahoehoe* after its Hawaiian prototype, this kind of lava is typically gas-rich and erupted at a high temperature. At other times, often from the same volcano and during the same eruption, the lava presents a rough, jagged crust resembling a heap of black slag or clinkers from a smelter. This type, called *aa*, is usually thicker and moves more sluggishly than the pahoehoe variety, probably because it is somewhat cooler and poorer in gas. On still other occasions, the lava crust breaks up into large angular blocks. It then seems to advance rather like a slow-motion avalanche as the blocks tumble over one another

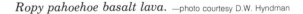

Ropy pahoehoe basalt lava. —photo courtesy D.W. Hyndman

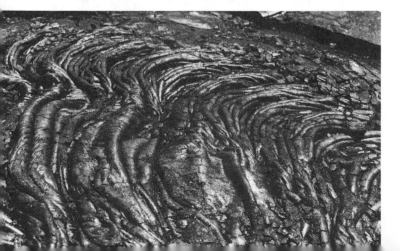

Devils Homestead basalt flow, a clinkery black aa flow. Looking down from partly vegetated older flow in Lava Beds National Monument.

—photo courtesy D.W. Hyndman

on their way downslope. Climbing over the chaotic surface of such a *blocky flow*, as on the upper reaches of Oregon's McLoughlin or Bachelor, is an exhausting experience.

Lavas are also catalogued by their chemical and mineralogical composition. The four most common types—all found in the Cascades—are *basalt, andesite, dacite,* and *rhyolite*. The constituents are virtually identical in all types of lava, silica (SiO_2 - silicon dioxide), and oxides of calcium, sodium, aluminum, potassium, magnesium, and iron, but they vary considerably in their proportion. It is the silica (SiO_2) content that determines the group to which a particular lava belongs. Basalts, rich in iron and magnesium—which gives them their dark color—are low in silica, less than 54 percent by weight. A high percentage of silica increases the viscosity of the lava. Conversely, the lower the silica content the more fluid the lava. Because their low silica content permits them to flow easily, basalts can travel great distances from their source and spread out in thin sheets over large areas. Shield volcanoes typically consist of basalt.

The most characteristic Cascade lava—that of which most of the high composite cones are built—is *andesite*, so named because it is plentiful in the Andes Mountains of South America. Andesite has an intermediate silica content of about 54 to 62 percent. Less fluid than basalt because of its higher proportion of silica, it flows shorter distances and thus piles up to form a moderately steep cone around the erupting vent. Varieties of andesite are generally named from the iron and magnesium bearing minerals they contain; the most common in the Cascades is pyroxene andesite, which is typically dark gray or brown in color. Some Cascade peaks, like Baker and Rainier, are constructed almost exclusively of this kind of lava. Others, like St. Helens, the South Sister, and Mazama (Crater Lake), demonstrate a much wider variety in the chemical composition of their lavas, erupting basalts or basaltic andesites and dacites as well as andesite.

When a lava's silica content reaches about 62 to 64 percent, it is dacite. Usually light colored, dacite has a low melting point, about 850 degrees Centigrade, and is pasty and viscous when molten. It tends to produce short, thick tongues of lava or bulbous, steep-sided domes, such as those of Lassen Peak and the nearby Chaos Crags. Intermediate between dacites and *rhyolites*, where the silica content is 72 percent or more, are *rhyodacites*, silica 68 to 72 percent. Although normally very light in color—light gray, beige, or pink—rhyodacites and rhyolites are sometimes unexpectedly dark. Perhaps the most striking form of rhyolite is *obsidian*, a glistening black volcanic glass. One of the world's largest obsidian flows erupted about 1400 years ago onto the floor of Newberry caldera. Another spectacular obsidian formation can be seen at the base of South Sisters, where a chain of rhyodacite domes stands like glittering black minarets along the Cascade Lakes Highway southwest of Bend, Oregon.

Magma Types and the Kind of Volcanic Eruptions Produced

The Cascade volcanoes have produced a wide variety of magma types with a corresponding diversity of eruptive behavior. The kind of lava a volcano erupts is significant because the chemical properties of the magma, along with its volume and gas content, largely determine how the volcano will erupt and therefore how it is likely to affect lives and property. In general, basaltic volcanoes are the least dangerous. Their low silica content makes basaltic magmas fluid enough to allow the gases dissolved within them to escape freely. Normally they do not produce major explosions, but erupt quietly, sometimes in enormous volumes. During Miocene time, beginning about 16,000,000 years ago, in parts of Idaho, eastern Washington and Oregon, tremendous quantities of basalt poured from long fissures, linear cracks in the earth's crust, and flooded many thousands of square miles, including much of the Cascade Range.

Smaller fissure eruptions have occurred in historic time; for example, at Laki in Iceland (1783) and in lateral eruptions along the flanks of Hawaiian shield volcanoes. During the initial stages of a fissure eruption, a long line of magma fountains spurts high into the air forming a "curtain of fire" as molten rock surfaces along narrow crustal fractures. During the second phase a line of several vents forms along the fissure, apparently because fluids move more easily through more-or-less cylindrical, than through sheet-like, conduits. Much lava flows directly from the vents, but other streams are fed largely from the fountaining showers of magma. The accumulation of still-molten fragments is called *spatter* or *agglutinate*. Finally, during the third stage, the fissure eruption centers at one major vent, flows from which typically bury products of the earlier activity.

Basaltic eruptions that build shield volcanoes are called *Hawaiian* because of their common occurrence in Hawaii. They differ from basaltic floods not only in producing smaller quantities of lava during single eruptions but also in generally being concentrated at or near a central vent so that over a period of time the flows accumulate to form a shield. No Hawaiian type eruptions have occurred in the Cascades during historic time; the most recent was about 1500 years ago at Belknap Crater northwest of Bend. The hundreds of basaltic shields in the range imply that we can expect future Hawaiian eruptions.

Gas-rich basalts, particularly when they encounter ground water, are explosive enough to produce quantities of tephra. But their explosions are tame compared to those of highly silicic volcanoes. Cinder cones of basalt or basaltic andesite are usually built by a kind of moderately explosive activity called *strombolian*, after Stromboli, a volcanic island off the coast of Sicily in the Tyrrhenian Sea, famous for its almost continuous but mild eruptions.

Volcanoes that erupt lavas with a high silica content, such as dacites and rhyolites, are much more dangerous. Sometimes the gases remain trapped in a dacitic or rhyolitic magma until all containing pressure of the overlying rock is removed, which happens the instant the magma reaches the surface. The sudden release of pressure allows the gas to expand explosively, shattering the magma into millions of fragments. Several such cataclysmic tephra explosions have happened in the Cascades during the last few thousand years, notably at Glacier Peak, Mazama (Crater Lake), and repeatedly at St. Helens, most recently in 1980. All involved a gas-rich dacite or rhyodacite that formed voluminous airfalls of light-colored pumice and ash.

Eruptions of moderate to strong explosiveness are known as *vulcanian*, after Vulcano, an island cone north of Sicily under which the Roman god Vulcan was thought to have his forge. A typical vulcanian event blasts large volumes of ash into the atmosphere, forming black "cauliflower" clouds, and hurls out old rock from the volcano's interior, with little new lava. Extremely violent explosive eruptions are called *Plinian*, commemorating the Roman statesman Pliny the Younger, who wrote two letters vividly describing the cataclysmic outburst of Vesuvius in A.D. 79. Whereas a vulcanian eruption may build a cone of blocks and ash, a Plinian event not only devastates a large area around the volcano but typically destroys much of the cone as well.

Immense quantities of fresh magma are ejected, both as tephra and as *pyroclastic flows*. A pyroclastic flow is a turbulent mixture of hot gas and incandescent rock fragments that moves along the ground like a liquid, at speeds known to exceed 100 miles per hour. The basal flow is often hidden by a billowing cloud of hot ash and dust, a *nuee ardent* or "glowing cloud" that spreads over wider areas than the pyroclastic flow itself, which is generally confined to valleys or other depressions. Plinian eruptions typically choke adjacent valleys with pyroclastic flow debris and deposit blankets of tephra that extend hundreds of miles from the volcano. Finally, a Plinian outburst may wreck the volcano's upper cone, leaving a large summit depression. This *caldera*, by definition a mile or more across, is not caused by the former summit being blown apart but by its collapse. Extensive summit collapse during or after Plinian eruptions occurred at Vesuvius in A.D. 79, Thera (Santorini), an island in the Aegean Sea, about 1475 B.C., Krakatau, an Indonesian volcano, in 1883, and Katmai in Alaska in 1912. A Plinian eruption 6900 years ago opened the caldera that holds Crater Lake.

Pyroclastic flows are a distinguishing feature of *Peléan* activity, which also creates thick, viscous flows or domes, typically of dacite or rhyolite lava. This kind of eruption was first recognized at Pelée, a dacite cone on the Caribbean island of Martinique. In May, 1902, it destroyed the city of St. Pierre, killing 30,000 people. After several weeks of warning activity, a sudden exposion at the base of Pelée's growing summit dome sent a glowing mass of lava fragments and superheated steam avalanching down the volcano's slopes. The pyroclastic flow was confined to a stream valley, but the accompanying gas and dust cloud swept over St. Pierre, flattening stone walls many feet thick, melting glass, and killing all but three or four inhabitants. Because of its suddeness and lethal effects, the pyroclastic flow is probably the most hazardous volcanic phenomenon, at least to persons living near the volcano. Two of the largest Cascade peaks—Hood and Shasta—have erupted domes and pyroclastic flows almost exclusively during the last several thousand years. Future

Pelean-type outbursts at Hood, Shasta, Glacier Peak, Lassen-Chaos Crags, and St. Helens are virtually certain.

Age of the Cascade Volcanoes

While most cinder cones are born, grow, and die within a decade or two, some Cascade stratovolcanoes have a life that stretches back several hundred thousand years or more. Recent studies suggest that many of the Cascade composite volcanoes are much younger than formerly supposed. Measurements of the magnetic polarity of the lavas that built Jefferson, the Three Sisters, Baker, and Mazama indicate that they have been emitted since the last reversal of the earth's magnetic field, within the last 700,000 years.

The Volcanic Evolution of the Cascade Range

Except for the North Cascades, a separate block of crustal material, today's Cascade Range consists of two parallel belts of volcanic rock of very different ages and appearance. The older belt, the folded and deeply eroded section known as the Western Cascades, extends from near Shasta in northern California into western Washington. Millions of years of erosion have totally obliterated all the individual volcanic cones, so that none of the ridges or peaks in the Western Cascades represent an original structure.

Immediately east of the Western Cascades lies the chain of young composite volcanoes known as the High Cascades, most of which have been built during the last million years. Erosion has had little time to modify these recently built edifices. In the High Cascades of central Oregon, the terrain is remarkably smooth, with few deep canyons cut into the coalescing shields and lava flow complexes on which the individual composite cones stand. Although the High Cascade volcanoes have buried most older formations, enough older rocks are exposed to show that volcanic activity on the site of today's mountains goes back many millions of years.

Cascade volcanic activity began in late Eocene and Oligocene time, producing many small cones composed mainly of andesite and rhyolite. This episode of Cascade volcanism began about 36,000,000 years ago and produced lavas compositionally similar to those of later periods. During Oligocene time the lavas grew increasingly silicic. Deposits in lower strata of the Western Cascades are mainly basaltic and andesitic, but grade upward into siliceous pyroclastic rocks of great volume. Despite the vast quantities of material erupted during

A Simplified Time Table of the Cascade Range

Approximate Age	Geological Events
Present (1980--)	Activity at St. Helens
60 - 200	Eruptions at Lassen Peak, Baker, Hood, St. Helens, Rainier, Chaos Crags, Cinder Cone, Shasta, South Sister, Glacier Peak
200 - 300	Eruption of tephra from Glacier Peak; lava flows at Cinder Cone, Lassen Park.
300	Volcanic avalanches create Chaos Jumbles, Lassen area.
350 - 450	St. Helens erupts ash, followed by lava flows and emplacement of former summit dome.
700 - 750	Pyroclastic flow and mudflows at Shasta, Hotlum Cone.
Less than 1100	Ash eruptions and obsidian flows at Medicine Lake Volcano. Pyroclastic flows at Glacier Peak.
1000 - 1200	Growth of Chaos Crag domes, pyroclastic flows, Lassen area. Pyroclastic flow at Shasta.
1300 - 1400	Eruption of tephra, pumice flow, Big Obsidian Flow, Newberry Caldera, Oregon.
1500	Latest lava flows at McKenzie Pass, Oregon.
1800	Pyroclastic flows, mudflows, floods at Glacier Peak. Several pyroclastic flows at Shasta, Hotlum Cone.
1900 - 2300	Ejections of tephra, followed by series of rhyodacite flows and domes, South Sister.
2000 - 2300	Pumice eruptions at Rainier. Lava flows build present summit cone, associated mudflows.
1,800,000-2,000,000 Pliocene Epoch	Growth of numerous overlapping shield volcanoes at site of High Cascades; stream deposition of volcanic debris along edges of mountain front.
7,000,000-8,000,000 Miocene Epoch	Late Miocene explosive volcanism at site of Washington Cascades, erosion of Mid-Miocene composite cones, deposition of sediments along flanks of range.
14-16,000,000	Floods of basalt spread over most of eastern Washington, Oregon, and much of present Cascade Range, partly inundating a chain of large andesitic stratovolcanoes. Meanwhile, Cascades silent.
	Intrusion of masses of molten granitic rock into older strata, forming Snoqualmie batholith in North Cascades, Tatoosh pluton at sight of Rainier, and Bumping Lake pluton east of Rainier.
26,000,000 Oligocene Epoch	Deposition of Keechelus volcanic group in central Washington and Little Butte volcanic series in Oregon; beds of light-colored ash (tuff), andesite flows and breccias, some interbedded sediments.
36,000,000 Eocene Epoch	Deposition of Colestin formation in Western Cascades, Oregon; lower units of John Day formation, Oregon; Ohanapecosh formation in Washington

See Peck and others, 1964; Fiske and others, 1963; Hammond, 1961; McKee, 1972 and McBirney, 1978.

this period and deposited along both the eastern and western margins of the range, few of the cones were high and the incipient mountain chain was low.

By mid-Miocene time, a series of large composite cones began to grow near the axis of the Western Cascades. Although the loftier peaks may have resembled the modern High Cascade stratovolcanoes, the range itself would have looked unfamiliar. At least during their early growth, the major cones were islands rising out of a shallow sea that then covered much of western and central Oregon.

The most significant volcanic event of Miocene time happened east of the Cascades. Floods of basalt lava poured from fissures in eastern Washington and Oregon, burying the older terrain and creating the vast Columbia River Plateau. Basalt flows streamed through the gap in the Cascades made by the ancestral Columbia River and buried parts of the southwestern Washington and the upper Willamette River valley in Oregon. Tongues of lava also extended westward between and around peaks in the central part of the future Cascade Range.

As the basaltic flood eruptions in the Columbia Plateau diminished in volume and frequency about 13 to 14 million years ago, the Western Cascades were broadly folded along axes that closely followed the trend of early mid-Miocene volcanoes. This folding of the older volcanic strata—which had attained thicknesses of thousands of feet—was followed by a renewal of activity in Late Miocene time, nine to ten million years ago. Andesites and basalts erupted, apparently from small cones built on the eroded remnants of the large mid-Miocene stratovolcanoes. Meanwhile, a westward-moving surge of silicic volcanism produced rhyolitic deposits across the plains of central Oregon.

The final major uplift of the range occurred during Pliocene time, two to eight million years ago. Most Pliocene activity seems to have produced only small basaltic shields and minor formations of andesite and rhyolite. Although the Cascades were by then high enough to create a climatic barrier that sharply reduced rainfall in eastern Washington and Oregon, no single large cones comparable to Hood or Shasta formed.

During late Pliocene and early Pleistocene time, approximately 1.5 to 2.0 million years ago, activity concentrated along a narrow north-south trending belt paralleling the Miocene volcanic chain of the Western Cascades. At first the easterly extension of the range was a series of broad, overlapping shield volcanoes with associated cinder cones and basaltic lavas. Later in Pleistocene time, the lavas diversified, producing larger volumes of andesites, as well as rhyolites and basalts; thus arose the procession of stately, glacier-bearing composite cones that now distinguishes the High Cascades.

As uplift continued, segments of the range were broken into great blocks. The Western Cascade region was raised and tilted to the west. Concurrently, the older rocks beneath the new chain of High Cascade volcanoes, particularly in Oregon, began to subside. Although this dropped section, about 20 to 30 miles wide, has sunk as much as 2,000 feet, the vigorous eruptions of the last million years filled the trough with such large accumulations of lava that the young cones, such as the Three Sisters and Jefferson, tower far above the surrounding mountainscape. The extraordinarily rapid growth of the Pleistocene stratovolcanoes coincided with similar activity around the entire Pacific rim.

Volcanoes as "Geologic Clocks"

Some of the more explosive Cascade peaks make extraordinary geologic clocks. By erupting widespread layers of ash, distributed over hundreds of thousands of square miles, several volcanoes have created distinctive "time markers" that enable the geologist to date the relative ages of other events. Bits of carbonized wood associated with a particular tephra deposit can be dated by the Carbon-14 method, if they are less than about 50,000 years old. The technique has dated a number of extensive ash layers from several Western volcanoes.

Discovering the age of rocks older than 50,000 years requires a different technique. Fortunately most volcanic rocks contain potassium 40, which breaks down into argon gas at a known rate. The potassium-argon dating method enables us to measure the ages of lavas erupted hundreds of thousands, even billions of years ago. This dating technique generally has confirmed the geologic youth of our Western volcanoes, most of which were born within the last half million years.

Very young eruption deposits can be dated by the tree-ring method, dendrochronology. Sometimes trees are damaged by an ashfall or by proximity to volcanic heat, slowing their annual growth rate. Using a sample core from a surviving tree, and counting back the annual growth rings to the point at which growth slowed, may make it possible to pinpoint the exact year of the event.

Close up view of a typical glacier. The giant cracks–crevasses–that split the ice stream are caused by tension as the glacier flows over an uneven bed. Winthrop Glacier, Mt. Rainier. —photo by Bob and Ira Spring

IV
The Ice:
How A Glacier Works

Among the most striking features of the high Cascade peaks are the masses of ice descended from their summits. These gleaming white ribbons are glaciers—flowing rivers of ice. The basic difference between an ordinary ice field and a true glacier is that the latter is *moving*. Because of this constant movement, glaciers play an important role in sculpturing the Cascade terrain.

How does a glacier form, and how can a snowfield be transformed into a stream of moving ice? The creation of a glacier does not necessarily require snowfalls of enormous depth, although they certainly help. All that is absolutely needed is a rate of snow accumulation that significantly exceeds the rate of melting over many years. The aging snow loses its light fluffy texture. Gradually the air is forced out as the tightly compacted snow slowly changes into granular ice. Impelled by gravity, it begins to slide downward over the underlying bedrock. If winter storms regularly supply more snow than can melt in the summer months, the glacier will continue to grow and advance. Conversely, if the annual precipitation decreases or the climate turns warmer, the glacier will diminish in size and may disappear.

In the Cascades of British Columbia, Washington, Oregon, and part of northern California, there is presently enough snow every year to maintain glaciers on the higher peaks. At Rainier more than 50 feet of snow falls annually, much of which does not melt above the 7000 to 8000-foot level. Large amounts of snow accumulate in the North Cascades; Baker is almost entirely sheathed in ice. Glaciers tend to be larger in the northern section of the range because the low sun angle at northerly latitudes provides less solar energy to thaw the ice. Even as far south as Crater Lake, the seasonal snowfall approximates 50 feet, although because of lower elevations and warmer summers most of it vanishes by mid-July. Throughout much of their length, the Cascades are white down to about the 3000-foot level from autumn until late spring.

The elevation above which snow persists throughout the year is called the *annual snowline*. On a glacier it is also known as the *firnline*. These elevations may vary with fluctuations in the weather from year to year. They also differ considerably in various parts of the range. On Lassen Peak, which rises 10,457 feet above sea level, only a few patches of snow on the shaded side of the mountain usually survive the hot California summers. Oregon's South Sister, approximately the same height as Lassen, but about 275 miles farther north, supports half a dozen glaciers and remains snowcapped all year long.

That portion of a glacier below the firnline where melting exceeds accumulation is called the *ablation zone*. In the California Cascades this zone lies much higher than it does farther north. The glaciers of Shasta, elevation 14,161 feet, do not descend below about 9000 feet. On Rainier, only about 250 feet higher, some glaciers extend into canyon bottoms as low as 3500 feet.

How Glaciers Work

On volcanic mountains, glaciers are particularly effective eroding agents because volcanoes are not very solid. Built of hundreds of individual lava streams and layers of fragmental material, they offer little resistance to glacial scouring.

As a glacier flows, it scoops up chunks of bedrock and transports them downslope. The underlying rock surface is also ground down by the abrasive action of the debris frozen into the glacier's base. The

Emmons Glacier toe on Mt. Rainier. Melting ice is completely covered with morainal debris —photo courtesy D.W. Hyndman

glacier thus sinks ever deeper into the trench it cuts for itself, polishing and smoothing some surfaces it passes over, gouging grooves and furrows into others.

The mechanics of its quarrying action are aided by numerous fractures in the rock, joints. These fractures form prior to glaciation and are enlarged by frost action. Ice moving over such a surface is able to quarry or extract these joint blocks and incorporate them into its base. Although the quarrying action at the base of the glacier cannot be observed, it may be intitiated when water enters these cracks, freezes, expands, and pries the joint block up where the moving ice can exert its tremendous force. Large boulders, as well as small stones, thus become embedded in the glacier; as it moves forward, the loosened rock is plucked from place and carried away in the moving ice stream. The rock subsequently becomes a tool of the glacier, used to scrape out still other fragments from the glacier's bed.

Although a clean, white glacier may be esthetically pleasing, a dirty ice stream is doing more work. Impregnated with sharp fragments of rock and grit, a blackened glacier is removing and transporting large quantities of material. Not only do active glaciers quarry and scour an area, they bulldoze their margins, undermining and steepening canyon walls. Avalanches of rock fall on top of the glacier's surface, adding to its weight and cutting power. When a glacier enters a winding stream valley, the ice typically broadens it by digging into the valley walls, planing off projections. Widened, deepened, and straightened, the valley becomes a glacially gouged

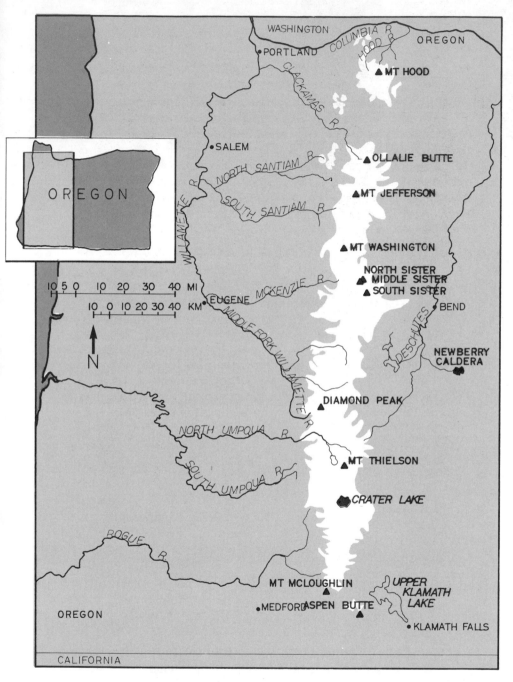

Extent of glaciation (white pattern) in the Cascade Range in Oregon.
Probably approximates the area covered during the maximum ice advance
of the Fraser Glaciation (25,000-15,000 B.P.). —After Crandell, 1965

canyon, with steep sides and a rounded bottom. One of the most typical and accessible of these glacier-scoured canyons is that occupied by the Nisqually Glacier on the south side of Rainier. Along the south rim of Crater Lake, the beheaded remnants of prehistoric glacial trenches—the broad notches of Munson Valley, and Sun and Kerr notches—scallop the caldera wall.

When a glacier descends to low elevations where melting exceeds replenishment of the ice from above, it dumps its load of rock. These deposits contain fragments ranging in size from large boulders to a gray flour-like substance. This unsorted aggregation of debris is called glacial *till*. Deposits of till are moraines. The long ridge of material thus formed at the glacier's front is called a *terminal moraine*. Those laid down along the edges of the glacier are known as *lateral moraines*. Such deposits, unmistakenly the work of glaciers, are often found many miles downvalley from the site of today's ice streams or in valleys throughout the Cascade Range where glaciers no longer exist.

Around Rainier, moraines extend as far as 65 miles away from the peak, yet the glaciers that deposited them clearly originated at high elevations on or near the mountain. One of the clues to the former existence of a great volcanic cone above Crater Lake is the presence of moraines 17 miles from the lake rim. From the regional distribution of moraines, it is apparent that in ages past the Cascade glaciers were vastly larger and longer than they are now. Several times during the past 2 million years glacier ice virtually buried much of the Cascade Range. During some of these glaciations, ice flowing down valleys in the northern Cascades merged with a stupendous ice sheet that blanketed the Puget Sound lowlands to a point somewhat south of Olympia, Washington. During the last of these glacial advances, the Fraser Glaciation, the site of Seattle lay beneath 4000 feet of grinding ice. This giant ice lobe melted only about 13,500 years ago. At the same time, a nearly unbroken ice cap smothered the high Cascades in Washington, and in Oregon from south of Hood to the base of McLoughlin.

In addition, thick ice sheets mantled local areas, such as mountainous regions bordering all the major peaks, including those in California. The full number and extent of these repeated glaciations are not yet fully understood, but their cumulative effects on the landscape were tremendous.

Most of the Cascade stratovolcanoes were built during and between episodes of intense glaciation. Consequently, during much of their history the cones of Baker, Rainier, Adams, Hood, Jefferson, the Three Sisters, Mazama, and Shasta were probably encased in ice to a far greater extent than they are today. During intervals of mild climate between glaciations, the growing cones of Rainier and its fellows may have been nearly as smooth and symmetrical as St. Helens was before 1980. But recurring glacial attacks deeply modified their conical forms, sculpturing trenches and bowl-shaped *cirques* into their slopes. Thus many of the older peaks are now highly asymmetrical, their original forms almost obliterated.

Pleistocene time, during which huge glaciers repeatedly developed, is popularly called the Ice Age. It began with a world-wide cycle of cold, wet weather about 2 million years ago and ended about 10,000 to 12,000 years ago. Since intervals between Pleistocene glacial advances often lasted many thousands of years, it is by no means certain that the ice ages are over. The short time since the last ice age has encompassed significant climatic fluctuations, which resulted in at least two episodes of renewed glacier growth within the last 3000 to 5000 years. A third *Neoglaciation*, the "Little Ice Age," ended only a century or two ago.

Today's Glaciers

The interaction of volcanic fire and glacial ice was dramatically illustrated during the 1980 eruptions of St. Helens. Hot rock ejected onto snow and icefields caused rapid melting of St. Helen's glacier mantle. Meltwater mixed with hot ash and rock fragments of all sizes generated massive mudflows that raced down all sides of the mountain. The largest stream of water-saturated rock debris poured tens of miles down the North Fork Toutle River, destroying bridges, roadways, and scores of houses and other structures. This mudflow emptied into the Cowlitz River, which carried sediment into the Columbia River to reduce the depth of the Columbia's shipping channel from 38 to about 13 feet and temporarily closed the river to ocean-going vessels.

The eruption-triggered floods and mudflows at St. Helens increased public awareness of the dangers posed by the voluminous icecaps on other Cascade volcanoes. Accordingly, the U.S. Geological Survey conducted a study to determine the quantity of ice present on five representative Cascade peaks. By measuring the ice volumes of 25 glaciers on Rainier, Hood, the Three Sisters, and Shasta, the Survey hoped to assess the potential threat from floods and mudflows during future eruptions of these volcanoes.

Rainier was already known to be the source of many mudflows of exceptionally large volume during the last several thousand years

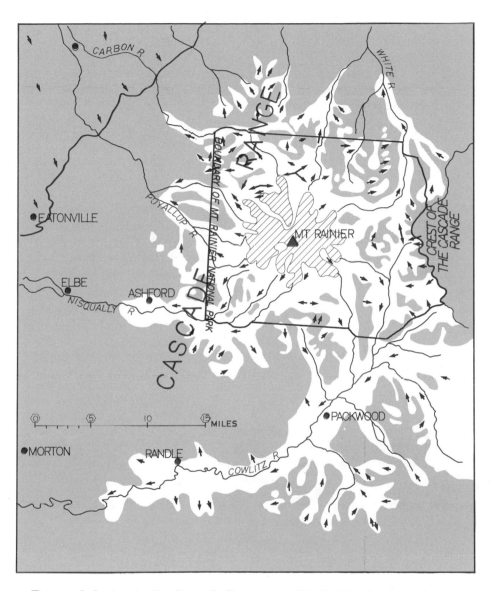

Extent of glaciers in the Cascade Range near Mt. Rainier between about 15,000 and 25,000 years ago. Arrows indicate the direction of ice movement; striped area represents modern glaciers on Mt. Rainier. —From Crandell, 1969

that have affected much of the surrounding terrain. It was not surprising to learn that Rainier bears by far a greater quantity of ice than that covering any other mountain in the 48 adjacent states—155.8 billion cubic feet. Rainier also has the most voluminous glacier, the Carbon, which extends to a lower elevation (3,500 feet) than any other U.S. ice stream south of Alaska. The Emmons Glacier, which occupies much of the volcano's east flank, has the largest surface area—120.2 million square feet. Fifty percent of the snow and ice lie between elevations of 6,000 and 9,000 feet, while 30 percent is above an altitude of 9,000 feet.

Rainier's snow-filled summit crater tilts east, indicating that this may be the first direction from which meltwater might flow. If future activity centers at the principal summit crater, the Cowlitz, Ingraham, Emmons, and Winthrop glaciers would be most affected, possibly triggering floods and mudflows down the valleys of the Cowlitz and White rivers. The ice mass poised above the Cowlitz drainage area is 20.2 billion cubic feet, while that above the White River drainage is 47.2 billion cubic feet. The glaciers that keep many of the Pacific Northwest's rivers flowing year-round could become sources of destruction for persons and property located on their valley floors.

The glaciers of Hood, Oregon's highest peak, feed several streams, all of which eventually empty into the Columbia River. The nine principal glaciers on Hood contain a total ice volume of about 12.2 billion cubic feet. The Eliot Glacier has the largest volume, 3.2 billion cubic feet, and the thickest measured ice, 361 feet. The Coe-Ladd Glacier mass has the largest surface area, 23 million square feet.

Although they may appear smooth from a distance, glacier surfaces are typically shattered and broken into colossal blocks of ice—seracs—which present a towering challenge to ice climbers. Ingraham Glacier, Mt. Rainier. photo by Bob and Ira Spring

Although it has a large ice cover, Jefferson has not produced a large-scale eruption since the last ice age, and was not measured. The Three Sisters in central Oregon, a cluster of three composite cones each exceeding 10,000 feet in height, have a total ice and snow volume of 5.6 billion cubic feet. The largest glacier, the Collier, descends from the north slope of Middle Sister across the west flank of North Sister. With an ice volume of 0.7 billion cubic feet, the Collier Glacier is also the thickest, with a measured depth of 300 feet. The South Sister has several small and medium-sized glaciers, including the Prouty, the Lewis and the Lost Creek, but those on the North Sister are relatively negligible. The Hayden and Diller glaciers cover much of the Middle Sister's steep eastern face. Future eruptions are likely to send meltwater and mudflows down several stream valleys heading on these peaks. Squaw Creek drains most of the Three Sisters' east flanks and flows through the town of Sisters before emptying into the Deschutes River, which passes through the center of Bend, the region's largest city.

The U.S. Geological Survey study found that Shasta, about 40 miles south of the Oregon-California boundary, has a glacier cover containing about 4.7 billion cubic feet of ice. Although its ice volume is comparatively small compared to that of Rainier or Hood, Shasta's glaciers are the largest in California and spawned major mudflows during prehistoric eruptions.

Severe winters can create a particularly thick snowpack on all of the Western volcanoes. Should an eruption occur in such a situation, significant floods and mudflows could be generated even from volcanoes without glacier covering. In the past mudflows have been secondary effects at virtually all the West's volcanic centers, including the Mono Lake-Long Valley area in central California.

In exploring the large Cascade volcanoes, we encounter the work of glaciers at every hand. The present shape and appearance of the high peaks is due almost as much to the power of moving ice as to the fiery volcanism that built them. The long contest between the constructive fires that raised the lofty cones and the erosive forces of ice which work to level them is not yet over. Glaciers will continue to sculpture the mountains that host them, while future eruptions will temporarily disrupt and melt the glaciers. Nature's two opposing powers—fire and ice—are still working on the great volcanoes of the West.

V
The Mono Lake-
Long Valley Region:
California's Potentially Most
Dangerous Volcanic Field

In May, 1980, as St. Helens produced its second major explosive eruption, the Mammoth lakes area of east-central California was shaken by a historically unprecedented series of earthquakes. Within one 48 hour period, four of the earthquakes scored a magnitude of six on the Richter Scale, triggering landslides in the Sierra Nevada Range and damaging buildings in the resort community of Mammoth Lakes.

Swarms of earthquakes, which had begun in 1978, continued intermittently through 1984, while new steam vents appeared near the Casa Diablo Hot Spring, about 1.5 miles east of the epicenter of recent quakes. A 1980 survey along U.S. Highway 395, which parallels the east front of the Sierra, revealed that a crustal bulge in Long Valley had grown by about 10 or 12 inches sometime during the previous two years. As of late 1985 the uplift measured about 18 to 20 inches over 1978 levels. Such surface deformations in volcanic areas are typically caused by an increase in pressure within a magma chamber at shallow depth in the earth's crust. Magma in the chamber underlying Long Valley is believed to extend within about three miles of the surface.

The Mammoth lakes earthquake sequence is particularly significant because the region has a long history of volcanic activity. Nestled in the eastern Sierra foothills, the Mammoth lakes lie inside the western margin of the Long Valley caldera, a huge collapse depression created by one of the most catastrophically explosive eruptions ever to occur in North America. The Long Valley paroxysm, which deposited a widespread ash blanket known as the Bishop Tuff, took place about 700,000 years ago. Eruptions of varying magnitude have continued sporadically up to and including the last few centuries.

Next to St. Helens, the U.S. Geological Survey considers the Mono Lake-Long Valley area as the site most likely to produce the next major eruption in the 48 adjacent states. The Survey issued a "Hazard Notice" of potential volcanic activity in May, 1982, much to the consternation of some local business persons, who loudly protested its alleged effect on the area's economy. In 1984 the Notice was modified through an official change in terminology. The original three-level system of Notice, Watch, and Warning, in ascending order of concern, was replaced by the single term, "Hazard Warning," which remains in effect. Federal geologists maintain a network of seismometers to record earthquake activity and devices to register swelling or deformation of the ground.

The Mono Lake-Long Valley Volcanic Field

Lying in the broad tableland between the White Mountains on the east and the crest of the Sierra Nevada on the west, the Mono Lake-Long Valley region encompasses three separate, but possibly connected volcanic complexes. Volcanic activity began about 3.5 million

Looking northeast along the Mono Craters chain, with Mono Lake in the distance. This geologically youthful line of obsidian domes, flows, and cinder cones has produced numerous explosive eruptions during the last 2,000 years.
—Photo courtesy of the California Dept. of Mines and Geology

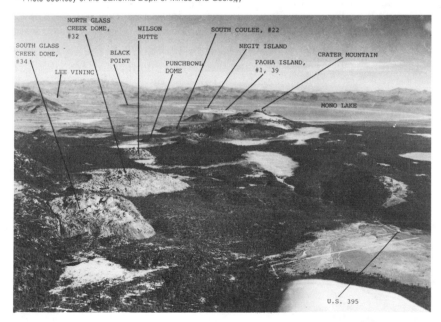

years ago and has persisted virtually to the present. The first eruptions produced extensive flows of basalt lava, eroded remnants of which exist in the Adobe Hills, around the periphery of Long Valley, and in the High Sierra. More silicic lavas also erupted, including andesites and quartz latites, another type of intermediate igneous rock basically similar to andesite.

A large silicic magma chamber evolved beneath the present site of Long Valley. Between about 2.1 and 0.8 million years ago, the first large volumes of rhyolite erupted, forming the Glass Mountain complex, along the northeast border of the present caldera.

The climactic outburst occurred about 700,000 years ago when the silica-rich, gas-charged rhyolitic magma rose to the surface through a series of circular fractures in the crust. The entrained gas could not escape from the viscous magma until all confining pressure of the overlying rock had been removed. The moment the magma reached the surface, the gas expanded explosively. Buoyed by the expanding gas, incandescent ash poured from the ring fractures in towering waves of seething lava fragments. The frothing magma rushed over the surrounding terrain at speeds exceeding 100 miles per hour, incinerating and burying everything in its path. So great was the volume of pulverized hot ejecta and so high its velocity, that one arm of the flowing mass surmounted the steep eastern face of the Sierra Nevada—an obstacle thousands of feet above the erupting vents—and, overtopping the crest, raced westward down the San Joaquin River drainage, perhaps as far as the Central Valley of California.

Another pulse of the pyroclastic flows traveled at least 50 miles southward down the Owens Valley past the present site of Bishop. The ash flow and airfall pumice deposits are known as the Bishop Tuff. Altogether at least 580 square miles of central California and southwestern Nevada were inundated by ash and pumice. Fallout from turbulent dust clouds generated by the pyroclastic flows darkened skies over most of the western states, leaving a recognizable ash layer as far away as central Nebraska. So much molten rock disgorged from the volcano's underground feeding chamber that its roof collapsed, causing the overlying section of the earth's crust to subside thousands of feet and creating the vast oval depression, 20 by 15 miles in diameter, known as the Long Valley caldera.

The volume of new material ejected during the Long Valley eruptions is phenomenal—140 cubic miles, enough to build a mountain larger than Shasta, the largest stratovolcano in the Cascade Range. This is nearly four times the quantity of fresh magma expelled when Mazama destroyed itself to form Crater Lake caldera almost 6900 years ago. The quantity of magma St. Helens erupted on May 18, 1980, 0.25 cubic mile, is minuscule by comparison.

After the roof of the magma chamber collapsed, activity continued intermittently on the caldera floor. Explosive expulsions of tephra were followed by flows of obsidian, the glassy form of rhyolite. Pressure from rising magma arched the caldera floor, creating a resurgent dome. Eruptions of rhyolite lava into the moat surrounding the resurgent dome occurred at intervals of roughly 200,000 years, beginning about half a million years ago.

Mammoth Mountain, one of the most conspicuous in the region, began to erupt about 150,000 years ago, building a massive cone of quartz latite at the southwest edge of the Long Valley basin. Banked against the east flank of the granitic Sierra, Mammoth Mountain consists of at least 20 overlapping domes and silicic lava flows. Over a period of 100,000 years it produced a major eruption every 5000 years, on average. Although no lava has erupted in the last 50,000 years, the volcano may not be extinct. Only 500 years ago a steam explosion blasted the north slope about two thirds of a mile west of the Mammoth Mountain Ski Lodge. Steam still issues from fumaroles at several localities on the cone.

The Mono Lake Volcanoes

When Mark Twain visited Mono Lake in the 1860s, he described the area as ". . .a lifeless, treeless, hideous desert. . .the loneliest spot on earth." Nearly twice as salty as the ocean and extremely alkaline, Mono Lake resembles the biblical Dead Sea in that it has no outlet, supports no fish life, and is the remnant of a once much larger body of water. During Pleistocene glaciation of the near-by Sierra, Mono Lake was as much as 900 feet deep and considerably larger than the ten by 14 mile area it presently occupies. The lake shrank drastically during this century as water from streams flowing into the lake basin was diverted to supply the demands of ever-growing Los Angeles. In 1985, protection was accorded the lake as a waterfowl and wildlife refuge when it was declared a natural preserve.

The gray desolation that fascinated and repelled Mark Twain has changed little since his visit. Small towns have sprung up along Highway 395 but the open country beyond remains almost as bleak and arid as ever. Shadowed by the high wall of the Sierra, the area receives little rain; many of its recently-erupted landforms remain little touched by erosion.

Eruptions related to the Mono-Inyo volcanic chain began about 200,000 years ago with flows of basaltic lava along a north-south zone

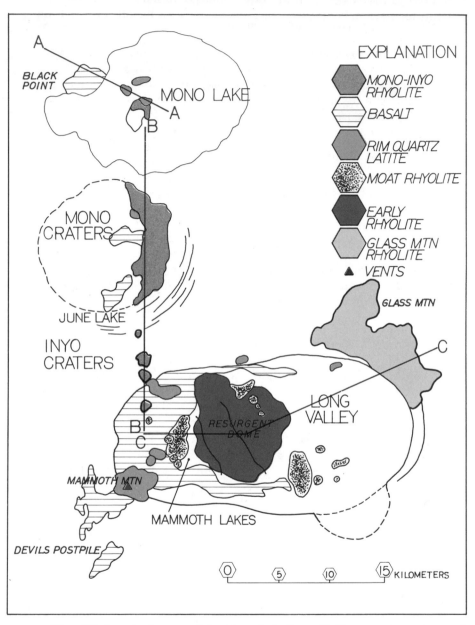

EXPLANATION

MONO-INYO RHYOLITE

BASALT

RIM QUARTZ LATITE

MOAT RHYOLITE

EARLY RHYOLITE

GLASS MTN RHYOLITE

▲ VENTS

BLACK POINT

MONO LAKE

MONO CRATERS

JUNE LAKE

INYO CRATERS

GLASS MTN

LONG VALLEY

RESURGENT DOME

MAMMOTH MTN

MAMMOTH LAKES

DEVILS POSTPILE

0   5   10   15 KILOMETERS

Simplified geologic map of the Mono Lake-Long Valley area in east-central California.

extending from the Devil's Postpile southwest of the Long Valley caldera to Black Point on the north shore of Mono Lake. The Black Point formation, composed of fragments of dark basalt, erupted underwater about 13,500 years ago.

The prominent chain of rhyolite cones and domes known as the Mono-Inyo Craters stretches from Mono Lake southward to the Long Valley caldera. Activity at the Mono Craters, and their southern extension, the Inyo Craters, started about 40,000 years ago and has continued almost to the present. During the last 2,000 years, at least 30 new vents opened, spewing ash and erecting an arcuate chain of cinder cones and rhyolitic domes from the center of Mono Lake to within about four miles of the town of Mammoth Lakes. Tephra from recent outbursts at the northern and southern ends of Mono-Inyo chain blankets a wide area, including the Sierra from northernmost Yosemite to Kings Canyon. At the northern end of the chain, recently active vents include the cones forming Negit and Paoha Islands in Mono Lake, and Panum Crater, a prominent rhyolite dome that partly fills an explosion crater 2,000 feet in diameter.

The Inyo Craters, which you can visit by driving a short distance off Highway 395, include steep rhyolite domes and funnel-shaped explosion pits that contain small lakes. The last eruptions occurred between about 500 to 600 years ago, with a rapid series of explosive eruptions at several sites along the chain.

Recent studies indicate that the rate of eruptions has increased significantly during the last 10,000 years; the past 2,000 years have brought eruptions every 200 to 300 years. If the frequency rate of past activity continues, an eruption from the Mono-Inyo Craters may be expected within the next 50 years. If the earthquakes and ground deformation that began in 1978 intensify, an eruption may indeed be imminent, as many geologists fear.

Hazards from Future Eruptions

Future eruptions in the Mono Lake-Long Valley region are inevitable, but geologists cannot predict their timing or magnitude. The worst case but perhaps least likely scenario would be a repetition of the voluminous pyroclastic eruptions that formed the Long Valley caldera. If the recent earthquakes, ground swelling, and steam emissions that trouble the Mammoth Lakes community were to culminate in another caldera-forming cataclysm, it would create havoc on an almost unimaginable scale.

According to C. Dan Miller and his colleagues with the U.S. Geological Survey, a similar eruption today could deposit an ash layer 50

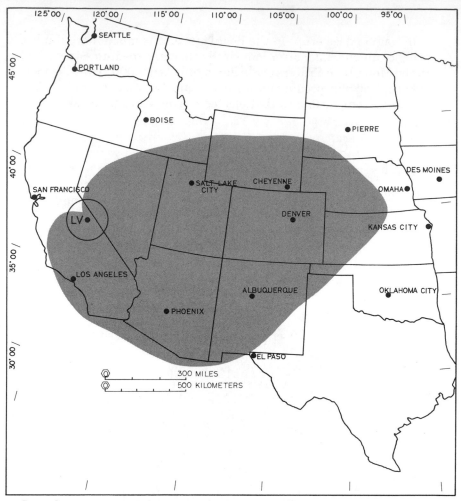

Distribution of ashfall from eruption at Long Valley Caldera (LV). Many feet thick near its source, the ashbed thins to about half an inch at its eastern extremity. Three feet or more of ash could be deposited within the 75-mile radius circle around Long Valley downwind from a future eruption with a magnitude similar to that which occurred 700,000 years ago. —After Miller, 1982, p.9

inches thick 75 miles beyond the erupting vents, an ash blanket 16 inches thick at a distance of 125 miles, and six inches 300 miles from the source vents. Destruction would be virtually complete within a radius of 75 miles of the volcano. Verdant Yosemite National Park, immediately west of Mammoth Lakes, could be reduced to a wasteland. In such an event, few parts of the American West could hope to escape unscathed. Fortunately geologists familiar with the area's volcanic history do not expect an imminent reenactment of the Long Valley paroxysm, certainly not without unmistakable warning signals.

If magma does erupt in the near future, it is much more likely to produce much milder eruptions similar to many medium-sized events in the Mono-Inyo Craters chain during the last 2,000 years. A moderate eruption may produce as much as a quarter of a cubic mile of lava, a volume comparable to that ejected during recent outbursts.

A Future Chain of Fire?

If the recent past is a guide to the future, the next eruptions along the Mono-Inyo volcanic chain may produce a whole series of new volcanoes erupting simultaneously along a zone several miles in

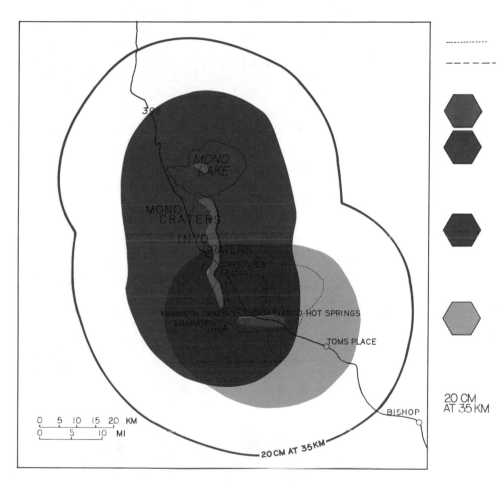

Potential hazard zones from future eruptions in the Mono Lake-Long Valley region. The hazard zones are based on the kind and size of eruptions that have occurred in this area during the last 10,000 years. —After Miller and others, 1982

length. Two recently completed studies of the latest activity at the Mono-Inyo craters suggest what kind of fireworks we may expect from this youthful volcanic chain in the relatively near future. Using tree-ring dating methods, geologists concluded that the last activity took place some time between A.D. 1325 and 1365, during a period of perhaps a few weeks or months.

Violent explosions occurred when a dike, a thin, sheetlike body of magma, broke the surface at the northern end of the Mono chain. Encountering ground water that flashed into steam, the gas-charged magma exploded with tremendous force, blasting open a four mile-long line of new vents and propelling columns of hot ash miles into the stratosphere. The explosive (Plinian) stage of the eruption occurred in a rapid succession of distinct pulses that deposited a series of airfall ash beds, extending north, northeast, east and south of the erupting vents. If a comparable eruption were to occur today, it would produce an ashfall eight inches thick 20 miles downwind, and as much as two inches thick 50 miles away. Depending on the location of the crater, wind direction, and the quantity of rock debris blown out, towns like Bridgeport, Lee Vining, Mammoth Lakes or other settlements could be endangered.

Composite cross-section of the Long Valley-Mono and Inyo Craters, and Mono Lake volcanic complexes, showing hypothetical form of their inferred magma chambers. —After Bailey, 1982, p. 21

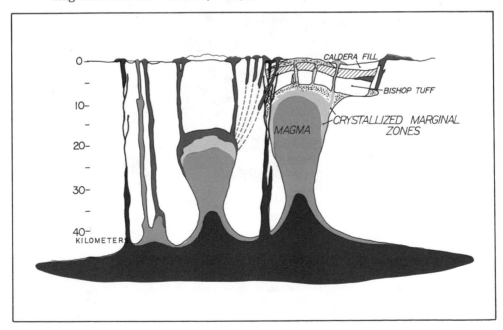

The second phase of the North Mono eruptive cycle produced pyroclastic flows and surges. Whereas the ashfall covered at least 3,000 square miles of the region surrounding Mono Lake, the pyroclastic flow and surge deposits are limited to about 38 square miles around the source vents.

The third and final stage of the eruption was much quieter. Most of the gas dissolved in the magma escaped during the ejections of tephra and pyroclastic flows. Following the explosive activity, thick tongues of rhyolitic lava squeezed out of the vents, like putty from a tube. Too stiff and pasty to flow far, the viscous magma formed short, steep-sided lava flows, called coulees, and massive domes. Five separate domes and coulees were emplaced. The northernmost is Panum Dome, which is surrounded by an oval-shaped ring of tephra about 250 feet high.

The largest lava extrusion forms Northern Coulee. Northern Coulee accounts for well over half of the total volume of material emplaced during the North Mono eruption. Upper Dome is the second most voluminous lava flow in the series, although Cratered Dome is also prominent. In addition to the thick obsidian domes, four tephra rings and ridges also formed during the final stages of activity.

The Inyo Craters Eruptions

Impressive as the North Mono eruptions were, they were only the first half of a two-part volcanic holocaust that afflicted the Mono Lake region during the mid-fourteenth century, A.D. Perhaps only a year or two after the North Mono explosive outbursts had ceased, another series of violent eruptions began only a few miles to the south, centered at the Inyo Craters. The sequence of events was similar. Again a dike of rhyolitic magma, about four and half to eight miles long, rose to within about 700 feet of the surface before erupting. A chain of new craters was blown open, generating towering plumes of ash that drifted tens of miles downwind, mantling the area with a new layer of pumiceous ash. Some of the new craters ejected a fresh magma in the form of tephra and pyroclastic flows, while others produced steam-blast eruptions. Following the explosive phase, degassed magma issued at several points along the rift, surging up in thick flows or domes at the South Deadman, Glass Creek, and Obsidian Flow vents.

Some geologists believe the current earthquake sequence and ground uplift centered near Mammoth Lakes suggests that a fresh dike, probably of rhyolitic composition, is now being intruded beneath the south moat of Long Valley. If this proves to be true, the area may experience activity similar to that of the recent eruptive cycles at the North Mono and Inyo Craters. The prospect of not one, but possibly two or more lines of explosive volcanoes erupting in the area between Mono and Mammoth lakes must give local residents and developers food for serious thought.

Should a series of explosive vents open simultaneously along two separate zones, each several miles in length, the effects on property and people living nearby could be devastating. Precisely how damaging future eruptions will be depends on a number of factors, some geological and some social or economic. The geological elements of the problem include such matters as the number of vents involved, the volume of the magma erupted, its chemical composition and gas content, and other physical properties. While none of these components can be known in advance, it is possible for civic leaders in the area to formulate contingency plans to deal with a major volcanic event. Land-use planning, community preparedness, and construction of evacuation routes to accommodate thousands of visitors during the winter ski season, are a few steps local authorities can take to mitigate the adverse effects of the next eruption.

Except for a short steam blast on the shores of Mono Lake in 1890, the volcanoes of east-central California have not erupted during historic time. With no eyewitness record, it is difficult to anticipate the course of events in a new eruption. Because the Mono Lake-Long Valley area has been the site of some of the most cataclysmic outbursts in geologic history, one cannot rule out the possibility, however remote, of a catastrophic event. The arcuate chain of Mono-Inyo Craters is part of the circular system of fractures that extends to the south and east. This system of concentric fractures may be the surface expression of a second magma chamber developing north of the Long Valley caldera. Although now believed to be much deeper and smaller than that which produced the caldera-forming outburst of 700,000 years ago, it is probably capable of generating eruptions that will affect thousands of people.

Many millennia may pass before another volcanic holocaust creates another great caldera. Meanwhile, lesser activity will surely modify the present landscape. Future travelers to this wind-blown waste of barren lava, saline water, and the hooting loon, will encounter newborn craters erecting new monuments to the ancient gods of fire.

A dark tongue of dacite lava, erupted in May, 1915, clings to Lassen Peak's western summit. The lava originally covered most of Lassen's summit and spilled down the steep northeast flank. Steam explosions that destroyed the east-side flow helped to generate the mudflow of May 19-20, more explosions on May 22 removed part of the summit lava. —U.S. Geological Survey photo by Robert Krimmel

VI
Lassen Peak:
California's Most Recently Active Volcano

The summer traveler driving northward on Interstate Highway 5 is likely to be disappointed by his first glimpse of Lassen Peak. As he approaches the town of Red Bluff, the motorist will see in the far distance on his right—usually through the haze of agricultural burning—a barren, earth-colored dome looming above what might be either a purplish cloud bank or a mountainous ridge. The rocky dome, standing 10,000 feet above the Sacramento River, is Lassen Peak, the second most recently active volcano in the conterminous United States.

Lassen is too far south to maintain glaciers, or even a year-round snowcap. By mid-July winter's snows have vanished, thus depriving the mountain of the glistening white mantle that beautifies the more northern Cascade peaks. Moreover, the mountain does not appear to rise much higher than its equally barren volcanic neighbors. For example, Brokeoff Mountain and Diller are more than 1000 feet lower in elevation, but from the Sacramento Valley they seem almost as high. Other nearby peaks, such as jagged Chaos Crags just north of Lassen, also diminish its visual impact.

If the traveler leaves the interstate freeway and drives eastward up the slope toward Lassen Volcanic National Park, he will discover that the smudgy ridge is actually a volcanic plateau that forms the eastern margin of the upper Sacramento Valley. In the lower foothills, it is a bleak, arid country whose thin stony soil supports little besides scrub pine and wild grasses. This is the Tuscan formation, an uninviting accumulation of ancient rock debris washed down from the volcanic highlands to the east. Farther upslope, the road winds among scattered cinder cones and lava flows. Southward, reaching almost to the Feather River Valley, lies a straggling collection of shield volcanoes and steep cinder cones—the southernmost extremity of the Cascade Range.

At about a mile in elevation, just west of Lassen National Park entrance, the landscape becomes more attractive. The forests thicken, and streams and lakes appear. Lassen Peak turns its best profile toward tree-shaded Manzanita Lake. Since its mountain competitors are not visible from this point, Lassen dominates the surrounding terrain. Framed by towering firs and pines, its image reflecting in Manzanita's still waters, Lassen's slopes reveal shades of pink, gray, slate blue, and warm earthen tints. Rising 4500 feet above the former visitor center, the volcano has at least a share of the majesty of the greater peaks to the north.

The 1000-foot-long tongue of coal-black lava that clings to the western summit reminds us that Lassen has one unmistakable distinction. It was the first volcano in the 48 adjacent states to erupt

Agglomerate at Lassen National Park. —photo courtesy D.W. Hyndman

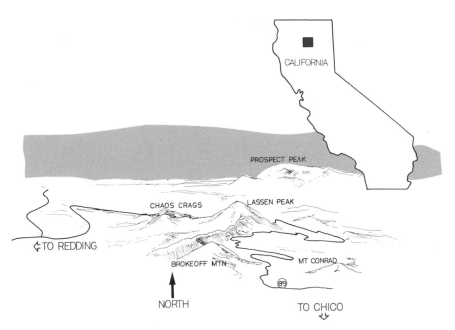

Drawing of Lassen Peak as seen from the southwest and showing the volcano's location in the Cascade Range.

during the 20th century. Lassen's reawakening in May, 1914, was sudden, and apparently unheralded by an earthquake or other warning. The first recorded outbreak occurred on May 30, when a local resident named Bert McKenzie happened to look directly at the peak. At about 4:30 or 5 p.m. Lassen suddenly belched forth a cloud of dense black smoke, actually steam laden with dark volcanic ash, which rose several hundred feet into the air.

Although McKenzie is usually credited with witnessing the first eruption, it may be that Lassen sent up some preliminary puffs of steam the day before. Anna Scharsch of Scharsch Meadows, a ranch near the present Park, was 19 years old and lived a few miles west of the peak. During the evening of May 29, she saw a column of smoke rising from Lassen's summit. She and other family members concluded that the newly constructed fire lookout must be burning. At 10 o'clock the next morning, they saw another, larger, smoke cloud rise from the peak.

The first person to investigate at close range was Harvey Abbey, a forest ranger who climbed through deep snow to reach the mountain's summit on May 31. Abbey found that a new vent—measuring about 25 by 40 feet—had opened high on the northwestern wall of Lassen's old summit crater. The old crater, which filled with lava in 1915, was then an oval depression about 1000 feet in diameter. A small lake occupied part of the crater floor, which lay about 360 feet below the highest point on the rim. Hot water from the vent poured downward into the lake, cutting a channel in the snow.

Abbey noted that the mountaintop was strewn with material blown out by the first eruptions. Stones up to 18 inches across and mud had fallen one or two feet deep over an area 200 feet in diameter.

Volcanic ash and sand covered an area about one mile wide, extending down the volcano's slopes. None of the solid material brought up was hot enough to melt the snow it fell on. Most of the rock debris ejected during the first year of eruptions was probably old lava that had plugged the volcano's conduit.

A second, or possibly fourth, explosion occurred at 8:05 on the morning after Ranger Abbey made his ascent. This was a heavier discharge and hurled out boulders "weighing all of a ton." The active vent, from which lateral fissures extended 100 feet in an east-west direction, was enlarged to 60 by 275 feet. As the steam explosions increased their duration and intensity, a new opening continued to expand until it occupied all of the pre-1914 crater.

On June 13 ashes fell for the first time on the village of Mineral, about 11 miles southwest of the peak. The first eruption to endanger human life occurred on Sunday, June 14. A party consisting of several local millworkers climbed Lassen to view the new crater, which had by then grown to 450 by 125 feet. The men were resting on the crater rim when:

> Without a warning, or explosion that could be heard, a huge column of black smoke shot upward with a roar, such as would be caused by a rushing mighty wind, and in an instant the air was filled with smoke, ashes, and flying rocks from the crater. They all ran for their lives. Mr. Phelps hid under an overhanging rock, which sheltered him from the rocks which brushed past him as they fell. Lance Graham was a few feet away when he was stuck by a flying rock, which cut a great gash in his shoulder and broke his collar bone. He was left on the mountain for dead [but was soon after revived by his companions].

> [Another member of the party, Jimmy Riggins, made a run for it]:. . . coming to a snowdrift, [he] slid down the mountain like a shot. The cloud of smoke kept pace with him, and the rocks from the crater came rolling down the mountain with him, and when he reached the bottom of the snowdrift he found a clump of bushes, and diving into it, buried his face in the snow to keep out the blinding smoke and ashes. The smoke is described as causing the blackest darkness, black as the darkest night. If it had lasted much longer some of them would have been smothered.

Close-up of a new crater, June 2, 1914. Note that the fragmental material thrown out was too cool to melt the snow it fell upon.
—Photo courtesy of Mary Hill

While Graham and his friends were being peppered with sharp rocks, a local businessman and amateur photographer recorded the June 14 eruption in a series of six remarkable exposures. B.F. Loomis, hoping to record the volcano in action, had set up his tripod near Manzanita Lake, about six miles west of the peak. As Loomis' photographs demonstrate, the eruption clouds were very dark—heavily charged with ash and dust. After shooting to a height of about 2500 feet above the mountaintop, they rolled downslope to enshroud the whole dome—as well as Phelps, Graham and the others—in inky black smoke.

Although it is famous because of Loomis' photos, the June 14 outbreak was not nearly so violent as many that followed during the summer of 1914. Beginning at noon on July 15, a series of detonations that lasted the entire afternoon hurled an ash cloud nearly 9000 feet above the mountain top. The crater remained in eruption intermittently throughout the next two days, scattering ash over an ever-widening area. The climax of this cycle of explosions was at 5:28 in the morning of July 18, when Lassen roared furiously, sending a black column of ash 11,000 feet above the mountain. One explosion followed another almost the entire morning. People living in nearby settlements began to calculate how quickly they could hitch their horses for a retreat from Lassen's lethal reach.

But the mountain had temporarily exhausted its energy. Attention turned again to a more sinister outbreak—military violence in Europe—which daily threatened to overwhelm the civilized world. By the time Lassen revived, on August 10, the European war had become a reality. However, local interest, plus the fact that the United States had not yet entered the conflict, helped ensure that the volcano would again monopolize headlines in northern California.

Four successive views of Lassen in eruption, June 14, 1914.

—Photos by B.F. Loomis, courtesy of the National Park Service.

63

The early August outbreak was moderate, but nine days later Lassen began a new cycle of particularly intense activity. For four and a half hours on August 19 the volcano sent up "huge clouds of ash" which rose 10,500 feet above the crater and were visible throughout the northern Sacramento Valley. Two days later the ash column, as measured by the Forest Service, towered 10,560 feet above Lassen's summit. On August 22 the volcano introduced a variation on its usual behavior. This time, instead of rising vertically as it had on all previous occasions, the eruption cloud "shot up obliquely" and rushed down the mountainside. This kind of laterally-directed explosion was a forerunner of what was to be the most dangerous aspect of Lassen's activity.

During the latter part of August and all of September, the eruptions lasted longer and produced heavier ash falls. From September 5 on, eruptions occurred almost daily, showering ash over Mineral, Viola, Chester, and other hamlets on all sides of the mountain. Eye-witnesses spoke of "terrific rumblings," "heavy vibrations," and the sound of rocks crashing against each other. Slides of rock debris tumbling down the mountainside must have added to the din.

On September 21, in what was described as, "probably the most violent eruption to date," ash darkened almost the entire sky over Mineral. A week later "luminous bodies hurled high into the air" and flashes of light from the crater led some observers to conclude that molten lava was now being erupted. It is likely, however, that electrical flares rather than hot lava fragments were responsible for the phenomenon. There is no evidence that molten lava was ejected until 1915.

Eruptions continued throughout the winter of 1914-1915. As soon as winter snows fell, they were often blanketed with black ash and cinders. Most of these eruptions were relatively mild, although that of October 22 lasted two hours and reportedly produced a fall of genuinely hot ashes. Charles Yori of Drakesbad climbed the mountain the next day and found "that the ashes were so hot he could not

West end of the new summit crater, June 28, 1914. The new vent is expanding along an east-west trending fault.

—Photo courtesy of Mary Hill

64

Close-up of ash cloud rising from Lassen's crater, September 29, 1914.
—Photo by Jack Robertson, courtesy of Mary Hill

remain long at one place and that his dog's feet were burned." By then the fire look-out atop Lassen had been demolished by falling rock.

New Year's eve, 1914, marked Lassens' 110th observed eruption. Between then and May 14, 1915, another 60 outbursts were recorded. The official reports, however, are not complete. During most of the winter, storms obscured the mountain, and most observers had departed. An unknown number of eruptions undoubtedly took place unnoticed.

E.N. Hampton, who climbed the mountain on March 23, 1915, noted that the new crater had expanded to a diameter of about 1000 feet. This enlargement is not surprising, since several of the observed winter eruptions had sent clouds more than 10,000 feet into the air.

The snowfall of 1914-1915 was one of the heaviest on record. As warm spring weather began to melt the snowpack, quantities of water undoubtedly trickled into the open fissures and vents of the volcano. Draining into the mountain's interior, some water may have reached hot magma and been transformed into steam. The rapid melting reached its height during May. In that month, almost exactly a year after the volcano reawakened, the climax occurred.

Clouds hid the peak from May 7th until the morning of the 13th, when brief clearings revealed almost continuous eruptions from what seemed to be a new crater. On the 14th, Alice Dines, postmistress at Manton, 20 miles west of the peak, began to telephone her friends that "fire" was visible for the first time. A glow was observed all that night and again on the 15th and 17th. This time the reports of hot material were accurate.

B.F. Loomis and Miss Dines were apparently the first to detect what appeared in daylight as "a small black mass pushing up into the cleft in the rim of the western summit." Miss Dines recalled in a letter of June 8, 1916, that the "dark formation began to rise above the level of the mountain about the 16th or 18th of May, black during a clear day." The "black mass" was the solidified crust of a lava column rising from the volcano's throat.

The night of May 19, 1915, brought the first of two spectacular eruptions. First, pasty lava oozed from the summit crater through notches in the eastern and western crater walls. To observers at the town of Volta, 21 miles west of the mountain, it looked as if the volcano were literally boiling over. In a letter to J.S. Diller, G.R. Milford described the most colorful fireworks Lassen had yet produced.

He saw a "deep-red glow" over the crater which illuminated "the entire outline of the mountain-top." Shimmering on "dense clouds of steam and smoke rising from the crater," the flow increased "in brilliancy" until:

> ... the whole rim of the crater facing us was marked by a bright-red fiery line which wavered for an instant and then, in a deep-red sheet, broke over the lowest part of the lip and was lost to sight for a moment, only to reappear again in the form of countless red globules of fire about 500 feet below the crater's lip. These globules, or balls of fire, were of varying size, the largest appeared at that distance [21 miles] about 3 feet in diameter, the smallest appeared as tiny red sparks. All maintained their brilliancy as they rolled down the mountainside until lost to sight behind the intervening range of hills.

Milford reported that these phenomena recurred about every 8 minutes for "two hours" so that it seemed as if he were "looking at a titanic slag-pot being slowly filled by molten slag in some smelter" until it repeatedly boiled over, breaking through the slag's crust to reveal "deep-red molten material."

The next day, in spite of a heavily overcast sky, Milford and some friends hiked to Manzanita Lake, where, through breaks in the clouds, they caught glimpses of the peak "steaming and smoking in great volumes." They also discovered a dark, almost black mass which to them appeared 1000 feet across and about 2000 or 2500 feet long extending down Lassen's western slope. They assumed this to be an unusually viscous volcanic mud, "as the surface still maintained its roughed, furrowed appearance."

The "dark mass" which Milford's group saw was not mud, but a genuine lava flow that had spilled over the western gap in the crater wall. Three-hundred feet wide at the crater lip, it had moved about

1000 feet downslope, where it cooled in position. Composed of glassy black dacite, its surface bristles with irregular spires and pumiceous blocks.

At about the same time, a second arm of the flow issued through a gap in the eastern crater wall and cascaded down Lassen's steep northeastern flank. Ploughing through a thick snowpack on the precipitous slope, the lava did not hold together as a coherent stream but soon broke apart into hundreds of incandescent fragments. It appears that the flow's disintegration triggered an avalanche of hot lava blocks that rapidly melted the snow. The melt water, mixed with the accumulated debris of more than 170 previous eruptions, was quickly transformed into a large mudflow—a mixture of water, mud and boulders—that rushed down the volcano's northeastern face.

An explosive event may also have accompanied the lava-generated mudflow. According to Dean Eppler, a geologist with the Los Alamos National Laboratory, heat radiated from the cascading blocks of lava caused melt water to flash into steam, producing an explosion that sent a ground-hugging blast cloud sweeping down Lassen's northeast side.

Carrying blocks of hot lava weighing up to 20 tons, the main arm of the mudflow traveled 20 miles down Lost Creek valley. Moving at a high velocity, part of the mudflow surmounted the 100-foot-high divide into Hat Creek, which was then occupied by several home-steaders and ranchers. This branch of the flow was far more dilute than that which inundated Lost Creek and is more accurately de-scribed as a turbid flood than a true mudflow. It nonetheless carried a significant load of rock debris and destroyed everything in its path: trees, fences, bridges, and farm buildings. Previously fertile meadows and grazing land were buried under a thick coating of mud, which was said to resemble wet concrete as it flowed.

The effect this volcanic flood had on the people living in the affected area is vividly described by Wid Hall, a rancher in Hat Creek. Hall's neighbor, Elmer Sorahan was homesteading in a tent about a mile and a half up Hat Creek, and was awakened by his barking dog. After dressing, he "peeped out" his tent to see "the mudflow coming like a wave about twelve feet high . . . The flood made a roar something like a gale of wind in the trees, with a crash and boom of the logs and rocks as they came tumbling along in the flood." Abandoning his posses-sions, Sorahan ran at top speed downstream to Hall's ranch, reaching it just five minutes ahead of the torrent.

He roused the Hall family in time for Mrs. Hall to telephone settlers on lower Hat Creek that "a flood" was moving their way. Meanwhile Sorahan ran across the creek to warn a man who was sleeping in Hall's barn, 150 yards distant, just before the bridge he crossed was washed away. All parties managed to reach high ground before the flood reached them, but communication was difficult: "The crash and roar of the flood was so intense that you could hardly hear one yell even at a short distance."

The flood arrived "about eleven or twelve o'clock," several hours after the lava flow had been observed by witnesses in the Sacramento Valley. About 3 a.m., when it began to rain, the Halls tried to return to their house, which they found "had moved 53 feet and lodged against a tree and the yard fence." The survivors spent a miserable night.

> We had some provisions in the house that were not affected by the flood, so we could eat. On leaving the house we left the door open and the flood entered about two and a half feet deep, then on the ebb flow part of the mud and water ran out, but when the mud had dried it was about sixteen inches deep in the house. A green log had floated in, a rough old thing, which was about all four of us could do to get it out of the house. The mudflow looked more like mortar than water, and where it ran over the ground it left it slick and smooth like pavement.

With daylight the Halls found that their barn had also been flooded, but still stood upright, so their horses were safe. Wading over to the barn, young Ellen Hall reported the water felt "something more than 'milk' warm, and that was in the afternoon of May 20th."

Other settlers along Hat and Lost Creeks had equally narrow escapes. Harvey Wilcox, who lived on Hat Creek below the Hall property, was also awakened by his dog. He had just time to run— barefoot—for high ground before the flood swept away his log house as well as two heavy iron stoves and other new furniture he had not yet placed inside. The mudflow destroyed four other ranches, including the prophetically named "Lost Camp."

Lassen's giant mushroom cloud of May 22, 1915, viewed from Anderson, 50 miles west of the volcano.

—Photo courtesy of the National Park Service

The "Great Eruption"

Worse was yet to come. Damaging as were the lava-triggered mudflows, Lassen had still to demonstrate its full destructive power. Although it was erupting regularly from what seemed to be new fissures and vents north and west of the main crater—now choked by the black dacite that had overflowed the night of May 19—tremendous steam pressure was building up within the mountain. It reached the bursting point about 4:30 on Saturday, May 22. The "Great Eruption" began moderately, but rapidly increased in force and volume. Before the eyes of thousands of persons throughout northern California, an enormous cloud boiled upward from Lassen's summit. Churning and rolling into a clear blue sky, the pale-gray to deep black cloud reached an estimated height of five to seven miles above the crater. As the giant umbrella spread out over the northern Sacramento Valley, ash rained down over a wide area. The prevailing winds carried the eruption cloud far to the east, where ash fell on towns as far as 200 miles from the volcano.

The Reno *Evening Gazette* of May 24 reported that the Nevada villages of Imlay, Elko, Winnemucca, Golconda, and Gerlach were "covered with a fine white ash." The article also noted that two Western Pacific overland trains arrived late in Oakland, California plastered with "ashes and a film of mud thrown out by Lassen Peak. . . Members of the crew reported they first noticed the baptism of ashes near Winnemucca, nearly 200 miles east of Lassen. When 100 miles from the mountain, they said, the trains were enveloped in an ashen cloud and were forced to decrease speed because the headlights could not penetrate.

The *Humboldt Star* of Winnemucca of the same date recorded that ashes began to fall about 8 o'clock in the evening of May 22 and continued for two and a half hours. ". . . people in the street noticed that their clothing was becoming coated with a grayish substance resembling soapstone. The moon was almost totally obscured. The wall of ashes continued until about 10:30 o'clock, when it began to subside, quieting the fear of those who fancied that Winnemucca might meet the fate of Pompeii . . ."

While columns of ash rose into the stratosphere, showers of a mixed andesite-dacite pumice blanketed areas northeast of the volcano and hot volcanic bombs catapulted down the north slope. As the photograph illustrates, Lassen's 1915 mushroom cloud bore a sinister resemblance to that of an atomic blast.

Lassen's northeast face before 1914, with Jensen Meadow in foreground.

Viewers west of the mountain saw only the upward-moving eruption cloud. The most devastating effects were produced by a ground-hugging cloud that simultaneously swept the northeast face. A seething mixture of hot gas, ash, and rock fragments of various dimensions traveled approximately 4.5 miles down Lassen's northeast flank into the heads of Lost and Hat Creek valleys. Described as a "pyroclastic surge," this swift-moving cloud mowed down thick stands of virgin timber, snapping off tree trunks as much as six feet in diameter and hurling them hundreds of feet from their stumps. Altogether, an estimated five and half million board feet of timber were destroyed. So great was the force of the surge that trees toppled along its margin defied gravity by falling uphill, away from the explosion source.

Lassen's northeast face after the mudflow of May 19-20. Note the wide swath which the mudflow cut through the forest on Lassen's flank. The large lava block in the left foreground is part of the dacite flow erupted on May 19.
—Photo by B.F. Loomis, courtesy of the National Park Service and Mary Hill

70

The pyroclastic surge followed the same route as the May 19-20 mudflow, but cut a much broader swath through Lassen's forest. It also melted more snow on the eastern slope, causing another flood in the valleys of Lost and Hat creeks. Following this second inundation in three days, even the tenacious Wid Hall family moved. Since most of the loose material mantling Lassen's northeast flanks had been transported downslope by the mudflow of May 19-20, that of May 22 was smaller and considerably more fluid. In any case, what the first flood had not buried the pyroclastic surge had scoured clean.

Heat from the May 22 explosion sent four additional mudflows streaming down the west and northwest sides of the peak. According to B.F. Loomis, all four reached Manzanita Lake, thus passing through or near the former site of the National Park Visitor Center and campground. Porous snowbanks of Lassen's lower flanks absorbed most of the meltwater so that only small amounts of mud and pumice actually washed into the lake.

Today a paved road runs through the heart of this four and half mile long region lying northeast of Lassen Peak, which has been aptly named the "Devastated Area." Natural reforestation is beginning to cover the old scars, but enough evidence of the 1915 eruption remains to give visitors an idea of Lassen's destructive potential.

Lassen's Declining Activity: 1915-1921

Lassen expended enormous energy during the blowoff of May 22, but was not yet ready to retire. The activity that followed during the next five and a half years did not in any single instance equal the force of the "Great Eruption," but it was sometimes of major proportions. The explosions of May 22 had opened, or considerably enlarged,

The devastated area on Lassen's northeast flank in 1939. Twenty-four years after the area had been swept clean by a hot blast or pyroclastic surge, the land had not yet begun to recover. Today reforestation is well underway.
—Photo courtesy of the National Park Service

71

a new crater northwest of that plugged by the May 19 lava. It, as well as two other new craters that were blasted out by subsequent eruptions, continued to send up impressive columns of ash and pumice; eruption clouds typically soared more than 10,000 feet.

J.S. Diller, the pioneer geologist who had mapped the Lassen region and written a geologic history of the volcano during the late 19th century, returned shortly after the hot blast to study its effects. He noted that on the night of July 13, 1915, "ashes fell at Drakesbad so that we could write our names on the porch railing in the morning." Other "heavy" eruptions lasting from one to two hours occurred on August 25 and 27, while billowing clouds of ash were visible for miles down the Sacramento Valley. Additional outbursts, accompanied by "rumblings" and sulphurous fumes, occurred spasmodically throughout the summer and fall of 1915.

On the evening of October 25, 1915, "flashes and bombs were seen shooting from the crater." Five days later the Forest Service reported a "heavy" eruption which produced a "glow over the crater and luminous bodies." One experienced observer, George W. Olsen, who lived at Chester, 20 miles east of the peak, also reported "large flashes of light and bombs" soaring "high over the top of the mountain."

During the fall and winter of 1916-1917, Lassen began another cycle of activity. White steam alternating with "black smoke" issued frequently during the first months of 1917. On April 5, George Olsen reported "the heaviest eruption" he had seen "since the big one of May 22, 1915." This was soon exceeded by a six-hour disturbance on May 18, 1917, during which clouds rose 10,000 to 12,000 feet above the peak and were "accompanied by loud rumblings." Activity was nearly continuous during May 19th and 20th.

Twenty-one eruptions were reported during June when the renewed activity reached its height. During this period, according to Day and Allen, who made a careful study of the disturbances, explosions were "of such violence as to displace large masses of material at the top of the mountain, and materially to change the appearance of the crater." Probably the major change was the blasting out of a new vent on the extreme northwest corner of the summit.

After June 29, 1917, no further action was reported until January 9 and 10, 1919, when Miss Dines observed "small smoke" issuing from the peak. If these were genuine eruptions, they were extremely mild and lasted for no longer than an hour each. Two more spurts of mild activity occurrred on April 8 and 9 of that year. Except for two days in October, 1920, when the volcano was seen to "smoke" for 10 to 12 hours at a time, the last eruptive incident occurred on February 7, 1921, when Miss Dines sighted "great clouds of white steam issuing from eastern fissures."

Even today, when humidity and temperatures are favorable, enough vapor escapes from tiny fissures in the northwest summit crater to produce an occasionally visible puff of steam. According to Paul Schulz, a former Lassen Park naturalist, about 30 of these steam vents were still active in the early 1950s. Thirty years later, the fumaroles are fewer and Lassen slumbers peacefully.

The Geologic History of Lassen Peak

Lassen Volcanic National Park is named after the region's most recently active volcano, but the Park encompasses a wide variety of other volcanoes with a history going back millions of years. The Park headquarters in Mineral, a small town near the south entrance, are on a wide plain called Battle Creek Meadow. This grassy flatland, surrounded by steep hills, is actually the floor of a caldera, a large oval depression formed by the collapse of an ancient stratovolcano called Mt. Maidu. One of several volcanic centers responsible for the numerous mudflows that created the Tuscan formation to the west, Mt. Maidu has been extinct for hundreds of thousands of years.

Highway 89 north from Mineral toward Lassen Peak passes through a somewhat younger and more spectacular caldera, a vast bowl cut into the interior of another ancient composite cone, the "Brokeoff Volcano," somtimes called Tehama. A hollow amphitheatre open to the southeast, the Brokeoff Volcano is in effect the parent of Lassen Peak, which grew on its north shoulder.

Old Brokeoff cone once towered above the present Brokeoff Mountain and Mt. Diller, the highest remnants of its vanished summit. Composed largely of andesite lava flows, the volcano began to form about 600,000 years ago. With a base covering 100 square miles, Brokeoff reached a maximum elevation of about 11,000 feet.

Late in its history, Brokeoff Volcano began to erupt more silicic magma. A violently explosive episode about 350,000 years ago produced the Rockland Tephra, a thick deposit of pyroclastic material that may have been associated with the collapse of Brokeoff's former summit. Quieter emissions of dacite and rhyodacite lava occurred later on the outer slopes of the volcano, building such landmarks as Bumpass Mountain, Reading Peak, and Flatiron Ridge. These formations are approximately 250,000 years old.

The Lassen Park region was repeatedly smothered in thick sheets of ice during Pleistocene glaciations. During an interval between glaciations, a new vent opened on Brokeoff's northeast flank, at or near the present site of Lassen's dome. From this crater poured copious

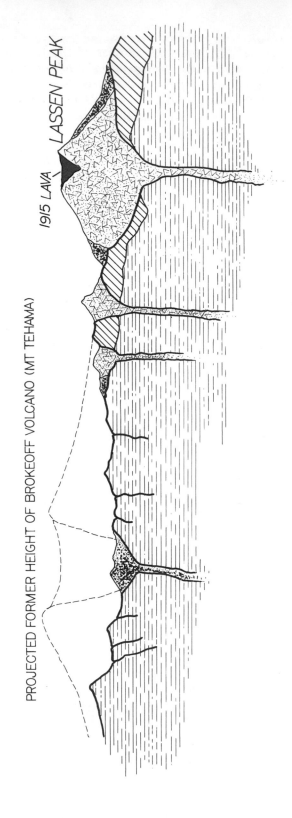

PROJECTED FORMER HEIGHT OF BROKEOFF VOLCANO (MT TEHAMA)

LASSEN PEAK

1915 LAVA

Diagrammatic cross-section of Lassen Peak, showing its relation to the deeply eroded Brokeoff Volcano. The dotted line indicates the former profile of the Brokeoff cone before erosion.

streams of rhyodacite lava, covering an area of perhaps 20 square miles. The late Howel Williams, a professor of the University of California, Berkeley, who made a pioneering study of the Lassen region, called these black glassy flows the "pre-Lassen dacites." In preparing a geologic map of the area for the U.S. Geological Survey, Mike Clynne dated the lavas at about 50,000 years and renamed this flow series the "Loomis Sequence."

Although lava flows and domes intermittently erupted on its slopes during late Pleistocene time, Brokeoff Volcano was repeatedly sheathed in ice and underwent intense glacial erosion. Long after the cone-building eruptions had ceased, the volcano continued to emit steam and corrosive gases. These gases gradually decayed the summit rocks through which they issued and hastened the disintegration of Brokeoff's upper cone. Heat and acidic solutions dissolved masses of rock in Brokeoff's interior, converting them into clay, opal and other vulnerable substances. Glaciers and the volcano's own residual heat conspired to destroy the cone. After glaciers had breached the hard outer shell of the mountain, the soft altered core was attacked and easily carried away by the moving ice. Even today considerable steam and thermal displays of Bumpass' Hell and the Sulphur Works continue to transform solid rock into soft clay. The headwaters of Mill Creek also pursue their work of cutting into Brokeoff's decayed interior.

The Pleistocene glaciers had largely melted when the lava dome of Lassen Peak formed about 11,000 years ago. The largest of the series of domes on Brokeoff's flanks, Lassen probably emerged through the vent that had produced the Loomis Sequence of rhyodacite lavas about 40,000 years earlier. Using the dacite dome that is now growing in the crater of St. Helens as an analogy, Lassen's dome may have grown by the repeated extrusion of thick, pasty tongues of lava. Too stiff and viscous to flow away from the vent as had the previous rhyodacites, the magma forming Lassen Peak oozed slowly upward, piling up around the vent to build a roughly cylindrical plug. As the putty-like lava was pushed up by gas pressure from below, the sides of the plug were polished and abraded by the sharp edges of the conduit. While rising, the dome probably bristled with spires and spines of hardened dacite. Some of these protrusions are still visible on the southeast flank of the mountain.

Relatively little of Lassen's solid core is open to view. As the plastic mass emerged, like toothpaste being squeezed from a tube, the surface rocks cooled and congealed rapidly when exposed to the air. As more magma was thrust up from below, the dome's brittle outer shell shattered and crumbled to form the banks of loose debris that now blanket most of the peak. Rock fragments avalanching down all sides

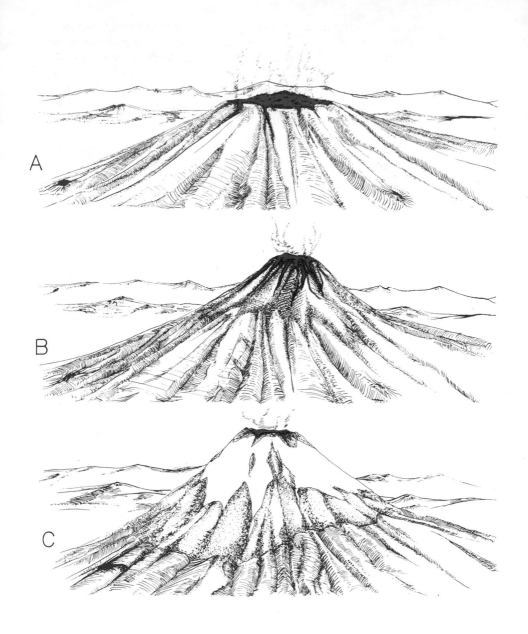

Evolution of Lassen Peak. A. Toward the end of the Pleistocene glaciation, a dome of viscous dacite lava rises into the old "pre-Lassen" crater. B. The dome eventually fills and overflows the original vent. C. Crumbling fragments from the sides of the plug-dome form aprons around the solid core. Hot avalanches from the sides of the dome create pyroclastic flows and small glaciers carve shallow cirques on Lassen's northeast flanks.

of the dome probably triggered pyroclastic flows that traveled down adjacent stream valleys. Williams estimated that Lassen's solid core, together with the talus, "crumble-breccia," that surrounds it, amounts to about one cubic mile of material. Lassen Peak is thus one of the largest plug domes in the world, certainly the largest of its kind in the Cascade Range. But it is minuscule when compared to the great composite cones of Shasta, Adams, or Rainier.

Dwight R. Crandell, an authority on Cascade volcanism, suggested that Lassen was emplaced during a brief interval near the end of the Pleistocene Epoch, after an extensive ice-cap glacier that formerly mantled the area had mostly disappeared. Had glaciers covered the site when Lassen appeared, they would soon have carried off the loose debris that blanket its slopes.

Glaciers did reform, however, shortly after Lassen was emplaced. Linear deposits of glacial debris, called moraines, now mark the limits of these recent glaciers, at least one of which extended three miles beyond the peak. Further evidence of their brief presence are the shallow cirques that indent Lassen's east and northeast slopes above 8000 feet.

In addition to its unusual size, Lassen's dome is distinguished by the presence of four summit craters. Most volcanic domes form during a single eruptive episode and do not ordinarily have an explosion vent or crater at their summits. Chaos Crags, on the northwest flank of Lassen Peak, and the conical Black Butte at the foot of Shasta, are typical large domes without craters. In most cases, explosions accompanying the rise of such domes burst from underneath their lower edges, as they did at Hood, Shasta, and St. Helens. The fact that a fairly large crater existed prior to 1914 suggests that Lassen may have had several cycles of activity similar to those of ordinary stratovolcanoes. Some explosive outbursts apparently occurred early in Lassen Peak's history, for pumice flow deposits are interbedded with late glacial sediments. Precisely how long the volcano had been quiet prior to 1914 is not known.

Chaos Crags and Chaos Jumbles

After Lassen Peak itself, and the steam-and-sulphur displays at Devil's Kitchen, Bumpass' Hell, and Boiling Springs Lake, two of the most remarkable sights in the Park are Chaos Crags and Chaos Jumbles. The Crags, which stand 1800 feet above their surroundings on Lassen's northwest flank, are, like Lassen itself, massive dacite plugs erupted into older craters. Explosive violence preceded their appearance. Not only were pumice and ash blown high into the air,

Chaos Jumbles and Chaos Crags. —photo courtesy D.W. Hyndman

but incandescent avalanches of pumice and fragments rushed westward along Manzanita Creek and northward along Lost Creek. The cinder cones built during this activity were later mostly destroyed by the rise of viscous dacite lava, which, like that forming Lassen Peak, was too stiff and pasty to flow away from the eruptive vent. As it emerged it broke into confused heaps of large boulders. The resultant masses are now known as Chaos Crags, whose magmatic kinship with Lassen is suggested by their similar—but fresher—pink and gray surface colors.

Recent studies indicate that the Crags formed about 1000 to 1200 years ago, so they are only about one-tenth the age of Lassen Peak. The Jumbles, a chaotic avalanche deposit two and a half miles long and covering about four and a half square miles, probably formed no more than 300 years ago, as the result of a rockfall triggered by a steam explosion that had undermined the northern dome, causing waves of shattered rock to rush northwestward at tremendous speeds. Another suggestion for the origin of the Jumbles is that a small dome was intruded among the Crags about 300 years ago. This possible late intrusion is thought to have toppled an unstable formation and triggered the rockfalls and avalanches. Whatever the cause, the series of avalanches must have been appalling. Moving at an estimated 100 miles per hour, they crushed and buried everything in their path.

Steam was reported still rising in large quantities from the northern dome as late as 1854-1857. This means that the volcanic life of the Crags extended over at least a thousand year period, and strongly suggests they may not be extinct. The park visitor center and campground, once located on the Jumbles, have been moved to safer ground.

Cinder Cone

The aptly named Cinder Cone stands about 600 feet above the flatlands near the northeast corner of the Park. Almost perfectly symmetrical, Cinder Cone is a consistent charcoal gray splotched with glassy black and rusty orange ash. The old Emigrant Trail connecting Nevada with the Sacramento Valley winds along its base and a sidetrail zigzags to its summit. The grade, about 35 degrees, is steep, but the summit is well worth investigating. Two nearly circular craters occupy the top, the smaller, deeper one within a broader depression. Remnants of perhaps two other crater rims are also evident, indicating that—unlike most cinder cones—this one experienced several distinct eruptive phases. In addition, extensive lava streams burst from the foot of the cone and dammed pre-existing drainage channels to form Butte and Snag lakes.

According to the San Francisco *Daily Pacific News* of August 21, 1850, an unnamed informant actually saw "burning lava . . . running down the sides" of the volcano. The witness had reputedly ". . . been as near to the base of the cone as the heat would permit, but was obliged to retreat." Light from the eruption, the article stated, ". . . has been distinctly seen at a distance of forty-five miles." At the time of publication, the ". . . mountain had been burning for about ten days."

In a paper before the California Academy of Sciences, H.W. Harkness summarized his interviews with "four different gentlemen" who claimed to have seen the volcano in eruption. A man named Wozencraft, who spent the winter of 1850-51 near Red Bluff in the Sacramento Valley, had "observed a great fire to the eastward of Lassen, which continued for many nights without change of position."

Another eyewitness, J.B. Trask, also observed this night display "for many nights in succession." His vantage point was on the north fork of the Feather River, about 40 miles from the cone. The eruption must have cast a bright glare, for a party of miners at Angel's Camp in the Sierra, about 160 miles from Cinder Cone, also reported the event. Tradition holds that lava was emitted during 1850-51, but the latest flows from Cinder Cone seem to be older than 150 years. In its latest outbursts, the volcano probably erupted only fragmental material.

Exploring Lassen Peak

Of all the Cascade volcanoes, Lassen Peak is the easiest and safest to climb. Take Route 89, the main road through Lassen National Park, to the point where it crosses the mountain's southeast shoulder, just north of the Sulphur Works thermal area. The summit trail begins at a parking lot on the west side of Route 89. Starting at 8500 feet, the well-graded trail is a relatively easy two and a half or three-hour hike for persons in fair condition. At the top, the trail branches: the right-hand, northeast, path leads to the highest crag, 10,457 feet; another, less well marked trail veers northwesterly across the jagged lava which filled the old crater in 1915. Past this lava, the hiker will find two bowl-shaped craters, the smallest and northwesternmost of which was opened in 1917. In clear weather, Shasta is visible 80 miles to the north.

The summit trail is usually open about July 4 and remains in use until the first heavy snows in October or November.

Chaos Crags and Chaos Jumbles, from the northwest. The rockfalls of Chaos Jumbles, center, slid about 300 years ago, damming Manzanita Lake. Lassen Peak looms above the younger Crags, with remnants of Brokeoff Volcano on the right. —U.S. Geological Survey photo by Robert Krimmel

Mt. Shasta from the northwest. Hotlum summit cone and Bolam Glacier appear at the left; Whitney Glacier occupies the saddle between the main summit and Shastina on the right. A blocky andesite flow from Shastina's north flank appears in the foreground. — Photo by Ed Cooper

VII
Mt. Shasta:
Mountain of Menace and Mystery

Physically, Shasta is as impressive as the myths created about it. At 14,161 feet above sea-level, it stands a full 10,000 feet above the wooded foothills at its base. One of the world's largest stratovolcanoes, it has a volume estimated at 108 cubic miles. The mountain's enormous bulk makes it visible for a hundred miles or more in every direction. Lassen Peak, 80 miles to the south-southeast, is a mere hillock by comparison.

"Lonely as God and white as a winter moon," so Joaquin Miller evoked Shasta's glacier-crowned, sky-piercing majesty. In equally poetic terms, native Indians once held that the mountain was home to Skell, a spirit chief who descended from heaven to its summit, where his presence was marked by fire and smoke.

Although other Cascade fire-mountains have their share of loyal boosters, none has attracted the bewildering variety of cults, sects and fanatical devotees as has Shasta. The village of Mt. Shasta, population about 5700, which lies at the volcano's western base, is the center for many of these off-beat religions. These groups seem to have little in common except for their conviction that Shasta is a "holy place." To such esoteric organizations as the Knights of the White Rose, the Rosicrucians, the Association Sananda and Sanat Kemara, the Radiant School of the Seekers and Servers, Understanding, Inc., and the I AM Foundation, Shasta is not merely another impressive landmark; it is a magic mountain.

Probably the best publicized fable about Shasta presents it as the home of the Lemurians, highly civilized refugees from the ancient kingdom of Mu, now submerged beneath the Pacific Ocean. These strange people, as described by those rare souls privileged to have seen them, are seven feet tall and display a walnut-sized sense organ in the middle of their foreheads. This cyclopean organ enables them to communicate by extra-sensory perception and probably contributes to their ability to appear and vanish at will. A few persons have claimed to have been taken on conducted tours of the elaborate tunnel system which the Lemurians have devised within the mountain.

The most famous instance of the Lemurian-related apparitions on the mountain allegedly occurred about 1930, when Guy W. Ballard, a Chicago paperhanger, encountered what he took to be a divine being high on the volcano's slopes. In his book *Unveiled Mysteries*, Ballard describes meeting with one he called St. Germain, "a majestic figure, God-like in appearance, clad in jewelled robes, eyes sparkling with light and love." Reacting to his mystic experience rather like Moses after his interview with God on Mt. Sinai, Ballard came down off the sacred mountain to found a new religion, the mysterious I AM movement, largest of Shasta's many sects.

But, at least according to other local cults, the Lemurians are not the only publicity-shy race to inhabit Shasta. The Secret Commonwealth, which dwells in the subterranean cities of Iletheleme and Yaktayvia also occupy choice real estate inside the mountain. According to "authoritative sources," these Yaktayvians have fashioned vast underground caverns for their cities by their cunning use of bells. Supersonic vibrations from mighty Yaktayvian bells and chimes not only hollowed Shasta's interior, but also supplied the citizens with heat and light. The eerie high-pitched bells also help frighten off intruders who might otherwise invade the Yaktayvians' highly valued privacy. How they managed to share quarters with the Lemurians, Atlanteans, and several other lost tribes who have also set up housekeeping inside Shasta is not known. Their mutual passion for solitude must severely conflict with the population explosion apparently taking place within the volcano.

Such beliefs represent only a part of Shastean mythology. According to other esoteric cults, Shasta is not only honeycombed by underground passageways, it is also a landing field for interplanetary travel. Flying saucers from distant points in the galaxy apparently make Shasta a scheduled stop. UFO reports, perhaps stimulated by the remarkable disc-shaped clouds that sometimes form over the summit, come in regularly from local observers.

The cause, if not the substance, of such wild surmise is understandable. For Shasta projects an unmistakable presence, an aura of combined physical power and alpine beauty. Silhouetted against a clear blue sky, its white glaciers shimmering in sunlight or reflecting the rose and deep purple of an autumn sunset, Shasta seems quite believable as a gateway to some mystical realm.

Mount Shasta as seen from the north.

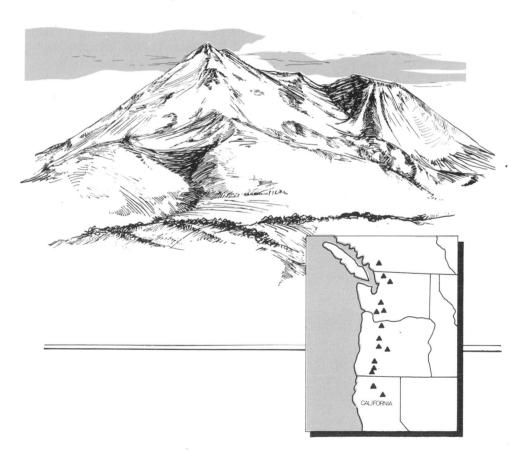

The Geology of Shasta

While Shasta is for some people a temple of the spirit, a lonestar for those looking for realms beyond ordinary human experience, to others it is an exceptionally attractive means of learning more about the forces that shape our earth. For the Shasta story, in purely scientific terms, is every bit as fascinating as any occult myth.

The most salient fact about the mountain is that it is not a single peak, but a compound structure. From the western flank rises a secondary cone—Shastina—that is large enough, if it stood alone, to rank as the third highest mountain in the Cascade Range. Only Rainier and Shasta itself exceed Shastina's 12,300 foot elevation.

Another striking aspect of the volcano is that the summit area is very little eroded. Compared with its more northerly neighbors—such as Hood or Rainier—much of Shasta's north and upper east surface has scarcely been scratched by glacier or stream cutting. Only on the south and to a lesser extent on the west and east flanks has the original constructional surface been dissected by erosion. Below the Konwakiton Glacier, Mud Creek has excavated a canyon about 1500 feet deep into the fragmental material that comprises much of the east slope. Oddly enough, erosion is much more advanced on the sunny south side, where the crags of Sargeants Ridge jut above the Avalanche Gulch, the largest glacial valley on the mountain, than it is on the glacier-covered north side.

Two factors contribute to Shasta's fine state of preservation. First, it occupies a southerly position in the Cascade Range where the annual rain and snowfall is considerably less than on Rainier or Hood. The nearby Klamath Mountains intercept the moisture-laden air moving inland from the Pacific before it can bring precipitation to Shasta. As a result, Shasta is much drier and has a thinner forest cover than either the Cascades to the north or the Sierra to the south. While the annual snowline at Rainier stands at about 8,000 feet, at Shasta it lies above 10,000 feet. Although Shasta supports five named glaciers—the largest in California—they are tiny compared to the 26 or more that bite deeply into Rainier.

In addition, Shasta has erupted significantly larger volumes of lava during the past several thousand years than have many of its northern counterparts. Although parts of the edifice were repeatedly immersed in glacial ice during Pleistocene time, the volcano has since covered its flanks, on all but its southeast side, with a veneer of fresh deposits. Post-glacial flows of andesite lava, pyroclastic flows, and debris from silicic domes have filled in most of the gullies and cirques that Pleistocene glaciers carved in the older portions of the mountain.

How Shasta Was Built

Shasta is a complex volcanic system in which numerous different vents were active at different periods over a long span of time. The magma has evolved chemically during many distinct episodes of cone-building, causing the volcano to change its eruptive behavior.

The oldest rocks known to have come from Shasta are andesites exposed low on the western flank near McBride Spring. These lavas, dated at about 593,000 years, record a Pleistocene volcano ancestral to the present Shasta.

The exact size and shape of this ancestral cone are unknown, but it was large enough to produce a world class landslide. Between about 300,000 and 360,000 years ago, the north side of the old volcano collapsed, sending an enormous landslide of andesite blocks and other rock fragments streaming northwestward into Shasta Valley, a broad depression between the Klamath Mountains on the west and the Cascade Range on the east. Today the Shasta River flows north over the hummocky surface of the avalanche deposit, consisting of hundreds of scattered mounds, hills, and ridges, some of which are 600 to 700 feet high. The avalanche extends about 28 miles northwest of Shasta and has a volume of approximately 6.5 cubic miles. This huge debris flow must have removed a large portion of the ancestral cone, in much the same way that the 1980 landslide at St. Helens destroyed the volcano's north flank.

Shasta includes many overlapping cones. Most of the main edifice consists of four large andesitic cones built during four distinct periods of activity widely spaced in time. The realization that Shasta is really four big stratovolcanoes of varying age piled atop and against each other helps explain some puzzling features.

The Sargeants Ridge Cone

The most thoroughly dissected part of Shasta, the south side is dominated by the long jagged buttress known as Sargeants Ridge. According to two U.S. Geological Survey geologists, Bob Christiansen and Dan Miller, this ridge is an eroded remnant of a large stratovolcano built some time after 450,000 years ago. The oldest of the four main overlapping cones, the Sargeants Ridge vent erupted lavas that overlie the Everitt Hill shield, which is exposed at Shasta's southern foot. The lavas last erupted from Sargeants Ridge were hornblende-pyroxene andesites and a hornblende dacite summit dome and flow. Subjected to at least two episodes of Pleistocene glaciation, its original form has been partly destroyed.

Representative Major Events at Shasta During the last 10,000 Years

Approx. Age	Event	Areas Affected
200 years	Pyroclastic flows, hot mudflows, several cold mudflows from Hotlum cone	East, southeast, north and northwest flanks
700	Block-and-ash flow from Hotlum cone, one hot and six cold mudflows	North and east flanks.
750	Hot and cold mudflows	East side.
1800	Several pyroclastic flows and mudflows from Hotlum cone	North, northeast, and east flanks. .
2000 - 3000	Pyroclastic flows, andesite lava flows, mudflows from Hotlum cone	Nearly all flanks.
3400	Pyroclastic flow from Hotlum cone	Northeast side.
4500 - 5000	Hot mudflow; andesite lava flows from Hotlum cone	Lava traveled 3.5 miles down north side.
6000	Pyroclastic flows; mudflows	North and south flanks
9000 - 10,000	Intense cone-building activity at both Shastina and Hotlum cone: andesite lava flows, domes, pyroclastic flows, mudflows. Red Banks pumice eruption and pyroclastic flows. Elevation of summit domes at Shastina and at Black Butte.	All flanks of the mountain. Pyroclastic flows from Shastina and Black Butte covered a wide area along Shasta's western flank.

(Summarized from Miller, 1980, Plate I.)

The next cycle of cone-building took place at a vent just south of the present summit, erecting an andesite pile on the eroded north flank of Sargeants Ridge and partly filling deep erosional valleys on the older structure. This large cone is called Misery Hill, for the upper part of its eroded surface is now the long slope of dark, hummocky material that climbers toil across on their way to the summit. The Misery Hill cone, built of pyroxene andesite flows, was later intruded by a dome of hornblende dacite that partly filled the summit crater. The cone formed before the last major glacial advance, probably 15,000 or 20,000 years ago.

Recent Activity

Since the last Pleistocene glaciers melted, Shasta has been frequently active, erecting both the large cone of Shastina on its northwest flank and the present summit cone. On average, the volcano has erupted at least once every 800 years during the last 10,000 years and about once every 600 years during the last 4500 years. The most recent significant eruption took place about 200 years ago.

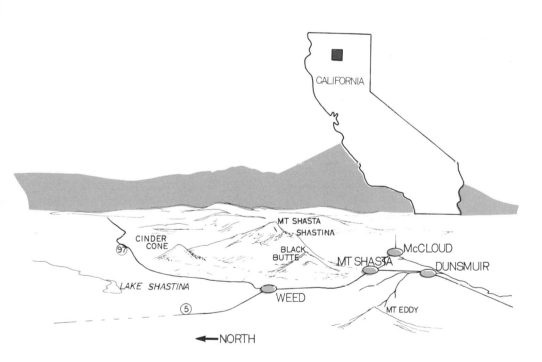

Map showing locations of towns near Mount Shasta.

Shortly after the end of the last ice age a small pyroclastic cone formed high on the west face of Misery Hill, about a mile and a half from the summit. This fresh-looking structure, which contains Sisson Lake in its crater, has been partly buried by the younger lavas of Shastina.

Shastina, which stands to the west of the other three cones about half a mile west of Sisson Crater, duplicated the eruptive history of its two predecessors. The bulk of the cone is pyroxene andesite lava flows, which extend as far as about 8.7 miles from the central vent. The broad summit area is composed of four or five overlapping domes of pyroxene-hornblende dacite. The steep upper slopes are heavily mantled with crumble breccia probably derived from collapse and shattering of the domes as they rose.

Shastina is the most precisely dated of Shasta's four principal cones. Its lavas lie atop the widespread Red Banks pumice deposit, erupted about 9700 radiocarbon years ago. Its last eruptions produced the summit domes and accompanying pyroclastic flows about 9400 years ago. Thus Shastina's entire cone formed in about three centuries, or less. Such episodes of rapid cone-building, lasting from a few centuries to a few thousand years, are probably typcial of the growth of each of Shasta's four cones.

Black Butte, a 2500 foot-high cluster of dacite plug domes at Shastina's western base, also dates from the period of Shastina's growth. Dome elevation and collapse at both Black Butte and Shastina produced a series of large pyroclastic flows. Pumice and block-and-ash flows travelled 6.8 miles south and about three miles north from Black Butte about 9500 years ago.

Only shortly before the Black Butte eruptions, dome emplacement and disruption at Shastina's summit had repeatedly sent avalanches of gas-charged hot rock sweeping down its west face. Deposits from the Shastina and Black Butte pyroclastic flows now underlie 43 square miles along Shasta's western base, including large portions of the townsites of Weed and Mt. Shasta. Explosions during the rise of Shastina's plug domes and the accompanying hot avalanches may have helped carve Diller Canyon, the sharp gash that heads in a notch in Shastina's crater wall and slices through the west flank of the cone. About a quarter of a mile wide and as much as 400 feet deep, this ravine may have been blasted out by downward explosions while the domes that now occupy Shastina's half mile-wide crater were rising. The abrasive force of pyroclastic flows, which can significantly erode surfaces over which they travel, may have helped excavate the canyon.

Aerial view of Shastina's summit crater. The crater rim is breached on the west (left) by a deep cleft that heads Diller Canyon. The snow-covered mounds partly filling Shastina's crater are plug domes blocking the volcano's central conduit. —Photo by Ernest Carter, courtesy of Mary Hill

Among the most prominent features on Shastina are a series of barren flows of blocky andesite extending along its northwest flank. The Lava Park flow emerged from a fissure below the terminal moraine of Whitney Glacier and spread nearly four miles downslope over a wide apron of pyroclastic flow deposits erupted about 9700 years ago. Despite its unweathered appearance, the flow is probably not much younger than the deposit it overlies.

Strewn with huge, angular boulders, the 360-foot-high terminus of the Lava Park flow can be seen bordering U.S. Highway 97 a few miles northeast of Weed. Inhospitable to vegetation, its jumbled surface supports only a few scrub juniper and clumps of sagebrush.

The Hotlum Cone

The youngest part of Shasta, the present summit cone, 14,161 feet, began to form some time before about 8000 years ago. Called the Hotlum cone after the glacier draped over its north face, it remained intermittently active until historic time. Following the Red Banks eruptions, the Hotlum vent erupted a succession of relatively short, thick, blocky pyroxene andesite lava flows. From a distance, these stumpy, high-sided flows, which compose the north and northeast flanks of the mountain, resemble eroded ridges. They are, however, original constructional features that form step-like terraces just below the summit. The longest of these Hotlum flows, the Military

Pass flow, originated near the summit and traveled 5.5 miles down the northeast slopes. About 500 feet thick at its snout, it overlies the Red Banks pumicefall and pyroclastic flow debris erupted about 9700 years ago.

As in the sequence of cone-building events of the Sargeants Ridge, Misery Hill, and Shastina eruptive cycles, a dacite dome invaded the Hotlum Cone late in its history. That dome is now the highest point on the volcano. Chemically altered by their exposure to intense heat and acidic gases, crag-like projections of the dome stand along the rim of Shasta's poorly defined summit crater. About 600 feet across and commonly snow-filled the year around, the crater is distinguished by a hot acidic sulphur spring which emerges near the foot of the topmost spires, staining the adjacent snow and ground surface a dirty sulphur yellow. A second small group of fumaroles occurs on the north side of the summit crags.

The Hotlum cone erupted at least eight or nine times during the last 8000 years, producing lava flows, pyroclastic flows, and mudflows. The latest eruptions, about 200 years ago, sent a pyroclastic flow, a hot mudflow, and three cold mudflows 7.5 miles down Ash Creek on the volcano's east flank. Another hot mudflow traveled at least 12 miles down Mud Creek, into areas now inhabited. The sand-like charcoal gray ash blanketing some open surfaces east of the volcano was probably deposited during the last few hundred to few thousand years by fallout from ash clouds that rose above pyroclastic flows erupted from the Hotlum cone.

Aerial view of three of Mt. Shasta's main cones. The largely uneroded Hotlum Cone (top) occupies the north flank of the older Misery Hill Cone (right). Whitney Glacier (center) descends between these two edifices and Shastina, whose wide crater is seen in the right foreground. —Photo by Austin Post, U.S. Geological Survey

90

Much of the Hotlum cone may have formed after the neoglacial period that began only a few thousand years ago. Its undissected lavas postdate early neoglacial deposits on the Misery Hill cone. No glacial deposits older than a few centuries exist on the Hotlum Cone. The total absence of well-developed cirques or erosional gullies and the lack of much soil oxidation on the flows and other Hotlum strata also suggest an age younger than a few thousand years. Continuing thermal activity indicates that the summit dome is still hot.

Shasta's last outburst may have been observed in 1786 by the explorer La Perouse, who was then sailing along the northern California coast, "only four leagues" from shore. At that time he "perceived a volcano [i.e., an erupting crater] on the top of a mountain, which bore east of us; its flame was very lively, but a thick fog soon deprived us of this sight. . ." It is uncertain whether La Perouse witnessed an eruption of Lassen or Shasta, but the fact that a mountaintop was visible suggests that it was the latter, which is high enough to be seen from the ocean.

A few feeble eruptions of steam or dust were reported as late as the 1850s. About 1855, one Nelson Harvey Eddy told of seeing "three puffs of smoke" rise from the summit of Shasta and drift southward. As he later described them, the "smokepuffs" were like those from the stack of an old-fashioned locomotive as it pulls away from a station. The son of a tribal chieftain from the Shasta area reported that his father, as a young man in the 1850s had also seen smoke issuing from the top of the volcano.

That Shasta's summit crater was hotter in the mid-19th century than it is now is confirmed by accounts of the first successful ascent of the mountain. In August, 1854, a climbing party from Yreka found "a cluster of boiling hot sulphur springs, about a dozen in number, emitting any amount of steam, smoke, gas, etc.," on the edge of a broad snowfield just below the summit pinnacle. Captain I.D. Pearce, who led the expedition, also noted that "the ground for fifty yards around. . . [was] considerably settled and completely covered with sulphur, and the rocks are hot enough to cook an egg in five minutes." Later, another group of hot springs was found on the north side of the peak. Today two small groups of fumaroles remain, while those lying next to the summit spire have been reduced to a single active spring, which, in dry seasons, provides only about a pint of sulphurous hot water per minute.

When the U.S. Geological Survey made aerial surveys to detect hot spots on Shasta, the infrared images taken clearly indicated thermal abnormalities in the summit area. These probably correspond to the sites of active vents reported by early climbers.

The Glaciers of Shasta

During Pleistocene time, Shasta was repeatedly enveloped in ice. Alpine glaciers, vastly larger than those of today, extended many miles downvalley. They may have merged to the east with the large icecap that radiated from the Medicine Lake highlands. Even on the south slope, glaciers rose to within 100 feet of the summits of the many domes, cinder cones, and other secondary formations that dotted the volcano's flanks. Although recent eruptions have buried many of the Pleistocene glacial deposits, on the southwest flank one can still find evidence of the glaciers' former extent. Several moraines form long ridges visible from Everitt Memorial Highway as it climbs the volcano's southwest side to a popular ski area.

The five named glaciers currently mantling Shasta—the Whitney, Bolam, Hotlum, Wintun, and Konwakiton—are puny compared to their Ice Age forerunners. They are not shrunken remnants of Pleistocene glaciers but probably formed during the Neoglaciation of the last 4000 years. None extend downslope beyond about the 9500-foot level.

A mid-1980s survey of ice volumes on several Cascade volcanoes revealed that Shasta bears an ice mantle of 4.7 billion cubic feet. More than a quarter of this is in the main lobe of Hotlum Glacier, which covers an area of 19.4 million square feet and has a volume of 1.3 billion cubic feet. The Whitney Glacier, in the saddle between Shastina and the Hotlum cone, extends over 14 million square feet and has a smaller volume, 0.9 billion cubic feet, but it contains the thickest ice measured on Shasta—126 feet. Ninety-four percent of the area covered by snow and ice lies above 10,000 feet, with the largest ice masses perched above the drainages of Mud, Ash, and Bolam creeks and the valleys of the Sacramento and McCloud rivers.

Destructive floods and mudflows, many caused by the melting of snow and ice during eruptions, have been common at Shasta. Because the volcano did not grow amid high ridges and deep canyons as did Rainier, many of the mudflows were not channeled away from the cone, but piled up in massive aprons about its base. Some mudflows, unrelated to eruptions, are triggered by heavy rains or rapid snowmelt caused by high summer temperatures.

During the summer of 1924 the length of the Konwakiton Glacier, which now terminates above a high cliff at the head of Mud Creek canyon was reduced by three-eights of a mile. Runoff from rapidly melting snowfields high on the mountain was especially large that season. Large torrents of meltwater poured into crevasses of the

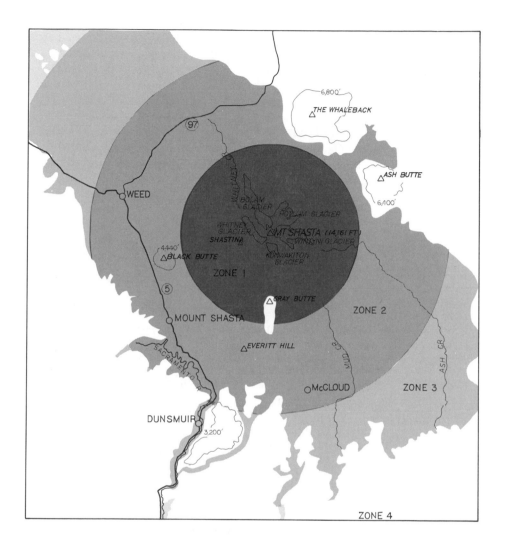

Volcanic hazards in the Mt. Shasta area reflect degrees of risk from future
eruptions involving pyroclastic flows and associated ash clouds, lateral blasts,
and mudflows. A. Zone 1: areas likely to be affected most severely and frequently during future activity. B. Zone 2: areas of intermediate potential risk
from pyroclastic flows. C. Zone 3: areas likely to be affected by mudflows and
ash clouds originating from pyroclastic flows in zones 1 and 2, or by exceptionally large pyroclastic flows. 3. D. Zone 4: areas (mainly valley floors and other
depressions) likely to be affected only by mudflows. Lava flows may erupt on
any side of the volcano but are not likely to reach more than about nine miles of
the summit. —After Miller, 1890

glaciers and burst with great force from its terminus. Flowing into the Mud Creek canyon below, carrying large blocks of ice and rock debris, the swollen creek undermined the loosely consolidated canyon walls. Avalanches momentarily dammed the flood, causing water to back upstream until it surmounted the temporary earthen dams. Bursting through these obstacles, it raced down the Mud Creek valley, finally spilling out as thick sheets of liquid mud and boulders on the plains near the town of McCloud.

Finer silt and mud were swept on into the McCloud River and hence into the Pit and Sacramento River. Because Shasta Dam did not then exist, flood-borne sediments were carried down the Sacramento all the way to San Francisco Bay. Similar floods occurred in 1926 and 1931.

The unusually hot summer of 1977 triggered another series of small mudflows on Shasta. After a light winter snowfall, rapid melting of glaciers generated mudflows that poured down almost every canyon heading at glaciers. Bouldery mudflows, some more than 16 to 18 feet thick, streamed more than 12 miles down Ash Creek, Mud Creek, and the Bolam and Whitney Creek valleys. Confined in deep ravines until they reached the mountain's base, the mudflows spread out over fans of older debris flows.

If a mere drought year can produce such results, one hesitates to contemplate the havoc that could be wrought should Shasta erupt during a spring thaw. This is the true riddle locked within the unexplored interior of California's "mystery mountain." It is not the forgotten lore of the Lemurians or the fabled prowess of Atlantean survivors. It is the geologic enigma characteristic of a very great dormant volcano. When will Shasta next erupt? And what will be the effects upon thousands of people living nearby?

Columnar jointing in dark gray basalt, weathering into rounded boulders, 37 miles west of Klamath Falls. —photo courtesy D.W. Hyndman

94

Lava dribbles three feet across on the flank of spatter cone, Fleener Chimneys, Lava Beds National Monument.
—photo courtesy D.W. Hyndman

The Medicine Lake Highland

Stretching eastward from Shasta are pine forested hills and youthful volcanic cones which culminate in the Medicine Lake Highland, 35 miles distant. This huge, sprawling volcanic complex—which holds the shallow waters of Medicine Lake—is generally regarded as an eastern promontory of the Cascade Range. Although never one of the Cascade skyscrapers like Shasta or Brokeoff, the Medicine Lake Volcano is perhaps the West's largest volcanic cone. Sprawling over 900 square miles, it has an estimated volume of 130 cubic miles.

It is commonly referred to as a shield volcano with a shallow caldera, but the 4000-foot high structure differs considerably from the ordinary basaltic shield. It has erupted a wide variety of magma types, ranging from basalt to rhyolitic obsidian. During its several hundred thousand-year-long history, it produced many explosive eruptions, emitting at least one large volume pyroclastic flow of dacite and andesite that mantled the cone. In its vast size and diversity of eruptive products, it closely resembles Newberry Volcano in Oregon. The summit caldera, approximately four by six miles across, has been partially filled in by recent flows, domes, and cones so it does not present as impressive a vista as Newberry's.

The Medicine Lake edifice is studded with bare, unforested volcanic piles emplaced since the last ice age. These include Paint Pot Crater, Burnt Lava Flow, Glass Mountain and Little Glass Mountain. For a spectacular panorama of much of the region, including a view of Shasta's seldom-seen east side, one can drive on a Forest Service road to the summit of Little Mt. Hoffman.

The last major eruptions in the Medicine Lake Highland occurred less than 1100 years ago. They began with violent ejection of rhyolite pumice from vents now buried by the massive flows comprising Glass Mountain and Little Glass Mountain. During the tephra eruptions, easterly winds bore the ash plume as far west as Shasta, where Dan Miller identified pumice fragments from the Little Glass Mountain vent. Glass Mountain is a classic example of a zoned lava flow, an early dacite and a later rhyolite from a single eruptive vent. Glass Mountain or another vent high on the volcano may have erupted as recently as 1910, when a minor ashfall dusted local vegetation.

Lavas from the Medicine Lake Highland extend northward to form the southern shores of Tule Lake and thus underlie the Lava Beds National Monument. Lava Beds encompasses a wealth of volcanic phenomena, including young spatter cones and lava ramparts formed as fountains of liquid basalt played along fissures spewing molten rock. Of particular interest is Schonchin Butte, a solitary, well-preserved cinder cone which rises above a bleak plain, formed by its flank lavas and flows from Mammoth Crater and other nearby vents. These and other pahoehoe flows are riddled with miles of lava tubes and tunnels, the exploration of which attracts many spelunkers and other visitors to the Monument.

The Modoc Indians, resisting a war of extermination by the U.S. Army, made their last stand here in 1872-73. For months "Captain Jack" and his men held out against superior numbers in the lava caves bordering Tule Lake. At a later period in our history, thousands of Japanese-Americans were interned in the Tule Lake area.

To See Shasta

If your car can endure narrow, rutted dirt roads, the best way to see Shasta is via the old Military Pass Road. The north end of the road leaves U.S. 97 about 15 miles north of Weed, beside a bronze tablet marking the Emigrant Trail. It circles the seldom-viewed east side of

Broken roof of lava tube at Lava Beds National Monument. —photo courtesy D.W. Hyndman

96

the mountain for 32 traffic-free miles and connects with route 89 at McCloud, a few miles east of I-5.

The Everitt Memorial Highway will take you to an elevation of 7703 feet on Shasta's southwest side. This 15-mile drive begins at the town of Mt. Shasta and ends at the Shasta Ski Bowl. Although once popular with skiers, the bowl area does not afford striking views of the mountain. Neither Shastina nor the actual summit are visible from here.

The most-traveled and least demanding route for climbing to the summit starts at the Sierra Club's Horse Camp on the southwest side. one must back-pack in about two miles from the Everitt Highway to the camp, at 7000 feet. The summit trail, well marked only for the first lap, leads up over weary miles of broken rock through "Avalanche Gulch," an empty glacial cirque, to the "Red Banks," 200-foot cliffs of crumbling pumice. Once over this obstacle, the only dangerous part of the climb, the route to the summit crags is direct.

Although the vertical distance from Horse Camp to the top is only about 7100 feet, it is a long and arduous climb requiring excellent stamina. Except in winter, crampons and ropes are necessary only in the Red Banks area.

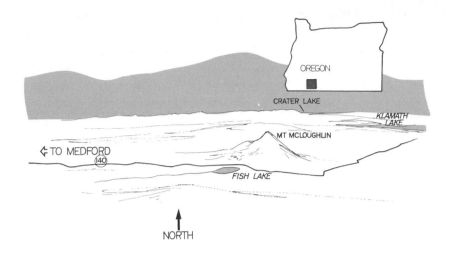

Relief map of the Mount McLoughlin area.

VIII.
Mt. McLoughlin:
Sentinel of Southern Oregon

About 70 miles north of Shasta stands the slim cone of McLoughlin, at 9493 feet the highest peak between Shasta and the Three Sisters. Little known even to most Northwesterns, except for fishermen who find excellent sport in the several lakes about its base, McLoughlin was first called Mt. Pitt, but renamed for John McLoughlin, the Hudson's Bay Company factor who generously helped American settlers at a time when both Britain and the United States claimed the Oregon territory.

The poetically christened Lake O'Woods is probably the most visited fishing spot near McLoughlin, although Four Mile Lake is the best point of departure for a summit climb. Those willing to exert the energy necessary to stand atop McLoughlin's narrow crest are rewarded by a sweeping panorama of the southern Cascades. To the west rolls a mighty sea of forested peaks and valleys, deep green close at hand, smoky blue where they meet the horizon. The most distant ridges belong to what geologists call the Western Cascades, a deeply eroded region of volcanic rocks which border and partly overlap its eastern margin. McLoughlin and its fellow stratovolcanoes are all in the most easterly High Cascade part of the range.

East of McLoughlin, somber forests extend to the curving shore of Klamath Lake, the largest body of fresh water in Oregon. Beyond it stretch the flat arid plains of Oregon's central plateau. To the south rises the white double cone of Shasta. Northward one can see the truncated mass which encircles Crater Lake. On an exceptionally clear day the snowy peaks of the Three Sisters appear dimly against the farthest horizon.

Spread out at one's feet, McLoughlin's cone provides a few surprises. To those who have previously seen the mountain only from the towns of Medford, Klamath Falls, or other settlements south of the volcano, it gives the impression of being a perfectly symmetrical cone. From the ridge that forms its present summit, however, McLoughlin is seen to be an asymmetrical victim of severe erosion. A vanished Ice Age glacier destroyed much of the northeast slope, leaving in its place a large semicircular hollow. This glacial cirque until recently was occupied by a stagnant ice field, although the ice has completely melted during the past few summers. Less than a century ago there was still an active glacier at the spot.

Until the early 1970s almost nothing was known about the geology of McLoughlin. Although a prominent landmark in southern Oregon, it has neither the size nor the grandeur—at least by Cascade standards—to attract much scientific attention. The only available literature devoted to McLoughlin was a brief monograph which Arthur B. Emmons published in 1886.

Fortunately for those who would like to learn more of McLoughlin's past, the late LeRoy Maynard, then a geologist associated with the University of Oregon Center for Volcanology, spent several summers researching its history. The results of his field work—which includes a detailed map of the area—have not yet been published, but Mr. Maynard had kindly permitted his preliminary findings to be used in this review of McLoughlin's story. The text which follows is based largely on Maynard's research.

Among Maynard's most interesting discoveries about the volcano is that it underwent at least three, and possibly four, distinct stages of growth, each involving a totally different mode of eruption. Because of the large basin cut into its northeastern flank, McLoughlin's past is more clearly discernible than that of some other Cascade peaks, such as Shasta. The presence of this large cirque indicates that McLoughlin cannot be one of the younger Cascade volcanoes. It must have reached maturity some time before the last major glaciation, at least 25,000 years ago. Its exact age is unknown, but it possibly has been in existence for 100,000 years.

The Stages of Cone Building

The earliest stage of growth is problematical, since its products lie buried beneath the deposits of later eruptions. In its first *authenticated* eruptive phase, McLoughlin was highly explosive and spewed forth quantities of cinders, bombs, and ash. Localized at a central vent, these outbursts built an unusually large fragmental cone which may comprise as much as a third of the mountain's total volume of approximately 1.84 cubic miles. Judging by McLoughlin's relation to the surrounding terrain, this remarkable cinder cone *seems* to have risen to a height of about 3000 feet!

Some flows of molten rock accompanied the erection of this exceptionally big pyroclastic structure, but they were apparently confined to the lowest parts of the cone. One lava stream followed the ancestral valley of Four Bit Creek, terminating at the site of Big Butte Springs, which emerge from beneath it. Although erosion has cut deeply into the original cinder cone, no lava flows appear among the upper layers of pyroclastic material.

In the second stage of McLoughlin's development, the volcano radically changed the nature and products of its eruptions. Instead of violently blowing out pyroclastics, it began to emit voluminous but quiet streams of lava—also from a centralized summit crater. As numerous thin flows of lava poured down every side of the cone, the original cinder edifice was completely encased in clinkery andesite. McLoughlin thus became an "armored cone," with a hard outer shell of congealed lava and soft inner core of loose fragmental material. If a glacier had not cut so deeply into its interior, we would have no way of knowing about its dual nature.

After activity at the summit crater ceased, McLoughlin began a third eruptive phase. This time copious floods of blocky andesite lava issued from vents below the summit, while fine-grained dark lavas poured from fissures along the base. Two conspicuous crags high on the west flank of the volcano, North and South Squaw Tips, now mark the vents from which exceptionally large blocky flows erupted. These two flows merged to cover much of the southwest slope below the 7800 foot level. A much smaller but prominent flow of the same kind, the Rye Spring Flow, emerged just south of the Squaw Tip effusions, but at a lower elevation. A fourth major stream of blocky andesite issued inside the northeastern cirque and spread over glacial moraines and landslide deposits.

The dark fine-grained lavas of this period are generally much smaller in volume than the blocky flows and are restricted to the south slope of the volcano. One notable exception is a large flow which issued from two closely-spaced vents at the northwest base of the cone. This late flow, which can be seen on the road to Butte Springs, is now bordered on the north by the South Fork of Four Bit Creek.

All of these third stage lava flows are thought to have occurred after the end of the last major glaciation. Some of them remain completely unweathered, their blocky surface as sharp and angular as the day they cooled. Many are devoid of vegetation, extending long arms of naked rock into the thick timberlands surrounding McLoughlin's base. Recent unforested avalanche and mudflow deposits, younger even than the lava streams, are conspicuous on the

Simplified cross-section of Mt. McLoughlin, showing the large pyroclastic cone that occupies much of the volcano's interior. McLoughlin's most recent eruptions have been andesite flows from vents near the base of the cone.

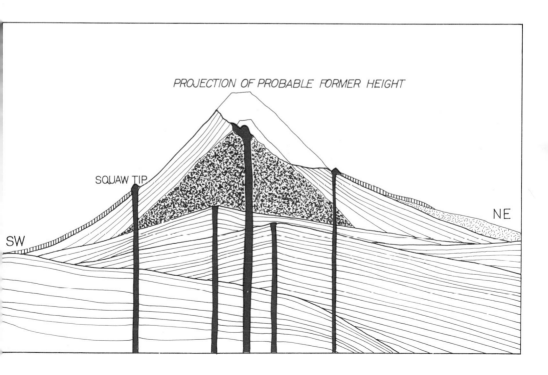

north and east sides of the volcano. Maynard estimated that McLoughlin's latest lava eruptions are probably contemporaneous with the last outpourings at McKenzie Pass, and are thus about 1500 to 2000 years old.

The Puzzle of McLoughlin's Stage One Activity

It was noted that the extraordinarily large cinder cone which forms much of McLoughlin's interior seems to stand about 3000 feet above its basement rocks. Since it is axiomatic that such loosely consolidated structures can not rise more than 1200 or 1500 feet without collapsing under their own weight, the McLoughlin cinder cone presents a special problem. It is possible that some extremely short, thin lava streams helped to cement the pyroclastic mass together, thus enabling it to reach an unusual height, but none have been found among the fragmental debris exposed in the upper part of the cone. Several lava plugs did intrude the original cinder cone, perhaps after it had already been buried under younger lava flows, and these now stand as monolithic protrusions at the head of the northeast cirque.

It may be that beneath the partly exhumed cinder cone there lies a steep-sided lava shield, the result of "pre-stage one" eruptions. This hypothetical shield, if it rose 1500 to 2000 feet above the surrounding land surface, could then have formed an elevated platform on which the cinder cone was later built. With a steep shield serving as its foundation, the cone would not have exceeded the usual height limits which are imposed upon such unstable structures.

Shield volcanoes capped by large fragmental cones are common in the Oregon Cascades. Some of McLoughlin's immediate neighbors, such as Brown Mountain and Pelican Butte, offer fine examples of steep-sided shields topped and partly buried by later cinder cones. Although it is presently impossible to prove, McLoughlin may, during its earliest stages of growth, have been similar in form and structure to Brown Mountain or other near-by volcanoes which closely resemble it. Alternatively, it simply may have been built over the glaciated ruins of an ancestral volcano which occupied the same site.

McLoughlin and the Cascade Icecap

The present crest of McLoughlin lies several hundred feet south of its presumed central vent. Since ice has removed not only much of the northeastern side but the former summit as well, there remains no trace of a crater. If one projects upward the original angle of incline of

McLoughlin's slopes, the lines meet at about 10,200 feet. Leaving room for a modest-sized summit crater, it is probable that the volcano originally attained an altitude of about 10,000 feet, approximately 500 feet above its present top.

According to recent studies, McLoughlin is near the southern terminus of an icecap which almost buried the High Cascades during late Pleistocene time. Apparently the glaciers that then mantled the volcano's northern and eastern slopes coalesced at the mountain's base with a continuous ice sheet which was there approximately 200 to 500 feet thick. Thus, except where recent lavas or landslides have moved downslope, both the northern and eastern flanks of the cone are blanketed with glacial moraines, outwash, and alluvium.

Exploring McLoughlin

Although not technically difficult, the six-mile trail to McLoughlin's summit, 9493 feet, should be attempted only by persons in good physical condition and then only in late July and August after winter's snows have melted. Be sure to carry water as there is none along the trail.

From Klamath Falls, take highway 140, west, for about 33 miles to the Four Mile Lake Road, No. 350, just west of the 31-mile post. Turn north here and proceed for 2.5 miles over rough road to the trail head located on the left hand side, where stands a sign reading Mt. McLoughlin Trail, No. 3716.

After crossing an open area and a clear stream, last source of water, the trail climbs through dense forest. It is a long 4.5 miles to timberline over increasingly steep grades and another 1.5 miles over large, angular boulders and loose rubble to the top. Above timberline, there is no real trail, only crosses and circles painted on lava blocks. At the top, there are fine views of Shasta to the south and Thielsen and the Three Sisters to the north.

If late snow drifts obscure the red "trail" markings on trees and/or boulders, mark your trail clearly; it is easy to lose one's way, particularly on the descent.

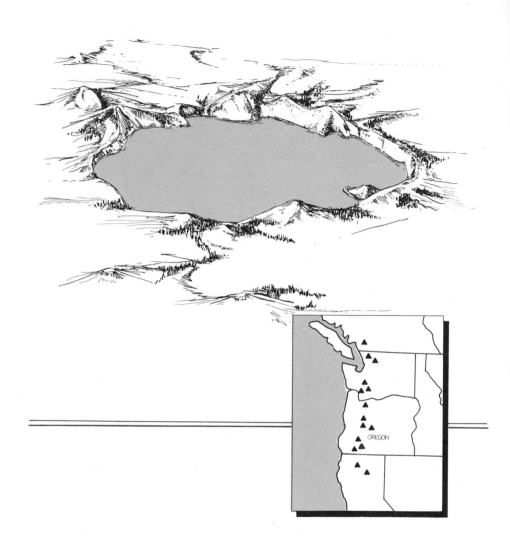

Crater Lake as viewed from the northwest.

IX
Crater Lake:
Beauty Born of Fire

The calm of Crater Lake belies the cataclysmic violence that led to its creation. The enormous lake basin, five by six miles across and almost 4000 feet deep, occupies the core of an ancient volcano, Mt. Mazama. Until about 6900 years ago, Mazama rose approximately a mile above the present lake level, making it one of the highest mountains in Oregon. Then a tremendous explosive eruption ejected a great volume of material from the magma reservoir beneath the volcano. That removed support from the former summit, allowing it to collapse inward, forming the present caldera.

Today, despite abundant evidence of the past violence, the intensely blue lake conveys a soothing impression of peace and permanence. To explore its geologic origins, the visitor can follow a one mile well-graded trail down the north caldera wall to the water's edge at Cleetwood Cove. From there launch trips convey passengers on a tour of the lake, permitting spectacular views of the interior structure of Mazama.

The classic study of Crater Lake is that by the late Howel Williams, formerly a geologist at the University of California, Berkeley. His work has recently been supplemented by Charles R. Bacon of the U.S Geological Survey, who modified some of Williams' interpretations of Mazama's eruptive history, particularly the timing and sequence of events that culminated in the volcano's catastrophic destruction. The following account of Crater Lake's development is greatly indebted to C.R. Bacon, whose work illuminates the forces that shaped one of our most beautiful national parks.

The Story of Mazama

Crater Lake has existed for only a few thousand years, but the mountain in which it rests began to form well over 400,000 years ago. Cone-building eruptions continued until about 50,000 years ago, when Mazama's activity produced smaller volumes of material. A long dormant interval that began shortly after the onset of the last major Pleistocene glaciation lasted approximately 20,000 years. About 5000 B.C., Mazama awoke again, only to destroy itself in one of the world's greatest volcanic eruptions since the end of the Ice Age.

Mazama's birth and growth were roughly contemporaneous with that of other large stratovolcanoes such as Shasta, Hood, Baker and Rainier. Like its Cascade neighbors, it consisted mainly of andesite lava flows, with relatively small quantities of pyroclastic material interbedded among the lava flows. In shape and form, however, Mazama differed strikingly from most other high cones in the range. Unlike Rainier or St. Helens, Mazama may never have been a single symmetrical cone. Instead, Mazama grew as an irregular cluster of overlapping composite cones and shield volcanoes. The oldest lavas exposed inside the caldera are those forming the Phantom Ship, a tiny island near Crater Lake's south shore. Potassium-argon dating has yielded an age of about 400,000 years for the Phantom rocks. The Phantom Cone was buried when larger nearby volcanoes in the Mazama complex spread over it and was subsequently exhumed when Mazama collapsed.

At 8926 feet, Mount Scott is the highest remaining cone of the overlapping volcanoes that composed Mazama. Despite its generally symmetrical form, it is also the oldest, 420,000 years. Pleistocene glaciers repeatedly formed on Scott, eventually carving away its western face and former summit, leaving a conspicuous cirque now facing lakeward. Like most of the different overlapping cones forming Mazama, Scott was probably built during a relatively brief cycle of intense activity and had become extinct long before Mazama's climactic eruptions.

During its early millennia, Mazama grew rapidly by voluminous eruptions from a number of related vents. Many of the early andesite flows, averaging only 20 to 30 feet thick, appear to have been emplaced in relatively quick succession, during eruptive episodes extending intermittently over a few years or centuries. Some of these old flows exposed in the west, south, and east caldera walls appear to be extensive sheets that covered large areas of Mazama's former surface. C.R. Bacon suggested that these fluid lavas erupted from fire fountains, jets of liquid rock that fed laterally wide-spread sheets of

Major Events in Mazama's Eruptive History

Approximate Age	Deposit or Event
Since 6845 B.P.	Latest activity on caldera floor: building of Merriam Cone; building of andesite lava platform east of Wizard Island, of Wizard Island, and of the rhyolite dome at the eastern foot of Wizard Island.
6845 B.P.	The climactic eruption in four main stages: (1) Violently explosive activity from a vent north of Mazama's main summit; (2) Formation of the first major pyroclastic flow, the Wineglass welded tuff; (3) Collapse of Mazama's former summit as many new vents discharged voluminous pyroclastic flows; (4) Steam explosions erupted the debris blanketing caldera floor.
6845 - 7015	Series of rhyolite eruptions at two vents along Mazama's north flank: (1) explosive eruption of pumice at Llao Rock vent, followed by massive lava flow blocking vent; (2) Cleetwood flow.
7015 - 30,000	Long dormant interval.
25,000 - 30,000	Eruption of Redcloud Cliff flow and the dome above Steel Bay; eruption of Grouse Hill flow and dome.
22,000 - 40,000	Building of Williams Crater cinder cone and eruption of associated flows, a complex mixture of basalt and more silicic magma.
30,000 - 50,000	Emplacement of Sharp Peak domes.
About 40,000	Eruption of silicic domes high on Mazama's southwest flank, triggers debris flows and pyroclastic flows in the Munson Valley and along the southwest slope.
About 50,000	Lava flows erupted on southwest and north flanks of the volcano.
About 70,000	Pumice Castle formed, followed by eruption of Scott Bluffs flow.
50,000 - 75,000	Lava flows, some probably under glacial ice.
About 70,000	Building of the Hillman Peak composite cone.
75,000 - 140,000	Lava flows, domes, and pyroclastic flow deposits of Meriam Pt.
110,000 - 185,000	Eruption of thin lava sheets of Llao Bay shield volcano.
200,000 - 300,000	Eruption of lava sheets of Cloudcap Bay.
About 320,000	Lava flows of Sentinel Rock volcano.
250,000 - 340,000	Lava flows of Dutton Cliff volcano.
340,000	Lava flow at Danger Bay.
400,000	Phantom Cone lavas and pyroclastic rocks.
420,000	Building of Mt. Scott composite cone.
About 600,000-700,000	Pre-Mazama rhyodacite flows and exposed domes east of Mt. Scott and south of Mt. Mazama.
More than 600,000	Basalt lava flows and related vents on the High Cascades.

Looking northeast across the mirror-still waters of Crater Lake. The Phantom Ship, the oldest rock exposed in the caldera, appears in the middle foreground; Mt. Thielsen's spire rises in the middle distance.
—Photo courtesy of the Oregon State Highway Department

agglutinated spatter. Similar kinds of eruptions occurred during Adams' early cone-building eruptions and are common today at the Hawaiian volcanoes.

These outpourings of molten rock built several shields, such as that exposed today beneath Llao Rock at Llao Bay. At other sites, some explosive activity accompanied the lava flows, forming a line of composite cones that stretched westward from Scott to Hillman Peak, the highest elevation on the caldera rim. Rising nearly 2000 feet above the present lake level, Hillman Peak was torn in two when Mazama collapsed, exposing its interior in perfect cross-section.

The rocks of Hillman Peak have been dated at about 70,000 years, but two of its conspicuous neighbors—Garfield Peak and Applegate Peak, east of Crater Lake Lodge, are significantly older and correspondingly more eroded. The popular trail leading from the lodge to the shattered summit of Garfield Peak affords not only spectacular views of the caldera but also good exposures of the volcanic bombs and other pyroclastic debris, interspersed with denser layers of andesite lava flows, that compose this part of the mountain. The highest remaining points of Mazama's primary south slope, Garfield and Applegate once extended an unknown distance toward Mazama's summit, but were beheaded when the old volcano collapsed.

Because it grew during and between repeated Pleistocene ice advances, Mazama may have supported glaciers for much of its existence. During prolonged dormant intervals between eruptions, substantial glacial erosion sometimes took place. At certain points on the caldera walls, as near Sentinel Rock, one can see glaciated canyons cut through layers of andesite. Ice-scratched and polished rock surfaces exist in many places in the caldera walls and along the southwest rim. Glacial trenches were typically refilled by thick flows of andesite lava or pyroclastic deposits when the volcano revived.

Evidence that large glaciers once descended from Mazama's summit can be seen in the three broadly gouged glacial valleys that indent the southern caldera rim. These canyons that scallop the volcano's southern flank—Munson Valley, Sun Notch, and Kerr Notch—were once the channels for mighty rivers of ice that flowed from Mazama's former summit area. Rock surfaces grooved or polished by the passing of glacial ice survive at various locations around the caldera, including sites just west of Crater Lake Lodge. Other clues to the vast extent of the icecap that accumulated on Mazama lie in the glacial moraines found as far as seventeen miles downvalley from the present caldera rim.

At the outbreak of the climatic eruptions, however, the canyons below the rim level were free of ice. Deposits at the head of Munson Valley and along the southwestern flank of Mazama that were formerly interpreted as glacial, are now known to have been left by pyroclastic flows.

During Mazama's development, numerous cinder cones erupted along its outer slopes. Crater Lake National Park contains at least 13 of these cones young enough to retain their original constructional forms; 11 others were built outside the Park borders. Williams believed that these small cones were parasitic vents that drew on Mazama's magma chamber, but Bacon suggests that they were general basaltic "background" volcanoes typical of the Oregon Cascades as a whole and not connected to Mazama's magmatic system. Williams Crater, formerly Forgotten Crater, a cinder cone about half a mile west of Hillman Peak, may be a partial exception. This partly eroded cone produced a variety of lava types, ranging from basalt to dacite, that indicates it may have tapped Mazama's magma chamber.

The mixing of different kinds of magma in the Williams' Crater complex is significant because it implies that by the time this cone was built a compositionally zoned magma body had evolved beneath Mazama. Far beneath the surface, profound changes were producing a large mass of highly silicic magma, a volatile-rich mixture that would eventually erupt catastrophically.

About 50,000 years ago the last series of cone-building lava flows erupted, sending streams of molten andesite down the north and southwest slopes of the volcano. Approximately 10,000 years later, viscous masses of dacite lava extruded high on Mazama's southwest face. The pasty domes formed during this episode collapsed or were shattered by steam explosions, precipitating avalanches of hot ash and angular blocks down Mazama's southern flanks, leaving conspicuous deposits at the head of Munson Valley and at various localities along the west rim as far north as the Devil's Backbone, the large dike exposed on the west caldera wall.

Mazama's magma reservoir continued to evolve an increasingly silicic lava. Between about 25,000 and 30,000 years ago, thick masses of rhyodacite magma oozed from the vent on the volcano's northeast flank, forming Redcloud Cliff. During the same period, another rhyodacite dome was emplaced above what is now Steel Bay.

The Redcloud flow looks like an inverted triangle, 600 feet high, atop the eastern caldera rim. Although at first glance it appears that this flow occupies an old stream valley, it actually fills a funnel-shaped explosion crater. Before the lava was extruded, explosions

Relief map of Crater Lake region.

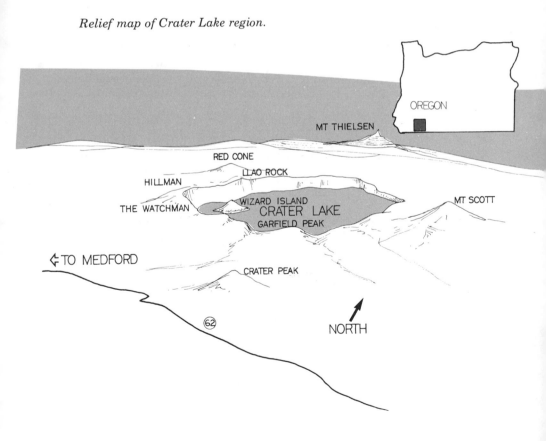

110

Pumice northwest of Bend, Oregon. —photo courtesy D.W. Hyndman

blasted open an almost vertically-walled crater, from which quantities of rhyolite pumice were ejected. The lava that later welled up to build a dome over the Redcloud vent was ultimately sliced nearly in two by the subsidence of Mazama's summit, transforming its southern face into a sheer precipice.

Following eruption of the Redcloud flow, Mazama apparently lapsed into a sleep lasting 20,000 years or more. Again, giant ice streams formed, deepening the cirques and canyons on the volcano's flanks. About 12,000 years ago, the last Pleistocene glaciers melted, their snouts shrivelling upslope above the elevation of the present caldera rim. To American Indians who inhabited the region during the post-glacial period, it must have seemed that Mazama had entered its final rest. Llao, the Klamath peoples' god of the underworld who was believed to dwell inside the volcano, was apparently at peace. Then in rapid succession, Mazama produced the climactic activity that, in tribal myth, decapitated Llao and almost leveled the mountain that housed him.

Mazama Awakes

Mazama's arousal about 7,000 years ago was abrupt and violent, a warning of what was to come. Again, activity centered on the northern flank of the mountain, possibly because this area, north of the main summit, was the region of lowest elevation above the magma chamber.

The first eruptions were at the site of Llao Rock. Explosions ripped open a new crater in the broad glacial basin that then occupied Mazama's north flank. Huge volumes of rhyodacite pumice were

111

hurled high into the air, where the winds carried the ashcloud many miles to the east and southeast, blanketing a large area with pumiceous ash. After the magma had blown off steam, the vent flooded with rhyodacite lava, which overflowed the crater rim and oozed a mile downslope. About 1.25 miles wide, this flow is as much as 1200 feet thick, with a volume approximating a quarter of a cubic mile. When Mazama collapsed, the head of the flow was bisected, revealing in cross-section its dark gray mass, now called Llao Rock. Rounded at the bottom, with lateral extensions near the top, it resembles a primordial bird of prey, crouched with wings outstretched atop the caldera rim.

Radiocarbon dating of charcoal twigs taken from beneath the Llao pumice deposit yielded an age of 7015 years, give or take 45, a scant century or two before the final outburst that destroyed Mazama. The Cleetwood flow, the last eruptive event before the culminating cataclysm apparently erupted only weeks or months before Mazama's destruction. The Cleetwood rhyodacite flow oozed from a vent near the present north caldera rim. When Mazama collapsed, the flow was still fluid enough near its vent to spill down inside the caldera wall to what is now lake level. This rare backflow phenomenon was first recognized by the pioneer geologist J.S. Diller in the late 19th century, but was then mistakenly interpreted as a feeding tube for the Cleetwood flow by later investigators. The Cleetwood flow was still hot during the culminating explosive eruptions; pyroclastic material blanketing the flow surface was chemically altered by gases rising from the lava.

Mazama's Appearance Before Its Destruction

Impressed by learning that a great mountain was suddenly transformed into an enormous hole in the ground, Park visitors sometimes try to make a good story even better. Greatly exaggerated estimates of Mazama's former height circulate freely. Informed by rangers that Mazama was once one of the loftiest peaks in the Cascade Range, tourists soon tell each other—and the folks back home—that it was 14,000, 15,000, or 16,000 feet high, taller than Shasta or Rainier. Stimulating to the imagination as these speculations may be, they are not supported by the available evidence.

To help determine Mazama's probable former height, Howel Williams compared the remaining slopes of the volcano's cone with those of other large Cascade volcanoes. Ignoring the secondary peaks and other formations along the caldera rim, Williams found that the primary surface of Mazama is now approximately 8000 feet high on the south side, 6000 feet on the north, dipping slightly below the current lake level, and at intermediate elevations along the other caldera walls. By hypothetically removing the summits of comparable Cascade peaks and imposing an imaginary caldera on them, Williams discovered that Adams, 12,286 feet, came closest to matching Mazama in size and girth. He concluded that Mazama, at its maximum height, stood about 12,000 feet above sea level.

By the time of the culminating eruptions, glacial erosion and subsidence caused by explosive activity may have significantly reduced the mountain's original stature. In a 1968 report, Williams revised his previous estimates downward, suggesting a final elevation of about 10,000 feet.

Mount Mazama begins its catastrophic eruption.

Crater Lake and Wizard Island. The caldera walls on the left rise almost 2000 feet above the lake surface. The dark, massive lava flow forming Llao Rock tops the caldera rim on the right. Mt. Bailey in the middle distance.
—Photo courtesy of the Oregon State Highway Department

C.R. Bacon's work indicates that Mazama's former profile was highly irregular, consisting of a line of variously eroded composite cones stretching across what is now the southern part of the caldera. Since the highest remaining cone, Scott, is still nearly 9000 feet high, it seems likely that some of the much younger stratovolcanoes built farther westward rose considerably higher.

The highest part of the Mazama complex, which was extremely asymmetrical, stood somewhat south of the present lake center. The eroded northern face of the volcano was apparently a large glacial basin. The size and extent of glaciers mantling Mazama just before its fall are unknown. If any existed during the climatically warm period of 7000 years ago, none extended below the present caldera rim.

The Climactic Eruptions

We may never know precisely what Mazama looked like just before its destruction, but we have ample testimony about the nature and power of the outbursts that beheaded Llao's mountain. By studying the composition and areal extent of the deposits laid down during the culminating blasts, geologists can reconstruct much of what happened. Besides the physical evidence the eruptions left behind, we have the example of similar eruptions that formed calderas during modern times. In 1883 the Indonesian volcano, Krakatau, staged one of the most violently explosive outbursts in recorded history. Like

Mazama, Krakatau collapsed after expelling such vast quantities of pumice that the peak foundered as its underlying magma chamber emptied.

The opening blasts that heralded Mazama's doom began as a crater somewhere north of the principal summit ejected a titanic mushroom cloud miles into the stratosphere. Winds carried the ash plume northeast, blanketing over 500,000 square miles. Near Mazama's base, pumice accumulated to depths of 20 feet, while 70 miles northeast of the volcano the initial ashfall was a foot thick. Thick sheets of rhyodacite pumice that fell from the plinian eruption column can be seen at the caldera rim from Hillman Peak clockwise to the vicinity of Sentinel Point. Most of the pumice deposits exposed in cuts along the Cleetwood Cove trail were derived from this stage of the climactic eruption and from pumice ejected slightly earlier from the Llao vent.

In 1883 winds carried ash from Krakatau around the globe, as they did from the much smaller eruption of St. Helens in 1980. The

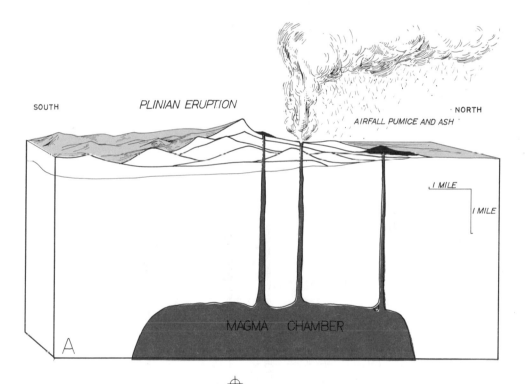

Formation of Crater Lake. A. Stage 1: A Plinian eruption column is ejected vertically from a single vent north of Mt. Mazama's main summit. Ashfall extends hundreds of miles to the northeast.

Mazama ashcloud was certainly larger than that of any historic eruption and spread ash over a vastly larger area, including virtually all of Oregon, Washington, Idaho, northern California, western Montana, as well as parts of Nevada and Wyoming. Much of southern Canada also was blanketed, including southern British Columbia, Alberta, and Saskatchewan.

The ash wafting through the stratosphere must have produced brilliant sunsets throughout the entire northern hemisphere. Ancestors of the Druids in England and Gaul may have observed this atmospheric phenomenon and wondered what it portended.

Native Americans close at hand experienced the cataclysm more directly. Their artifacts have been found buried beneath Mazama ash and their myths interpreting the great eruption as a battle between Llao and his rival the sky god Skell preserved memories of the event for many generations. In the late 19th century an aged chief of the Klamath tribe recounted the tale, including the collapse of Llao's mountain, to a soldier at Fort Klamath.

Although Mazama's rain of fire created an ashen desolation for hundreds of miles north and northeast of the volcano, the southwesterly winds allowed only a thin sprinkling of pumice to litter areas

Formation of Crater Lake. B. Stage 2: Overburdened with fragmental material, the Plinian column collapses, triggering the first major pyroclastic flow (the Wineglass Welded Tuff).

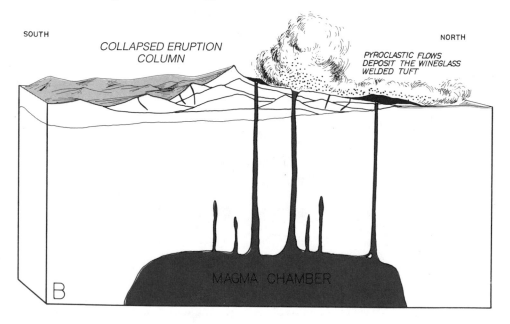

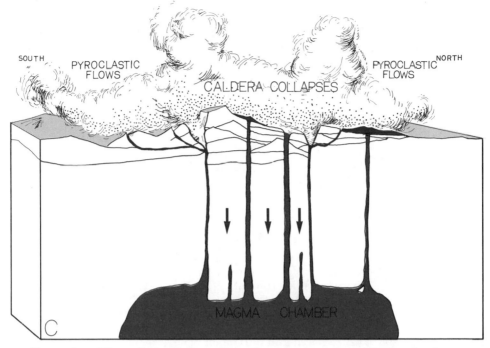

Formation of Crater Lake. C. Stage 3: As the underground magma chamber is drained, Mt. Mazama's former summit collapses. New vents open along "ring fractures," discharging voluminous pyroclastic flows that sweep down all sides of the volcano.

west of the peak. An observer standing near Union Peak, only five miles away, would probably have enjoyed a relatively safe, if terrifying, view of the event.

The eruption abruptly changed in character during the next phase of the activity. The towering column of gas and ash, issuing from a vent north of the main summit, collapsed as the vent apparently widened. This collapsed eruption column, burdened with the enormous volume and weight of erupting pumice, deposited huge quantities of incandescent pumice fragments on the north slope, forming a pyroclastic flow that poured northward between Llao Rock on the west and Redcloud Cliff on the east. Emplaced at high temperatures that welded the glassy pyroclastic fragments together, the initial pyroclastic flows—the first of many—now form a conspicuous deposit on the north caldera wall. Known as the Wineglass Welded Tuff, it contains large blocks of orange-brown to gray rock, and is well exposed near the top of the Cleetwood Cove trail.

The Wineglass pyroclastic flows were but a prelude to what followed. So much magma had been ejected from Mazama's subterranean reservoir that much of Mazama collapsed. As the mountaintop foundered, a series of concentric fractures opened around the summit block, creating a chain of new vents that encircled the volcano. Explosions began anew at these ring fractures, tearing out fragments of old rock from deep within the cone and spewing huge volumes of frothy pumice. This third stage of the eruption produced massive pyroclastic flows that rushed down all sides of the volcano at dizzying speeds. Triggered also by a collapsed column overburdened with rock fragments, these glowing avalanches surmounted high ridges, even the summit of Scott, and then plunged headlong into the glacier-cut valleys radiating away from Mazama. Divided into many branches beyond the base of the volcano, they raced outward through forested valleys.

The abrasive pyroclastic flows eroded most of the surfaces over which they passed, removing much of the airfall pumice from Mazama's upper slopes. Although they scoured all sides of the mountain, below Mazama's base the pyroclastic flows were largely confined to topographic depressions. Turbulent ashclouds broiled upward along the flow fronts and margins, projecting searing ash over neighboring ridgetops. Following the twisting valley of the Rogue River, one

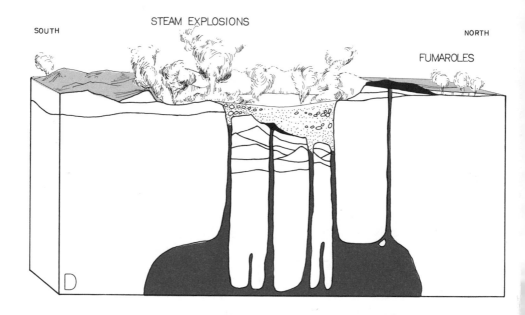

Formation of Crater Lake. D. Stage 4: During the final phase, steam explosions deposit large quantities of ash, partly filling the caldera.

pyroclastic flow traveled 40 miles from its source, mowing down and incinerating thick stands of timber before coming to rest near McLeod. Another pyroclastic flow containing pumice and dense rock fragments descended to the north, sweeping across Diamond Lake and depositing its burden of pumice bombs and fine ash into the valley of the North Umpqua River. One arm of this flow was deflected westward down Lava Creek and the Clearwater River, leaving a stratum of pumice 20 to 30 feet thick.

Pyroclastic flows that moved east sped over 25 miles of flat ground beyond the base of the volcano. Pumice blocks six feet across were carried 20 miles from their source; at least one lump 14 feet long was deposited at Beaver Marsh, near U.S. Highway 97. Similar pumice flows traveled south down Annie and Sun Creek canyons, filling them to a depth of at least 250 feet. The basal deposit of these flows, consisting of old andesite fragments blown from vent walls, has not yet been fully exposed. Southeastward, the pyroclastic flows raced down Sand Creek, continuing for more than ten miles across the flatlands beyond Mazama's base. Some flows reached into Klamath Marsh, from which masses of floating pumice washed down the Williamson River into the Klamath Lakes.

As this paroxysm ended, Mazama apparently tapped ever-deeper levels of its magma chamber. The final pyroclastic flows carried volumes of smoke-gray scoria, which formed a contrasting layer atop the pale buff and yellowish pumice erupted earlier. Although these andesitic scoria flows did not spread so far as those transporting the lighter colored pumice, on the north and northeast flanks of Mazama they created the arid pyroclastic field known as the Pumice Desert and inundated the wide valley of Desert Creek. Some scoria flows reached out toward the plains bordering Klamath Marsh. Especially in the center of the valleys down which they traveled, these scoria flows laid down a conspicuous dark band capping the more voluminous deposits of rhyodacite pumice. The dark upper layer is particularly prominent where streams have since cut narrow gorges into the valley fills.

Long after the pyroclastic flows came to rest, they remained extremely hot. Gases seething from the accumulations of pumice, some of which attained a thickness of 250 feet or more, produced cylindrical vents or fumaroles in the deposits. The walls of these natural flues were cemented and hardened by the hot vapors so that they became relatively resistant to later erosion. When rainfall and streams finally cut through the soft pumice beds, they left the hardened pipes standing as columns and spires. Outstanding examples of such fossil fumaroles now rise prominently along the upper walls of Annie and Sand Creek canyons.

Immediately following the explosive eruptions, each of the pumice-choked canyons extending from Mazama must have resembled a Valley of Ten Thousand Smokes. Writhing plumes of acidic gas issued from hundreds of fumaroles in the deposits; when rain fell, immense columns of steam rose from the slowly congealing masses of pumice and scoria. The rapidly settling ashclouds that had accompanied the pyroclastic flows added to the volcanic haze. In some places, fallout from the ashclouds accumulated to depths of many feet. In the meantime, Mazama—in a series of weak, dying explosions—deposited a final layer of pumice lapilli and crystal-rich ash around the caldera rim. These last ejecta form well-bedded drifts up to 50 feet thick along the caldera rim.

When all this volcanic smog finally dissipated or drifted away, the native American survivors of the holocaust must have rubbed their eyes in wonder. Not only was the once-green land transformed into a sea of dirty-gray pumice, but the huge mountain responsible for the catastrophe had virtually disappeared. Where a snow-capped peak once towered, there was now only a colossal depression, five to six miles wide and nearly 4000 feet deep. The ash-crusted Indian of 6850 years ago must have asked himself the same question that the visitor to Crater Lake does today. Where did Mazama go? Was it blown apart in the earthshaking detonations? Or did it subside into a subterranean pit of its own making?

And what of the devastating pyroclastic flows? What enabled them to travel so far from their source? How were they able to move so fast, at speeds exceeding 100 miles per hour?

According to the present understanding of Mazama's pyroclastic flows, they achieved their astonishing mobility through a combination of several factors: (1) the momentum derived by their falling from the eruption column and plunging down the slopes of Mazama; (2) the continuous discharge of hot gas from the millions of molten particles they contained; and (3) the heating and expansion of air trapped within the flow. As it condenses from an eruption column, a pyroclastic flow apparently overrides a wedge of air, incorporates what it can't push ahead or aside, and heats the entrained air, causing it to expand and thus keeping particles in suspension and giving further impetus to the flow's outward movement. Buoyed up and lubricated by trapped gas within the flow, propelled by the myriads of miniature explosions within them, the pumice flows were able to travel many miles, even after they reached level ground or had overrun forests.

Considering the enormous volume of material evacuated from the magma chamber beneath Mazama, it is no wonder that the mountain collapsed. In calculating the difference between the volume of new

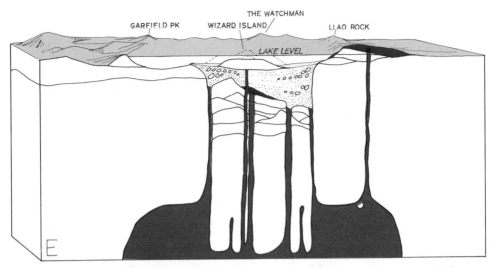

SOUTH

NORTH

THE WATCHMAN

GARFIELD PK   WIZARD ISLAND   LLAO ROCK

LAKE LEVEL

E

Formation of Crater Lake. E. The present: Post-caldera eruptions have built Wizard Island and other submerged cones on the caldera floor. Rain and snow melt have half filled the basin, forming Crater Lake.

magma expelled and the volume of the vanished mountaintop, it is necessary to remember that magma in its liquid state, before being erupted, is much less voluminous than when discharged as frothy, gas-charged pumice. C.R. Bacon estimated that about 11 to 14 cubic miles of fresh magma erupted during the precursory and culminating eruptions. This figure agrees closely with Williams' estimate of the volume of Mazama's missing summit. The amount of material erupted from the magma chamber roughly equals the assumed bulk of Mazama's vanished cone.

The Caldera Floor

Immediately after its creation, the caldera must have been an awesome sight. Enclosed by almost vertical cliffs towering up to 4000 feet above the chaotic jumble of giant blocks that littered the caldera floor, the basin probably seethed with escaping gases and bubbling pits of sulphurous mud. Yawning fissures crisscrossed the block-strewn floor, and avalanches of loosened rock thundered down from the precipices high on the caldera walls.

Today, soundings of Crater Lake indicate that parts of the caldera floor are remarkably smooth. Several geologic processes may have

121

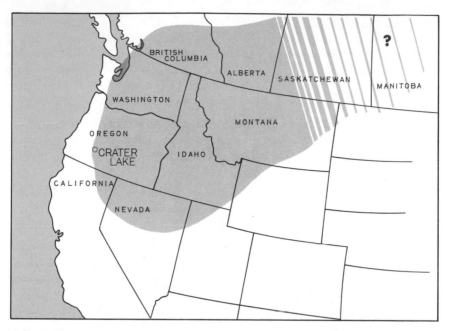

Map showing the areal extent of Mt. Mazama's climactic ash fall, which blanketed virtually all of the Pacific Northwest and extended as far northeast as Saskatchewan, Canada.

played a part in creating this condition. Molten rock may temporarily have formed a lava lake similar to Kilauea's famous fire pit, which filled in many chasms and depressions on the basin floor. Or perhaps the smoothness of the present lake bed may reflect thick layers of ash. Large quantities of rock, pumice, and ash probably fell back into the caldera during Mazama's collapse. Deposits of ash settling from the pyroclastic flows also may have created a layer of tuff, composed of densely compacted fine ash. In addition, a 1979 U.S. Geological Survey seismic-reflection study, which included taking core samples from the lake bottom, revealed that about 350 feet of sediment fills depressions in the caldera floor. The sedimentation occurs as rock debris cascades from the surrounding cliffs and spreads over the lake floor.

Mazama's Post-Caldera Activity

Since Mazama's collapse, all volcanic activity has been restricted to the caldera interior. Some of the first eruptions after the caldera formed built a broad platform of andesite lava that rises 1200 feet above the basin floor. Following this phase, mild explosions built the almost perfectly symmetrical cinder cone that forms Wizard Island, so called because it suggests a sorcerer's hat.

The youngest formation may be a small rhyodacite dome emplaced underwater at the northeastern foot of Wizard Island's cone. A larger subaqueous structure, age unknown, is Merriam Cone, which stands close to the north caldera wall. Approximately a mile across the base and 1320 feet high, this andesitic pile is almost precisely the same size as Paricutin, the famous Mexican volcano.

The eruptions that built the lower part of Wizard Island seem to have occurred underwater. It seems that the lake surface had already risen to within about 300 feet of its present level when the cone formed.

Judging only by the fresh appearance of Wizard Island's charcoal-gray, orange and rust-streaked slopes, the cinder cone would seem to have been created only a few centuries ago. Despite its youthful appearance, however, it is probably about 6000 years old.

An easy climb up the 763-foot cone rewards the hiker with an opportunity to inspect one of the most interesting and best preserved craters in the Cascades. Near the crater lip is a miniature lava flow apparently formed by semi-molten rock being sprayed into the air and accumulating to form a small agglutinated stream. Extremely thin and narrow, this tiny flow forms a black clinkery splotch on the southwest crest of the cone. Another mass of dark lava, forming a small humped dome on the crater floor, may represent an incipient lava flow that failed to rise high enough to stream over the crater rim. Instead, it became a conduit filling and now plugs the eruptive vent.

Large lava blocks perched atop the narrow crater rim are six or more feet in diameter, and occur in clusters at several points along the eastern rim. It is difficult to believe that such large blocks were blown from Wizard Island's relatively small crater, only 300 feet in diameter and 90 feet deep, especially since most of its ejecta, with the exception of a few bombs and blocks, is smaller than a man's fist. But there seems no other way to account for their presence except as the product of final, weak explosions that discharged fragments of lava already solidifying in the volcano's throat.

Perhaps some time after the young cones and flows were emplaced, water from rain and melting snow began to fill the basin. At first a network of small pools and connecting channels must have formed. But eventually the expanding waters joined and even rose about 50 feet higher than the present level of the lake. How long it took to fill the enormous caldera to its present depth is not known. The water level has probably fluctuated greatly during the past several thousand years. During the first part of the 20th century, the water dropped about 40 feet. Today, the National Park Service reports that water added by precipitation almost perfectly equals that lost by

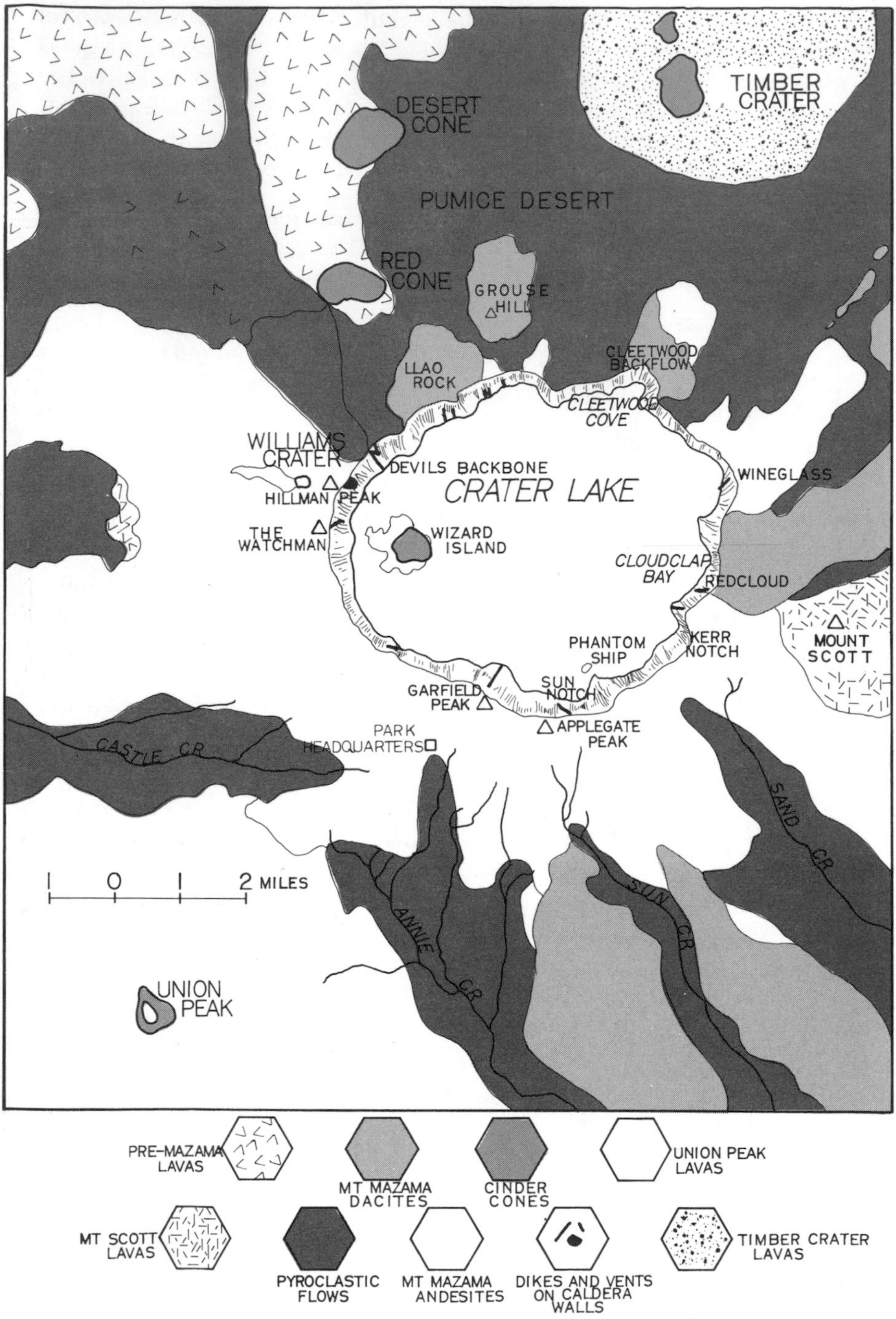
TIMBER CRATER
DESERT CONE
PUMICE DESERT
RED CONE
GROUSE HILL
CLEETWOOD BACKFLOW
LLAO ROCK
CLEETWOOD COVE
WILLIAMS CRATER
DEVILS BACKBONE
CRATER LAKE
HILLMAN PEAK
WINEGLASS
THE WATCHMAN
WIZARD ISLAND
CLOUDCLAP BAY
REDCLOUD
MOUNT SCOTT
PHANTOM SHIP
KERR NOTCH
GARFIELD PEAK
SUN NOTCH
APPLEGATE PEAK
PARK HEADQUARTERS
CASTLE CR
SAND CR
SUN CR
ANNIE CR
1 0 1 2 MILES
UNION PEAK
PRE-MAZAMA LAVAS
MT MAZAMA DACITES
CINDER CONES
UNION PEAK LAVAS
MT SCOTT LAVAS
PYROCLASTIC FLOWS
MT MAZAMA ANDESITES
DIKES AND VENTS ON CALDERA WALLS
TIMBER CRATER LAVAS

Map of the Crater Lake area. Dacite flows and domes form the north caldera rim: Llao Rock, Grouse Hill, and the Cleetwood flow, all of which were erupted shortly before the climactic events that beheaded Mt. Mazama.

subterranean drainage and evaporation. Except for minor seasonal changes, the lake level now remains almost constant. With its maximum depth officially measured at 1932 feet, Crater Lake is the second deepest in North America, ranking next to Canada's Great Slave Lake. The extreme depth and purity of the water allows all colors of the light spectrum to be absorbed, except for blue, which gives the lake its incomparable indigo tone.

The Future of Crater Lake

The tranquil beauty of Crater Lake makes the viewer wish it never to change. Nature, however, does not permit even her greatest triumphs to remain unaltered for long. The future will bring changes of unknown kind and extent. Even if old Mazama were never to erupt again, erosion will eventually breach the caldera walls and drain the lake. Prolonged drought could diminish the lake or perhaps dry it up. New glaciers like those of the great ice ages could grind down the encircling cliffs and level even the ruins of Mazama's cone.

Before any of these eventualities occur, it is possible that Mazama may reawaken. It has been active sporadically for perhaps half a million years. It has seen ice ages come and go and has revived to deposit new lava atop its ice-carved slopes. The date of its latest activity only a few thousand years ago implies that Mazama has not finished its volcanic career.

Future eruptions are likely to occur inside the caldera, erecting new cinder cones like Wizard Island or emitting lava flows from fissures on the lake floor. Some eruptions will probably take place entirely underwater, while others may build new islands, shooting clots of red hot lava over the caldera rim or splashing them into the lake. Eventually vents may open on Mazama's outer flanks, creating additional cinder cones or lava domes.

Comparison of Mazama's caldera with that of Newberry Volcano suggests that Crater Lake ultimately may be divided into two or more bodies of water, separated by ridges of volcanic cones and flows. Or a new series of cone-building eruptions may create a cluster of andesite and dacite stratovolcanoes that will tower above the caldera rim, gradually burying it beneath their ejecta and restoring Mazama to its former lofty status.

Recent studies of other Mazama-like volcanoes around the world suggest that more violently explosive eruptions are possible in the future. Since the climactic outburst about 6900 years ago, the volcano has produced a wide range of lava types, the latest a return to the highly silicic rhyodacite erupted shortly before and during the caldera-forming debacle. Some volcanoes produce a number of caldera-forming paroxysms; Mazama may prove to be one of them!

Although geologists can not forecast how the volcano will behave in the future, the intrusion of even relatively small volumes of gas-rich magma into the water-filled caldera may produce some highly explosive events. When Llao next stirs from his subterranean hibernation, we can be confident only of being surprised.

Exploring Crater Lake

Much of Crater Lake can be enjoyed from your car. Take the 33-mile Rim Drive and stop at the roadside observation points.

A short uphill hike eastward from Crater Lake Lodge to Garfield Peak, 1.5 miles, offers a fine vantage point almost 1900 feet above the lake surface.

A somewhat longer, 2.5 miles, and steeper trail leads from the eastside Rim Drive to Mt. Scott, 8926 feet, highest elevation in the park. From the deck of a fire lookout there, one can see 100 miles in every direction, including the entire extent of Mazama's caldera.

Cleetwood Trail, one mile, descends the northern caldera wall to Cleetwood Cove, from which launch trips to Wizard Island and around the lake run during the summer season, beginning in late June or early July, depending upon snow conditions. Signs mark a parking lot on the north Rim Drive where the trail starts. Although perfectly safe if followed cautiously, these trails have been the scene of several fatal accidents involving persons who ran blindly or tried to take short cuts.

Both the Lodge and the Park Headquarters distribute maps and trail guides to numerous other points of interest in the Park.

Several of central Oregon's "Matterhorn" peaks, Mount Washington is in the foreground, Three Fingered Jack in the middle distance, and Mount Jefferson on the horizon.

—Photo by Delano Photographics

X
Mt. Thielsen:
Lightning Rod of the Cascades

Unique to the Oregon Cascades is a series of high pinnacles whose sheer pointed summits remind travelers of Switzerland's Matterhorn. Two of these spires can be seen from many points along the rim of Crater Lake: Union Peak, 7698 feet, to the south, and Thielsen, 9178 feet, 12 miles north. Still farther north in the range, between the Three Sisters and Jefferson, are the strikingly similar peaks of Washington and Three Fingered Jack. Because they all rise from broad, gently sloping bases to extremely steep summits with no trace of a summit crater, these mountains appear unlike any of the other volcanoes in the High Cascades.

Several geological processes are responsible for the difference between such peaks as Thielsen and the more conventional looking cones of Hood, McLoughlin, and the Three Sisters. These include the manner in which Thielsen and its fellows were constructed, their relatively greater age, and their more advanced state of dissection.

Thielsen, Washington, Union Peak, and Three Fingered Jack not only belong to an older period of Cascade volcanism, they apparently ceased cone-building sufficiently far back in Pleistocene time that no fresh material erupted to repair the ravages of repeated glaciation. All these volcanoes were probably extinct before the last two or three major ice ages, so they are more extensively eroded than younger Cascade volcanoes.

Oregon's unusual pinnacles did not receive their peculiar form solely through glaciation, however. The nature and sequence of the eruptions that raised them also contributed to their obelisk-like profiles. Because each of the cones in question apparently underwent similar stages of growth, what holds true for one also seems to be valid for the others. Thanks to pioneer investigations by J.S. Diller in

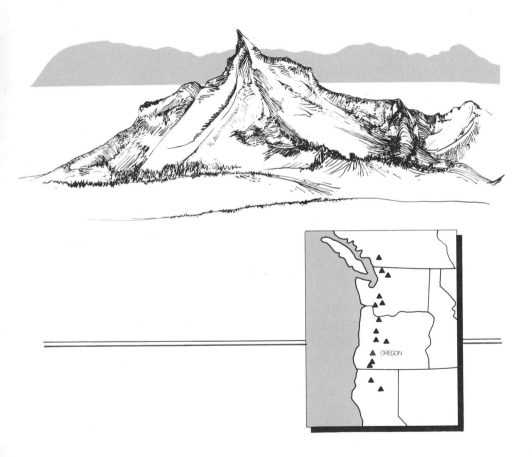

Mount Thielsen as seen from the south.

128

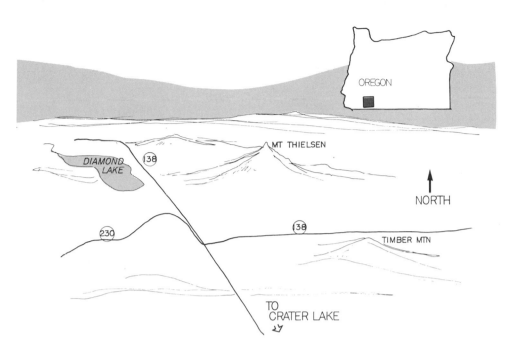

Relief map showing the Mount Thielsen region.

1902, Howel Williams in 1933, and David Sherrod from the University of California, Santa Barbara, Thielsen is the best studied example, and makes a good representative choice.

Thielsen rises above the east shore of Diamond Lake; directly opposite, dominating the western shore, stands Bailey, a younger stratovolcano that is much less eroded. Thielsen's eastern slopes extend down to the basaltic plateau of central Oregon; southward they end against a complex of basalt volcanoes.

Whereas most of the geologic literature refers to Thielsen as a basalt shield surmounted by a pyroclastic cone, it is actually a small composite cone standing on an older shield volcano. Like the majority of prominent cones in the High Cascades, Thielsen's lavas have a normal magnetic polarity, which probably indicates that they erupted less than about 700,000 years ago, although it is conceivable that the volcano might have formed during much earlier periods of normal polarity, such as the short-lived Jaramillo event between about 900,000 and 970,000 years ago.

Thielsen also resembles several other Cascade volcanoes in that its initial eruptions were explosive and produced a large pyroclastic structure that comprises the core of the volcano. Pleistocene glaciers that cut deeply into the cone of McLoughlin, Crater Peak, a few miles

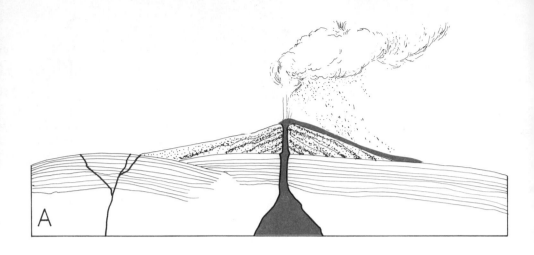

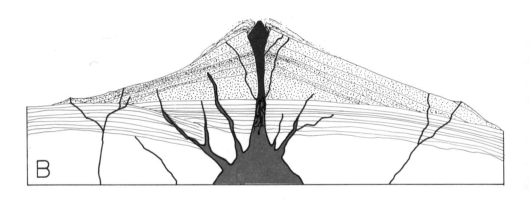

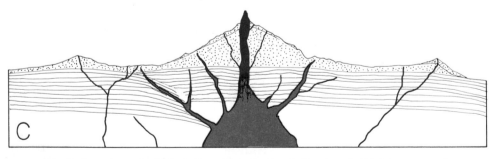

Evolution of a typical High-Cascade volcano: A. During the Pleistocene, explosive eruptions erect a medium-size cone on top of a late Pliocene or early Pleistocene shield volcano. B. After lava flows from the central and peripheral vents have largely covered the pyroclastic cone and partly buried the older shield, a swarm of dikes and a massive plug invade the cone. C. Glacial erosion cuts deep into the volcano, demolishing much of the summit cone and exhuming the central plug as a free-standing spire.

A close view of the massive character of the lava plug forming Mt. Thielsen's summit. Repeatedly scarred by lightning, it is known as the "Lightning Rod of the Cascades." —Photo by Ed Cooper

northeast of Lassen, Three Fingered Jack, and Jefferson, exposed their interior structures. All are revealed to have begun by erecting sizable edifices of pyroclastic material. In most cases, the original fragmental pile was buried by later lava flows as the eruptive sequence moved from eruption of pyroclastic material to lava flows. Thielsen apparently has a similar history, although the central cone was never entirely sheathed in lava.

According to Sherrod, Thielsen first built a pyroclastic cone at least 1000 and perhaps as much as 3000 feet high. Although very few flows were exposed in the main core, peripheral eruptions at vents half a mile from the central conduit simultaneously discharged thin flows of basaltic andesite. The presence of cinders and scoria among these generally thin and fluid lavas implies that fire-fountaining accompanied their emplacement. Sprays of magma from numerous satellite vents fed many flows that traveled several miles from their source.

The final act in creating Thielsen was the intrusion of a massive plug of basaltic andesite, half a mile in diameter, into the central conduit. A second, lower plug stands about a mile east of that filling Thielsen's main vent, but it belongs to an older basaltic andesite volcano that Thielsen's eruptive products partly bury.

Scored or grooved surface of viscous lava rising as a plug in the throat of a volcano west of Bend, Oregon. —photo courtesy D.W. Hyndman

Thielsen's summit was brought to its present needle-like state by erosion that stripped away much of the friable pyroclastic material surrounding the central plug, exhuming the precipitous summit pyramid. Glaciers scoured large basins into all sides of the cone, leaving broad cirques separated by thin rocky spurs radiating away from the mountain's apex. As moving ice demolished the upper part of the main cone, and also bit into the plug, meltwater and streams carried away the debris to give the mountain its distinctive Matterhorn appearance.

Thielsen's lofty spire attracts countless lightning bolts. It has been struck so often that the mountain is known as "the lightning rod of the Cascades." Over the years repeated electrical charges have formed on the summit rocks a peculiar substance called fulgurite. Derived from the Latin term for thunderbolt, fulgurites form coatings on rock surfaces or small carrot-shaped tubes inside the rocks. These lightning tubes form as intense heat fuses the volcanic rock.

On Thielsen, the fulgurites appear as brownish-green glass that resembles greasy splotches of enamel paint. Sometimes the fulgurites also line small holes and crevices in the rock. According to climbers, most are in the top five or ten feet of the highest pinnacle.

Mt. Bailey

Facing Thielsen across the shallow waters of Diamond Lake is broad-shouldered Bailey, 8362 feet. Although younger than its pyramidal neighbor, Bailey has been modified by glacial erosion,

132

which scooped out large cirques and gullies along its flanks. In a 1978 report, Calvin Barnes described the 2000-foot high main cone as resting on a basaltic andesite shield, elongate to the north. Most of the eruptions that built Mt. Bailey produced flows of andesite about seven feet thick, although the steep upper part of the cone consists of stubbier tongues of scoriaceous lava. A large summit crater, over 1600 feet across and about 350 feet deep, later filled with reddish orange scoria and other pyroclastic material. A secondary vent north of the present summit covered the volcano's crest with black scoria, while several scoriaceous flows issued from another lateral vent on the southern flank. Apparently the last of Bailey's summit eruptions formed an explosion crater, approximately 350 feet wide and 100 feet deep, about 3500 feet south of the present summit. Late activity along the volcano's flanks include formation of Rodley Butte, a cinder cone which produced at least ten flows of olivine andesite.

Diamond Peak

Like Bailey, Diamond Peak, approximately 20 miles to the north, is a composite cone of basaltic andesite, but much larger. It consists of at least two overlapping cones. The older edifice underlies the volcano's lower northern summit. A second major vent then built a somewhat younger cone to the south, forming the true summit, 8750 feet. The broad, serrated crest has been significantly modified by glaciation and the mountain still supports small glaciers.

Mt. Washington

Motorists crossing the barren lava wastes at central Oregon's McKenzie Pass cannot fail to notice the towering summit pinnacle of Washington. From some angles Washington's spire is sharp enough

Vesicular edge of black basalt flow along the Rogue River Gorge, Union Creek, Oregon. —photo courtesy D.W. Hyndman

to be another Cleopatra's Needle; from others it bears a general resemblance to Sugarloaf Mountain which guards the famous harbor of Rio de Janeiro.

Like Thielsen, Washington is a composite cone consisting of a central pyroclastic edifice and attendant lava flows into which a huge basaltic andesite plug intruded. Washington has apparently been extinct for a long time, so that glaciers have largely demolished the upper part of the mountain, leaving only the sheer obelisk of its central plug. Narrow ridge-like outcrops of the ravaged summit cone radiate from this volcanic neck much as they do at Thielsen.

Three Fingered Jack

Northernmost of the Matterhorn volcanoes, Three Fingered Jack, 7841 feet, is as picturesque as its name. Williams' reconnaissance of the mountain concluded that its development almost exactly parallelled that of Oregon's other eroded volcanic necks which it so closely resembles. Unlike those of Union Peak or Washington, however, its summit—a long saw-toothed ridge with a generally north-south axis—does not consist of a central plug, but of loose tephra deposits underlain by a ten-foot thick vertical dike. Three Fingered Jack's rapidly disintegrating summit pinnacles are so precarious that climbers say the mountaintop shudders in the wind.

Glaciers have cut so deeply into the volcano that its eruptive history is clearly displayed. A 1980 study found that Three Fingered Jack, built on older shield lavas, consists of several overlapping cinder and composite cones which, with their associated flows, cover an area of about 34 square miles. The main edifice, about 1000 feet west of the initial tephra cone, is massive, light-gray flows of basaltic andesite interbedded with layers of multicolored pyroclastics. Late in its history, the volcano became more explosive, producing a larger proportion of tephra so that the upper part of the cone is chiefly fragmental material, interspersed with progressively thinner lava flows, the last averaging only three feet thick. Another cone rose about 350 feet south of the main vent, while several secondary craters erupted both lava and pyroclastics, extending the long axis of the volcano to the north and south.

The summit cones were repeatedly invaded by radial dikes and viscous plugs which form the volcano's exposed core. Perhaps after the central conduit had been permanently sealed, olivine basalt lava flows poured from vents on the north and south flanks.

Cone-building eruptions had probably ceased before the earliest of the last three Pleistocene glaciations occurred, about 200,000 years ago. Glaciers excavated large basins on all sides of the structure, particularly the east and northeast flanks. The northeast face presents an almost vertical wall of rotten, crumbling rock.

The degree of erosion can not always correctly reveal the relative ages of adjacent volcanoes. Three Fingered Jack and Washington, although thoroughly dissected, are younger than their almost perfectly symmetrical neighbor, Black Butte, 6415 feet. Rising 3200 feet above the High Cascade plateau at the southern end of the Green Ridge fault zone, this basaltic composite cone stands in the rain shadow of mountains farther west. Not enough snow accumulated on its slopes during the ice ages to form glaciers. Despite its youthful profile, Black Butte's slopes are etched by deep ravines, its surface rock bears thick weathering rinds, and its lavas possess reverse paleomagnetic polarity, indicating that it was built sometime before about 700,000 years ago.

Volcanic breccia or agglomerate.
—Photo courtesy of D.W. Hyndman

This view from Paulina Peak on the south rim of Newberry Caldera shows, left to right, Paulina Lake, Central Pumice Cone, and East Lake. A light snowfall emphasizes the wrinkled surface of the Big Obsidian Flow (right center).
—Photo by Delano Photographics

XI
Newberry Volcano:
New Sources of Geothermal Power

Newberry Volcano, approximately 20 miles southeast of Bend, lies about 40 miles east of the Cascade chain. Nevertheless, it so closely resembles its western neighbors that it seems to belong to the Cascade Range.

If Crater Lake did not exist, Newberry might have become Oregon's only national park. For this enormous shield volcano, 20 miles in diameter and approximately 80 cubic miles in volume, possesses some of the same attributes that make Crater Lake so famous. Like Mazama, Newberry has lost its former summit, so that the top of the mountain is now a huge depression. Although slightly smaller than

Mazama's caldera, that of Newberry contains both Paulina and East lakes. Instead of a single volcanic cone in the caldera, Newberry can boast of many cinder cones, as well as lava flows and domes of glistening black obsidian. Judging by the 400 parasitic vents and flows on its flanks and the abundance of fresh-looking structures inside the caldera, activity at Newberry volcano during the past few thousand years has been far more frequent and varied than at Crater Lake.

The best place to get a comprehensive view of Newberry's attractions is the crest of Paulina Peak, 7985 feet, the highest elevation on the caldera rim. To the south, Paulina Peak's slopes descend to the basaltic flatlands of central Oregon; its northern face is an almost vertical cliff that plunges 1500 feet to the caldera floor. Except for the latest eruptive deposits, which remain stark and unweathered, the caldera is heavily timbered. Four by five miles in diameter, it forms an irregular oval with its major axis trending east-west. Hemmed in on most sides by cliffs rising up to 1500 feet, the caldera is drained by Paulina creek, which has cut a narrow gorge through the sunken western wall. East lake has no known outlet and maintains a water level about 40 feet higher than that of Paulina Lake.

The two lakes are separated by very young flows of basalt and obsidian, amidst which rises the appropriately named Central Pumice Cone, literally the central feature of the caldera. This mile-wide, 700 foot-high cone is not, as some visitors speculate, the former peak of the volcano which sank to its present position. It formed only about 6700 years ago, shortly after Mazama ash blanketed the Newberry shield. The presence or absence of Mazama ash on recent caldera formations is a useful means of determining their relative ages.

The caldera's prominent features include several large flows of obsidian, which look like masses of coal spilled over the caldera floor. Easily the most impressive of these is the Big Obsidian Flow, which

Streaky obsidian layers of Big Obsidian Flow, Newberry Caldera, Oregon.
—photo courtesy D.W. Hyndman

137

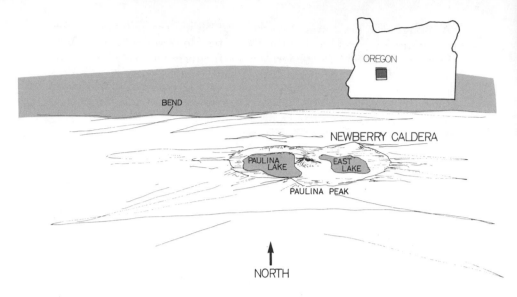

Location of Newberry Caldera.

erupted about 1400 years ago and is the product of Newberry's most recent activity. Issuing from a fissure near the south wall of the caldera, it spread over a square mile before erecting a hummocky dome over the erupting vent. Its blocky surface furrowed by giant "wrinkles" formed as it surged forward, the Big Obsidian Flow terminates in abrupt margins typically about 100 feet high. It is quite a feat to scale the front of this lava, whose sharply broken crust challenges even the toughest bootleather. Other post-glacial obsidian flows, such as that which poured from crevices on the north caldera wall about 6700 years ago and partly encircled the Central Pumice Cone, are only slightly less formidable. After these glassy masses had cooled, the prehistoric Indians of central Oregon found that the obsidian made excellent projectile points, which they apparently bartered with other Northwest tribes.

On the northern caldera wall is The Fissure, a deep gash that rises from the shore of East Lake. It marks the end of the Northwest Rift Zone, a 29-mile-long series of surface fractures that extends from Newberry northwest to Lava Butte, on U.S. Highway 97. The East Lake Fissure may have erupted fountains of molten rock, known as "fire curtains," such as are common on the flanks of Mt. Etna and the Hawaiian volcanoes. Along its margins are heaps of scoria, fused particles of congealed lava, and thin glassy flows, which may have formed from the falling spray of magma. Roughly 6100 years ago, this fissure system produced a sequence of basaltic flows along Newberry's northwest flanks.

Although largely forest covered, the Newberry shield includes areas created so recently that they seem too bleak and inhospitable to belong on this planet. Some aerospace officials must have shared this impression; sections of the volcano's outer slopes were chosen as a training ground for astronauts. The stark, pumice-dusted surfaces were thought to offer a rough approximation of what lunar explorers might expect when they landed.

The Growth of Newberry Volcano

Geologists have studied Newberry for the better part of a century, but only in the last few years have its great age and complexity been well understood. Recent investigations by Norman MacLeod, David Sherrod, and others have radically changed many formerly accepted views of the volcano's history. Their research demonstrates that not only is Newberry much older than previously supposed but has held a caldera for perhaps as long as 500,000 years. Several large tephra and pyroclastic flow deposits underlying broad areas of the 500 square-mile shield suggest a revised concept for the caldera's origin. Howell Williams, who studied Newberry volcano in the early 1930s, proposed that caldera formation resulted from ". . .drainage of the underlying reservoir (magma chamber) either by subterranean migration of magma, or, more likely, by copious eruptions of basalt from flank fissures. . .", with summit collapse occurring along ring fractures. Later studies suggested that Newberry's upper cone subsided along tectonic fault zones that intersect at the volcano's summit. However, other geologists were unable to confirm the existence of such fault zones on Newberry's shield.

Recognizing the abundance of tephra on Newberry's flanks, a fact unknown to Williams, MacLeod concluded that the caldera formed in much the same way as that at Crater Lake. The volcano's summit collapsed following violent Plinian outbursts that produced large quantities of pyroclastics and rapidly emptied the magma chamber, thus withdrawing support from the mountaintop above. The main difference is that Newberry caldera collapsed much earlier, formed a deeper caldera, and happened several times. Today's vast summit depression actually consists of several roughly concentric collapse areas, each somewhat smaller than its predecessor. They are a nested set of calderas of different ages.

The oldest evidence for a Plinian event that may have initiated summit collapse and caldera formation is a widespread rhyolitic ash flow, named for its exposure in Teepee Draw, that covers much of Newberry's entire shield. Estimating its volume at "much more than

10 cubic miles," MacLeod believes that it represents material ejected during the earliest caldera-forming eruptions, perhaps half a million years ago.

On Newberry's west side, the oldest tephra unit visible is gray to black lapilli tuff that covers about 30 square miles up to a depth of 200 feet. Look for it along the Paulina Lake Road, about 9.6 miles east of Highway 97. Deposited at extremely high temperatures, it was probably laid down by a hot pyroclastic flow or series of flows. Its large original volume of 5 to 10 cubic miles suggests that it may record another caldera collapse.

A younger series of ash flow deposits mantles the upper slopes of Newberry's west flank, forming a smooth, gently dipping surface that extends about 2 miles from the caldera rim. Composed of numerous thin to thick andesite pyroclastic layers, the flows and ashfalls were hot enough when erupted to weld their bases and form agglutinates in which molten or plastic lava fragments fuse together, near the caldera lip. The agglutinated material, which grades downslope into ash flow deposits, probably erupted from ring fractures bordering the western caldera wall. Scenic Paulina Creek, the only stream on Newberry's shield, has incised a narrow ravine through this formation, which underlies the cliffs at Paulina Falls.

The large volume of pyroclastic material interlayered with lava flows and domes makes it clear that, in spite of its shape, Newberry is not an ordinary shield volcano of the Hawaiian type. It staged violent Plinian and Peléan eruptions throughout history, and has also mixed typical basalt flows with great quantities of highly silicic lava. Cores taken from an exploratory drill hole 1200 feet deep on the upper cone showed that Newberry produced more dacite to rhyolite lavas than it did basalt. Ash flow tuffs also exist. Rocks at the core's base were 350,000 years old; midway down they were 250,000.

The present caldera floor, most of which consists of rhyolitic lava flows, domes, pumice flows, tephra, and other explosion debris, represents only a fraction of the caldera fill. The U.S. Geological Survey drilled over 3000 feet beneath the caldera, penetrating numerous layers of solid and fragmental rock. The upper 950 feet of fill contains airfall pumice and obsidian flows, as well as basaltic and rhyolitic ash that erupted underwater—indicating that lakes ancestral to Paulina and East lakes occupied the caldera thousands of years ago. Between 950 and 1180 feet down were lake-bottom sediments, beneath which were thick layers of pumiceous ash and breccia, 1180 to 1640 feet, rhyolitic to dacitic flows, 1640 to 2449 feet, and basalt to basaltic-andesite flows and breccia, 2485 to 3057 feet. Newberry's caldera has clearly been the scene of extremely varied volcanism for a long period of time.

Geothermal Resources

Hot springs in both East Lake and Paulina Lake are the only surface manifestation of Newberry's internal heat. Confirmation that very high temperatures exist at a relatively shallow level beneath the caldera floor was made in the summer of 1981. Drilling revealed temperatures of 510 degrees F at a depth of 3057 feet. These are the highest temperatures yet recorded at a dormant Cascade volcano and are higher than those at The Geysers of California, which are the world's largest producer of geothermal power.

Eruptions of the Outer Shield

Activity has been as varied and intense on Newberry's outer slopes as in the caldera. Approximately 400 cinder cones, domes, flows, and fissures are distributed over the shield, more than on almost any other known volcano in the world. So numerous are these miniature volcanoes along Newberry's flanks that local residents refer to them as the "Paulina Mountains" as if they constituted separate ranges in themselves. These parasitic vents, which range in age from several hundred thousand to about 6100 years, are concentrated on the three broad zones that join on the upper part of the shield. The eastern zone is apparently a continuation of the basaltic volcanism that characterized Oregon's High Lava Plains to the east. Most cones in this alignment appear relatively old. The second, northwestern zone of vents parallels a series of faults on the lowermost flank that extends to Green Ridge in the Cascades. The latest basaltic eruptions occurred along this fault zone. The third, a southwestern belt, parallels the Walker Rim fault zone. Some aligned cinder cones and fissures near Newberry's summit follow an arcuate pattern paralleling the caldera rim, and may lie along ring fractures associated with caldera collapse.

Most of the cinder cones range in height from 200 to 400 feet, although a few rise 500 feet above their bases and have diameters exceeding half a mile. The majority have shallow, saucer-like summit depressions, but some, such as Lava Top and Kawak Buttes, retain steep-sided craters 200 to 300 feet deep. Flank eruptions either concentrated at a single vent and formed cinder cones, or occurred along fissures and built long ridges, such as the Devil's Horn.

Streams of andesite or basalt typically emerged from the bases of these cones and flowed between and around them, leaving a network of intermingling flows. Both pahoehoe and aa lavas erupted, some of which retain extremely fresh surfaces and sharply defined margins. In spite of their youthful appearance, some of these flows are many tens of thousands of years old.

In addition to hundreds of cones and fissure eruptions of lava, parasitic activity produced about 20 rhyolite domes or flows on Newberry's western, southern, and eastern flanks. Two large rhyolite and obsidian domes at the extreme eastern base, East Butte and China Hat, are, respectively, 850,000 and 780,000 years old. McKay Butte, on the west flank, dates to 580,000 years ago, but some small rhyolite vents on the upper northwest side may be less than 10,000 years old.

Also of a surprisingly great age is the massive rhyolite dome underlying Paulina Peak, highest point on the volcano. This elongate dome, a mile wide, extends southwestward from the caldera wall for three miles. It may have extruded from the northeast-trending fissure or fault about 240,000 years ago.

The Latest Eruptions

The wide variety of lava types and eruptive behavior that characterized Newberry during the Pleistocene Epoch has also been typical of the last few thousand years. Most basaltic and basaltic andesite features within the caldera are somewhat older than Mazama ash, about 6900 years. These include the Interlake flow, the Sheeps Rump cinder cone and flow on the northeast side of East Lake, and a flow from the east-rim fissures that extends to East Lake.

Also predating the Mazama eruptions are several rhyolitic formations: two domes along the south shore of Paulina Lake, a large obsidian flow in the northeast corner of the caldera, an obsidian dome near the caldera wall south of East Lake, with an obsidian flow that reaches northward from the dome to East Lake. Some rhyolitic flows and landslide debris also locally underlie Mazama ash.

The first cycle of post-Mazama eruptions built the Central Pumice cone between Paulina and East lakes, followed soon after by the Interlake and Game Hut obsidian flows to the north and south, all about 6700 years old. A widespread deposit of pumice, ash, and mud-coated pumice lapilli in the eastern part of the caldera may have been erupted slightly earlier from vents in East Lake. It is locally as much as 14 feet deep and is overlain by the East Lake obsidian flow, which erupted about 3500 years ago. The Pumice Cone Crater is now blocked by a series of flows dated at about 4500 years.

Newberry's most recent eruptions are associated with the now buried vent that produced the Big Obsidian Flow. Activity began violently with a Plinian explosion of pumice that blanketed the southern half of the caldera and eastern flanks of the volcano. The pumicefall forms a relatively narrow lobe extending to the northeast.

It is 12 feet thick about 5.5 miles from its source and at 40 miles decreases to 10 inches. As increasingly fine ash, it probably continues hundreds of miles farther. Radiocarbon dating indicates that the airfall pumice erupted between about 1550 and 1720 years ago.

It was followed by an extensive ash flow from the same vent. Bearing pumice bombs in a matrix of fine, slightly pinkish ash, it extends from the Big Obsidian flow to the shores of Paulina Lake. Carbon 14 tests of organic material beneath the ash flow yielded ages of about 1350-1400 years. The climatic event was emplacement of the Big Obsidian flow, which issued from a vent or fissure near the southern caldera wall and flowed northward over the earlier ash flow deposit, partly engulfing the Lost Lake pumice ring. A rough trail ascending the blocky obsidian flow offers views of its various features, including exposures of flow-banded obsidian, pumiceous obsidian, brown streaky obsidian that was pumiceous before the bubbles collapsed, and indications that the flow behaved in both a plastic and brittle manner as it oozed over the caldera floor. The airfall pumice, ash flow, and obsidian flow all have about the same chemical composition.

The East Lake Fissure (Northwest Rift Zone)

Extending for 20 miles northwest from the Newberry Shield is the East Lake Fissure. Beginning at the vertical scar on the north caldera wall above East Lake, it terminates near Lava Butte, 10 miles south of Bend on Highway 97. Newberry's most recent basaltic eruptions have followed this line of crustal weakness. Between about 5800 and 6400 years ago, at least eight flows of basalt or basaltic andesite poured from the fissures, overwhelming nearby timberlands. The Lava Cast Forest flow, so called because the molten rock engulfed standing trees, incinerated them, and formed hollow molds of their trunks is a favorite stopping-place for tourists. Despite its unweathered appearance, the flow is 6150 to 6380 years old. Several other lava tongues emerged from the Northwest Rift Zone at about the same time. These include the Lava Cascade, Gas Line, and Forest Road flows. All these young flows may have erupted during a much briefer time than the age uncertainty of radiocarbon dates suggests, perhaps within a few weeks or years. All other Newberry basalt flows have Mazama ash on their surfaces. That means they are older than about 6900 years, some as much as several hundred thousand years.

The most voluminous outpouring from the 20 mile-long fault occurred near its northern terminus. The site is marked by 500-foot-high Lava Butte, which rises at the edge of the main highway into

Bend. Steep-sided, beautifully symmetrical, with a deep, well-preserved summit crater, Lava Butte is one of the most accessible examples of recent cinder cone in the Cascades. A paved road spirals to the summit, where the Forest Service maintains a museum and you can peer from your car window into a crater where lava once bubbled and sprayed fountains of molten rock.

Lava Butte received its name from the floods of lava that poured from its southern base. This flow, or series of flows, spread over 10 square miles, extending northward for 5 miles and westward for 3 miles. The flows dammed the Deschutes River, forcing it to cut a new course around the flow margin, creating spectacular rapids and waterfalls. Charcoal found beneath a tephra layer northeast of Lava Butte yielded a date of about 6160 years for the cone. The associated lava flows are probably about the same age.

Some geologists believe that before erecting its present cone, an earlier vent at the site of Lava Butte erupted streams of fluid basalt. If so, the cinder cone formed during a second stage of activity, a pyroclastic episode between two distinct phases of effusive volcanism. Another geologist suggested that if Lava Butte had not squandered its resources by pouring lavas over territory so far removed from the main crater, it might have built a volcano to match Mt. Bachelor.

The Three Sisters, looking south. Collier Glacier, center, still has a maximum thickness of 300 feet although it has shrivelled drastically during this century. In about 1900 the ice tongue surmounted the crater wall of Collier Cone, bottom center, a cinder cone formed about 2500 years ago.

—U.S. Geological Survey photo by Austin Post

XII
The Three Sisters:
Oregon's Volcanic Playground

Throughout most of the length of the Cascade Range, large volcanic peaks are spaced about 40 to 80 miles apart. This distribution of the higher peaks allows a particular mountain to reign visually over a wide domain, of which it is the chief peak. In central Oregon, however, the usual pattern is broken. Instead of a single snowy cone dominating the immediate countryside, a whole cluster of closely grouped volcanoes creates an impressively crowded skyline.

Viewed from the hills near Bend, the prominent elevations, from south to north, include the symmetrical Bachelor, craggy Broken Top, the South, Middle, and North Sisters, and Belknap Crater. Most of these peaks remain snowcapped even in summer; the Sisters, all over 10,000 feet, bear at least a dozen active glaciers. Farther north, are the pinnacles of Washington and Three Fingered Jack.

No part of the range has seen a greater number and variety of recent eruptions than the vicinity of Oregon's Three Sisters. From the oldest, highly eroded spires of Washington and the North Sister to the smooth, almost unmarred summit cones of Bachelor Butte and South Sister, the skyline view offers a lavish display of volcanoes, old and new.

Because of its scenic beauty and protection by the Federal Government, the Three Sisters Wilderness is well known as a year-round recreational area. A ski resort on Bachelor attracts a throng of winter visitors, while summer sports enthusiasts find excellent camping and hiking opportunities along the 240 miles of developed trails that criss-cross the region. Scented pine forests, flowering meadows, and mossy-banked streams add to the charm of this natural playground. West and south of the Sisters are 37 alpine lakes, ranging in size from small tarns to respectable bodies of water suitable for fishing, swimming and boating.

Motorists following the Cascade Lakes Highway southwest from Bend can stop conveniently at many of these scenic glacial ponds,

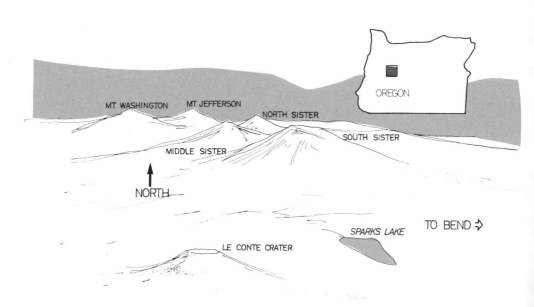

Location of the Three Sisters in central Oregon.

146

some of which are bordered by cliffs of glassy dacite or flows of treeless blocky lava. Nature here presents a vivid contrast: luxuriant alpine parklands surrounded by bleak monuments to the earth's past violence.

The Geology of the Three Sisters

When he published the first geological reconnaissance of the Three Sisters region in 1925, E.T. Hodge concluded that these mountains were but remnants of a once vastly larger volcanic cone that had collapsed in Miocene or early Pliocene time. The broad arc of peaks that runs from Broken Top through Devil's Hill, the Wife and Sphinx, Husband, Little Brother, and North Sister seemed to outline an ancient caldera rim. Observing the long slopes leading outward from the ragged arc, one might easily imagine them projected upward to an enormous central cone. Hodge called this theoretical ancestor of the Sisters Mt. Multnomah. He concluded that it had been engulfed, in much the same way that ancient Mazama had collapsed to form the caldera in which Crater Lake lies.

In the 1940s the late Howel Williams, then dean of Cascade volcanologists, examined the Three Sisters and associated peaks. He concluded that each of the edifices in question is a totally separate volcano. While several cones may be parasitic to the larger peaks, none represents survivors of a demolished older structure. Multnomah never existed!

The North Sister

The oldest mountain is the steep, rugged North Sister, 10,085 feet. Considered the most difficult climb of the three chief peaks and commonly termed "the Black Beast of the Cascades," its summit consists of almost sheer cliffs of disintegrating rock. Its original surface has eroded away, leaving no trace of a crater.

With a base 15 to 20 miles in diameter, the North Sister was once one of the mightiest volcanoes in the Oregon Cascades, at least 11,000 feet high. Williams' investigations also revealed that the cone was built in three distinct stages, almost duplicating the three-part development of Thielsen, Washington, and similar early-Pleistocene volcanoes of Oregon. First, the North Sister formed a wide basalt shield that eventually rose about 4000 feet above its foundation. Then, when it had attained a height of some 8000 feet above sea level, quiet eruptions of fluid basalt gave way to explosions, which erected a large fragmental cone in the summit crater. This fragmental summit

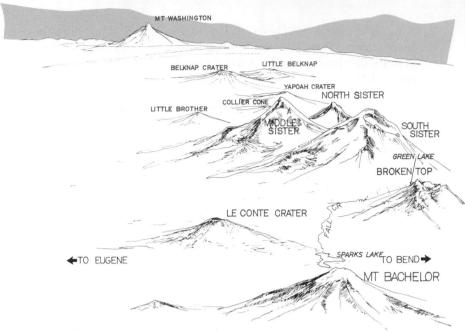

Map of the Three Sisters region, looking north. The volcanoes erupted sporadically from the Pleistocene to the recent past. The North Sister, Mt. Washington, and Broken Top are among the oldest cones, while the Belknap shields and Collier Cone are among the youngest.

cone stood about 3000 feet above the underlying shield. Glacial erosion later scooped out large sections of this cone, so that its composition is clearly displayed. It consists of shattered lava blocks, derived from older plugs and flows, as well as pumice, scoria, bombs, and cinders. In addition, slaggy flows of vesicular basalt issued from both the central vent and from fissures along the sides of the cone. Compared with those that built the shield, most of these later flows are notably thin. An exposure south of the Thayer Glacier, for example, shows that 50 superimposed flows make a cliff only about 120 feet high.

In its last stage of growth, the North Sister also resembled Thielsen and Washington. The summit cone was invaded by swarms of dikes, more numerous than those on any similar Cascade peak. Now exhumed by glacial erosion, dozens of vertical or oblique lava walls ranging in width from a few inches to 25 feet radiate away from the summit. Finally, a massive "steep-sided plug of . . . brown-crusted, greenish lava" welled into the main conduit, thrusting aside and displacing many of the earlier dikes and fragmental deposits. Completely filling the volcano's throat, this 300-yard-wide plug now forms the two summit pinnacles, known as the Middle and South Horns.

148

As in the cases of Thielsen, Union Peak, and other mountains of comparable age, glaciation removed a quarter to a third of the North Sister's original volume. During maximum glaciation, ice entirely covered the peak, extending north and east to elevations of about 4000 feet; some glaciers stretched 15 miles west to an elevation of 1000 feet. Judging by its extremely denuded state, North Sister was probably extinct before the last major glacial advance.

At least some of the many irregular mounds and protrusions that clutter the North Sister's flanks are probably the eroded remains of various satellite cones. One of the most prominent is Little Brother, which reproduces in miniature the development of its parent.

Other conspicuous and deeply eroded volcanoes, likewise miniature replicas of the North Sister, abound. West of the Middle and South Sisters is The Husband, notable for the unusual fluidity of its early lava flows, some of which reached as far as the Western Cascade boundary. The Husband is also remarkable for its two enormous summit plugs. The southern plug measures 600 by 200 yards; the northern one is three quarters of a mile along its major axis and 300 yards along its minor and stands 800 feet high. Although these protrusions may have been conduit fillings, they probably rose above the original vents as steep-sided lava domes.

The Wife, Sphinx, and Burnt Top also form secondary outcroppings along the central Cascade skyline. All belong approximately to the North Sister's main period of activity and all have had their original features effaced by glacial cutting.

Broken Top

One of the largest and most picturesque volcanoes in the Three Sisters constellation is Broken Top. A Pleistocene composite cone of

The Three Sisters from the southeast. The wrinkled surface of the Newberry lava flow, erupted about 2000 years ago, is in the lower right. A half a dozen active glaciers are visible. —Photo by Delano Photographics

The North Sister (foreground) has been deeply carved by glaciers. Collier Glacier, the largest and thickest in the area, originates on the Middle Sister (right background). Mt. Bachelor is seen to the southeast (left) and South Sister appears directly behind North Sister's summit (center).
—Photo by Austin Post, U.S. Geological Survey

basaltic andesite that once stood appreciably higher than its present 9175 feet, Broken Top is now a glacially steepened, deeply eroded amphitheatre, open to the southeast. Its former summit, entire southeast slope, and much of its interior have completely disappeared, leaving jagged rocky spurs standing between glacial cirques, where small ice streams, including the Bend and Crook glaciers, continue their work of laying bare the volcanic core. In its exhumed interior one can see brilliantly colored strata: bands of red, purple and black scoria alternating with layers of yellow, brown, and orange welded pumice and ash flow tuffs. A conspicuous stratum of white pumice is interbedded with fragments of black lava, giving a striking salt-and-pepper appearance to the deposit.

A recent study of the Broken Top area showed that the volcano grew by first building a large pyroclastic cone atop a platform of older, glaciated basaltic andesite lavas. The cone of scoria and ash was repeatedly invaded by dikes and sills that strengthened the edifice and probably fed lava flows that formed an armor of solid rock covering the original fragmental structure. During this cone-building stage, Broken Top produced a variety of lava types, alternating streams of basaltic andesite with more silicic flows of dacite and rhyodacite. Explosive eruptions ejected tephra and pyroclastic flows.

Part way through its growth, the volcano's former summit subsided or collapsed, creating a summit depression about half a mile in diameter. Thin flows of basaltic andesite later filled the crater, then spilled down the outer slopes of the cone, mantling it with a thin veneer of porous lavas. During its final stages of development, magma congealed in the volcano's central conduit to form a plug of micronorite, similar to that in Three Fingered Jack. The central plug and adjacent deposits were then partly altered by circulating hot water and steam.

Vents near Broken Top were once thought to be the source of large pyroclastic flow deposits that covered more than 200 square miles on the east flank of the Cascade Range near Bend. More recent studies indicate that these deposits are about two million years old, much older than Broken Top volcano. The pyroclastic flows and airfall pumice apparently erupted from vents south of Bachelor Butte, perhaps near Sitkum Butte and Lookout Mountain. Somewhat younger pumice flows and tephra did originate at a now buried source in the Broken Top vicinity, perhaps from a site near the modern Triangle Hill. The wide distribution and great volume of the pumice flows and accompanying tephra deposits, totalling about 10 cubic miles, show that the Three Sisters area has produced truly cataclysmic explosive eruptions. If such eruptions were to occur today, the timberlands and settlements of central Oregon could be devastated.

The Middle Sister

Smallest of the trio of volcanic siblings, Middle Sister, 10,047 feet, has few distinctions. Were it not for its height and glacial covering it would merit little attention. In general outline, the Middle Sister is a regular cone, the eastern face of which has been stripped away. It thus resembles a half-dome rising above the thick lava accumulations of its predecessors in the area. The Hayden and Diller glaciers continue to enlarge the amphitheatre excavated in its eastern slope and probably obscure the site of its former central conduit. The Middle Sister thus has neither a crater, like the South Sister, nor impressive summit pinnacles, like the North.

Collier Glacier, which descends from Middle Sister's north shoulder, flows north-northeastward directly across the North Sister's western flank, into which it has cut a broad channel. The glacier has large lateral moraines, which seem to have been formed in late post-glacial time. After several decades of shrinking, the glacier seems to have stabilized.

The South Sister

The highest and best preserved of the three major peaks, the South Sister, 10,358 feet, retains an almost perfectly circular summit crater, about 1300 feet in diameter. In spring and summer melting snow forms a small crater lake, the highest body of water in Oregon. Recently christened Teardrop Pool, on a sunny day its turquoise surface offers a dazzling contrast to the encircling snowfields.

Although its form is generally conical, the South Sister has a complex history. Ed Taylor found that its lavas have a normal palomagnetic polarity, indicating that the whole mountain is less than 700,000 years old, but much of the structure has clearly been glaciated. According to Howel Williams, the volcano's base consists of a broad basaltic shield, perhaps contemporaneous with that of the North Sister. It now seems probable, however, that the shield lavas underlying the volcano are much older than South Sister.

The steep main cone is mostly andesite and dacite lava flows, along with small quantities of basalt. After the edifice had been partially eroded by Pleistocene glaciers, mildly explosive activity at the central vent erected a symmetrical summit cone of thin dark flows and deposits of red and black scoria. This basaltic andesite cone, formed in latest Pleistocene time, records the most recent activity at the main conduit.

South Sister has had only two significant eruptive episodes during the last 10,000 years, both from vents along its flanks. These eruptions occurred only a century or two apart between about 2300 and

The South Sister, from the southeast. The chain of rhyodacite domes and flows (foreground and center) was erupted about 2,000 years ago when a dike of silicic magma intersected the South Sister's cone. —Photo by Austin Post, USGS Tacoma

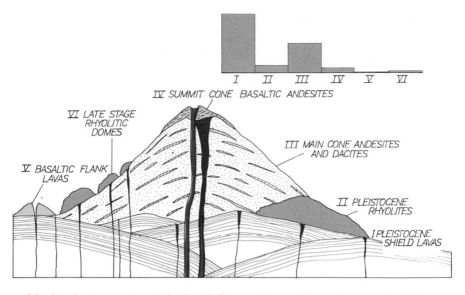

Idealized cross-section of the South Sister volcano. The main cone of andesites and dacites sits atop an older formation of Pleistocene shield lavas. The basaltic-andesite summit cone was probably formed during very late Pleistocene or early Holocene time, while the rhyodacite domes and flows on the south flank were erupted about 2000 years ago. —After Clark, 1983

1900 years ago. Both events, known respectively as the Rock Mesa and the Devil's Hill episodes, began explosively, ejecting moderate quantities of pumice, followed by quiet eruption of lava flows and domes. Hikers along the Fall Creek trail to the Green Lakes basin, which occupies the saddle between the South Sister and Broken Top, find abundant light gray and buff-colored pumice bombs and lapilli lying on the ground surface. Similar ejecta lie on or near the summit.

The first eruptive cycle, which took place about 300 B.C., centered at two vents at the southwest and south flanks of the cone. Pulsating columns of ash distributed tephra to the south and east, to a maximum distance of about 20 miles. This vent was then buried by the massive flow and dome that formed Rock Mesa. After a short dormant interval, the South Sister erupted again, opening a linear series of 16 new vents along its northeast and southeast slopes. This was the Devil's Hill episode, which occurred about the time of Christ. Its order of events duplicated those of the Rock Mesa activity: Plinian eruptions showered pumice over a wide area, after which viscous masses of rhyodacite lava surged from the active vents, forming a series of domes and thick, stubby flows. The highest vents on the southeast flank produced the Newberry Flow, a blocky stream that crept down the east flank to the base of the cone, probably damming Fall Creek

153

and helping to impound Green Lake. A chain of steep-sided domes with black, glassy, deeply furrowed crusts also erupted along the southeast slope. Beginning at an elevation of about 8000 feet and moving progressively downward for a distance of three miles, the obsidian domes extend as low as the 5500-foot level. Bristling with sharp spires, huge angular blocks, and other chaotic surface features, these massive structures rear bleakly amid thick stands of evergreen timber. Look for the lowest obsidian dome from the Cascade Lakes Highway, near the shores of Devil's Lake.

The two eruptive episodes together produced more than a fifth of a cubic mile of new rock, 95 percent of it as lava flows and domes. The activity also generated mudflows and pyroclastic flows, some of the latter extending as far as two miles from their source. A new analysis of these late Holocene eruptions suggests that they were produced when a dike of highly silicic magma intruded the South Sister's cone.

The rising sheet of magma apparently bisected the volcano, creating the linear chain of nearly 20 vents when it reached the surface. If this hypothesis is correct, the South Sister's latest eruptions followed the pattern of those at the Mono and Inyo Craters in California. If, as several geologists have suggested, the recent rhyodacite eruptions indicate a developing body of silicic magma beneath the volcano, the South Sister may eventually produce an eruption similar to the one that destroyed Mazama and created Crater Lake.

James Clark compiled estimates of the kinds and volumes of magma erupted during each stage of South Sister's evolution. He found that the underlying basaltic shield lavas represent about 14 cubic miles of material, while the younger rhyolites account for only a tiny fraction of the total. The main cone consists of approximately seven cubic miles of andesite and dacites, whereas the later summit cone of basaltic andesite has a volume of only about 0.65 cubic mile. As at Jefferson and some other Cascade volcanoes, the proportion of cone-building andesites and dacites to earlier basalts diminishes with time. The later the eruption, the smaller the quantity of magma.

Glaciation

South Sister underwent at least three distinct glaciations. Little evidence remains of the oldest, but during the last glacial advance, which ended approximately 10,000 to 12,000 years ago, a vast icecap buried the High Cascades, smothering all but the loftiest ridges and peaks and merging with alpine glaciers descending from the Sisters. Ice streams flowing down South Sister's southern slope filled adjacent valleys to a depth of 500 feet. Neither the central ice sheet nor its related glaciers, some of which were 19 miles long, extended beyond the High Cascade platform; in the Sisters region few of the glaciers reached below 3600 feet.

A more recent glacial episode deposited fresh moraines and outwash between elevations of 7000 and 9000 feet on the high peaks. Radiocarbon ages of associated lavas and ash deposits indicate that these moraines formed less than 2500 years ago and belong to what is sometimes called the "Little Ice Age." The last minor glacial advance culminated near the end of the last century.

The small, generally shallow cirques cut into the south and southeast flanks of the South Sister's present summit cones probably formed during the neoglacial period. A long ridge of reddish scoria separates the south-side Lewis and Clark glaciers, which have significantly steepened but not yet breached the crater's outer walls. Lewis Glacier occupies a hollow bordered on the south by a steep youthful moraine; the northern headwall, incised into the summit cone, reveals layers of relatively thin, dense basaltic-andesite flows just below the crater lip. Along the inside crater rim one can see strata of black and red scoria, almond-shaped bombs as much as a yard long, and flows of dark, scoriaceous basaltic andesite. When snowcover is low, two depressions on the crater floor, one east-southeast of the other, appear. These may mark the vents of the last eruptions.

Glaciers cut more deeply into the north face of the volcano, where 1200 feet of layered andesites, dacites, and basaltic andesites are exposed just below the summit. Three principal terraces of lava, perhaps belonging to an older summit cone's eruptive cycle, separate hanging icefields. Existing glaciers seem too small to carve this slope, which probably dates from late Pleistocene time.

Considering the fresh appearance of the volcano's summit area, it is not surprising that we have at least one report of historic activity. In July, 1853, James P. Miller, a Presbyterian missionary then stationed in the Willamette Valley, wrote to his home board that on the 9th of that month he saw "one of the Three Sisters belching forth from its summit dense volumes of smoke."

> The smoke appeared to arise in puffs at intervals, which continued until the mountain was hid from view by the intervening clouds, between sundown and dark; since which, constant cloudy weather has hid the range from view. . . .There can be but little doubt that a crater existed in the peak of the Three Sisters, which issued smoke on Saturday. . .

Miller did not specify from which of the three peaks the apparent ashcloud originated, but if he observed a genuine eruption, the South

The northeast face of Mt. Bachelor, a basaltic composite cone built during the late Pleistocene and early Holocene Epochs. Except for the small glacier-filled cirque just below the summit, this symmetrical volcano is little altered by erosion. A young parasitic cinder cone is in the lower right. A chair lift now carries skiiers and sight-seers all the way to the summit.
—Photo courtesy of the U.S. Forest Service

Sister is the probable source. Miller's description of the "smoke" "issuing in puffs of intervals" recalls the mild eruptive behavior recorded at St. Helens and Hood during the 1850s. Like many of the 19th century reported volcanic events, this was probably a steam explosion that scoured old ash out of the volcano, leaving too scanty a deposit to be recognized 130 years later. Both Taylor and Clark, among the geologists most familiar with South Sister's eruptive history, expressed considerable skepticism about the report.

Mt. Bachelor

Bachelor, formerly called Bachelor Butte, is a remarkably symmetrical lava cone standing approximately 3000 feet above the surrounding woodlands a few miles south of Broken Top. It is familiar to thousands because of the popular ski resort located on its northern slope. Except for a small glacial cirque on its shaded northern side, the cone is little modified by erosion.

This Bachelor, however, is somewhat older than he looks. Most of the volcano was constructed shortly before the end of the Pleistocene Epoch, between about 18,000 and 14,000 years ago. At least three distinct stages of activity have been recognized. The first built a broad shield, the second erected a steep composite cone of basaltic andesite on top of the shield, and the third produced streams of blocky lava from vents on Bachelor's north and northwest flanks. About 12,000 years ago, the last brief advance of Pleistocene glaciers formed moraines on the north slope that were later partly buried by lava erupted from the north-side vents. The last activity built a prominent cinder cone on the north flank. The lava that issued from this cone looks deceptively fresh, prompting Williams to remark that "no one who sees the barren flows of basalt which poured from fissures on the northern flank and spread in branching tongues into Sparks Lake can doubt that they must have escaped only a few centuries ago." The latest Bachelor flows, however, have traces of Mazama ash on their surfaces and were thus erupted more than 6900 years ago.

Climbing Bachelor along the edge of the shallow trench that its one surviving glacier excavated reveals in miniature the glacial processes that sculptured the larger Cascade volcanoes. Although this un-named glacier is tiny by Cascade standards, it has significantly modified Bachelor's northern face. Lavas bordering the margins of the glacier's bed appear as fresh as if they had just congealed. But where the glacier has flowed over the lava and cut beneath the surface, the ice has ground massive lava blocks into a fine, charcoal-gray flour, easily carried away by streams issuing from the glacier's snout. The transformation of solid rock into gritty dust is particularly dramatic at the glacier's terminus, where stands a large semi-circular ridge, concave toward the glacier-front. Some climbers mistake this formation for a cinder cone; it is actually the glacier's terminal moraine.

During recent years, the Bachelor glacier has noticeably shrivelled and no longer shows the crevasses that indicate movement. If this trend continues, the now-stagnant icefield will disappear. A future series of heavy winter snowfalls and cool summers, however, may increase the glacier's mass and stimulate a new advance.

A chair lift transports skiiers and sightseers to Bachelor's summit. You can rise effortlessly to an elevation of 9000 feet, where a sweeping panorama of the central Oregon Cascades and high lava plains unfolds at your feet. Marked foot paths encircle the summit area, allowing one to see far beyond the Three Sisters, with their sparkling icefields, to Jefferson, Hood, and the humped top of Adams in Washington state.

No large crater occupies Bachelor's summit, but several more or less circular depressions amid the summit lavas may mark the sites at which magma sank back into the supplying conduits as the last eruptions ended. Some fine examples of smooth pahoehoe lava surfaces exist near the vent areas; a short distance downslope from their source, the undulating, ropy crusts grade into rough lavas. The transition from smooth to rough flow surfaces reflects the lava's loss of heat and steam as it flowed down the slope.

Although Bachelor formed during a relatively brief period during latest Pleistocene and recent time and has been quiet for at least seven millennia, it may only be resting between cone-building episodes. The volcano's youthful appearance and the reported presence of the fumaroles on the northern slopes may indicate the presence of magma beneath the cone. If it does erupt again, it may eventually become one of the larger volcanoes in the region.

Cones North of the North Sister

In a north-south alignment between the large cones of the North Sister and Three Fingered Jack, at least 125 separate eruptive centers have blazed and died in post-glacial time. These recent vents built cinder cones several hundred feet high, covered the surrounding countryside with thick mantles of ash and pumice, and—most impressively—poured out immense streams of lava whose surfaces are so fresh and unweathered that they look as if they had congealed only yesterday. Lava issuing from a cinder cone near Sand Mountain dammed the McKenzie River to form Clear Lake, now a favorite Cascade fishing spot. Ghostly forests, denuded of branches, still stand in the depths of the lake. The cold waters must act as an effective preservative, for, according to radiocarbon dating from wood samples taken nearby, the forest drowned about 3000 years ago.

Remarkable volcanic displays greet the traveler over McKenzie Pass, between Bend and Eugene. Following the winding but generally east-west trending McKenzie River valley, the highway crosses lava fields covering an area of 85 square miles—a forbidding desert of black, jagged basalt. Several major vents contributed lava; the most important were Belknap Crater, Little Belknap, and the Yapoah cinder cone. From the Dee Wright observatory, built of and atop the

The rubbly surface of a fresh lava flow is no place for bicycles. This is the Yapoah flow near McKenzie Pass.
—photo courtesy D.W. Hyndman

basaltic flood, one can enjoy a sweeping panorama of the several volcanoes involved. To the south stands the white massif of North Sister, its base dotted with recent peripheral cones. One of them, Yapoah Cone, erupted the flow on which the observatory stands. Northward rises the eroded volcanic neck of Washington, on whose glaciated lower slopes the Belknap volcanoes rose.

Most of the visible lava erupted from Belknap Crater and its satellites, Little Belknap and South Belknap. The main Belknap cone, 6869 feet, stands in the wide depression between the opposing slopes of North Sister and Washington, whose lavas probably underlie Belknap's broad shield.

Belknap's lava shield is topped by a 400-foot high cinder cone, indented by three summit craters of varying size, aligned in a north-northeasterly direction. The northernmost and smallest vent apparently erupted some lava, for part of the crater wall is breached by a small flow. The much larger south crater emitted large quantities of ash that drifted over an area exceeding 100 square miles.

Later, fissures split Belknap's cone, and issued copious streams of basalt, which inundated the entire shield. Most of this lava is of the blocky or scoriaceous type, although some pahoehoe crusts also formed. The collapse of lava tubes, and the breaking up and "peeling back" of cooling lava crusts as the fluid interior of the flow moved forward, make these basalt surfaces unusually chaotic, virtually impossible to cross.

Despite their fresh appearance, some of the Belknap lavas are surprisingly old. About 2900 years ago basalt lava erupted from Little Belknap, a vent about a mile east of the main crater. These hummocky, virtually treeless flows surround two prominent islands of older volcanic rock, leaving them isolated amid a sea of black basalt. Williams described these Little Belknap effusions, with their crusts broken into jumbled piles of sharp rock, as "reminiscent of a shattered ice jam." The final lavas from Little Belknap filled the crater, leaving a craggy mound of red clinkery rock, from which collapsed lava tubes diverge radially.

The next major overflow occurred about 1800 years ago at South Belknap, a subsidiary cone on the south flank of Belknap Crater. Moving over the older Belknap flows, the South Belknap lava streamed southward to McKenzie Pass, where it banked against the margins of a solidified flow from Yapoah Cone, which is about 2900 to 2600 years old.

The largest and last eruption happened about 1500 years ago, from fissures near the base of the main Belknap shield. Simultaneously,

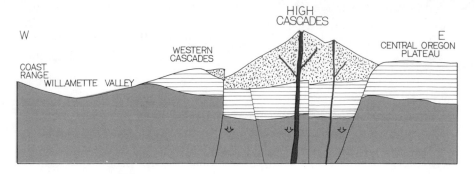

Simplified cross-section of the Cascade Range in central Oregon. The High Cascade composite cones are built atop a down-dropped block in the center of the range. —After McBirney, 1978 and Taylor, 1981

explosions blew quantities of ash from the northern of the two principal summit craters. Blocky basalt flowed 12 miles west, then plunged onto the McKenzie canyon floor. Lava spread across the riverbed, damming an extensive swamp—Beaver Marsh—upstream. Today the McKenzie disappears into the permeable sediments along the flow margins and runs underground until it reappears at Tamolitch Falls.

Other Recent Eruptions

Because such an abundance of new cones and lava flows characterize the area north of The Sisters, only two of the most notable will be mentioned. Four-in-One-Cone, as its name implies, consists of a row of four closely-spaced, almost identical cinder cones, which, about 2600 years ago, had their northeast crater walls breached by extensive lava flows. It must sit astride a fault zone, for it is part of an alignment of 19 similar cones.

Collier Cone, on the flanks of the North Sister, may be the most recently active volcano in the group. Its eruptive history is long and complex. Not only did its explosive eruptions create a mile-square plain of desolation, known as the Ahalapam Cinder Field, it also produced a series of lava flows which extended three miles to the northwest and eight and a half miles to the west.

Standing in the path of Collier Glacier, Collier Cone blocked that ice stream's advance during several recent glacial resurgences. As late as 1924 Collier ice rode high on the side of the cone. At an earlier period, when the glacier was at least 200 feet thick at its terminus, water from its melting surface flowed into the crater of Collier Cone and covered much of the crater floor with outwash. Stream gravels were simultaneously deposited in the lava gutters and other flow features surrounding the cone. Today, Collier Glacier has retreated far up the flanks of the North Sister, leaving behind a conspicuous lateral moraine that borders the cinder cone.

The Volcanic Future of the Three Sisters Area

The number and variety of post-glacial eruptions in The Three Sisters vicinity make it seem unikely that all the recent vents are extinct. In the past few thousand years, old volcanoes have reawakened and many new ones have been built. The brief period of quiescence during historic time, which in this part of Oregon goes back less than a century, is probably only a hiatus in a long history of intermittent volcanic activity. Which one of the many cones, fissures, blow-holes, or craters will next break into new life? Perhaps an entirely new mountain will one day rise before our eyes, like Paricutin, which was born in a Mexican cornfield in 1943. The Three Sisters region is as likely as any in the whole range to stage a spectacular volcanic renaissance.

Exploring the Three Sisters

Enjoy a panoramic view of the Three Sisters and their satellites from your car along Highway 97 for miles north and south of Bend. The short paved road to the summit of Lava Butte, a few miles south of Bend, affords an unobstructed view of the entire region. Watch for the roadside marker.

The Cascade Lakes Highway leaves 97 at Bend and leads west-southwestward to skirt the base of South Sister before rejoining Highway 97 near LaPine. Although this route provides access to the many scenic lakes that dot the sparsely timbered highland the road does not actually take you *on* any of the Sisters. The closest approach is to the south flank of South Sister at Sparks and Devil's Lakes. For an enjoyable summit climb take the Fall Creek trail from a meadow near Sparks Lake to its terminus at the Green Lakes basin, at the volcano's eastern foot. From there to the crater rim it is a steep walk.

Energetic mountaineers in good condition have been known to scale all Three Sisters, plus Broken Top and Bachelor, in a single day. A trail leading from the McKenzie Pass Highway via Sunshine Camp conveniently approaches Middle and North Sister.

Bachelor, an inviting climb for the beginner, has become almost too accessible. Follow the Cascade Lakes Highway 22 miles westward from Bend and turn left at the sign indicating the Bachelor Ski Area.

If you choose to ignore the chair lifts, a new trail offers hikers a brisk walk to the summit. From the south side of the parking lot east of the lodge, the trail climbs about 2600 feet in about 4.5 miles of switchbacks to the top.

The Three Sisters viewed from the east. —photo courtesy of D.W. Hyndman

Whitewater Glacier on Jefferson's eastern flank has cut deeply into the volcano's interior. —U.S. Geological Survey photo by Austin Post

XIII
Mt. Jefferson:
Guardian of the Wilderness

Jefferson, the second highest peak in Oregon, lies within a rugged wilderness that makes access difficult. A popular hiking trail leads to Jefferson Park on the mountain's northwest side; the closest gravel logging road (No. 1044) connects with Highway 22, and comes within four miles of the peak. Most people see Jefferson from a distance; the stretch of U.S. Highway 97 between Bend and The Dalles affords good views of the heavily glaciated east side.

Despite its inaccessibility, Jefferson has frequently appeared, without being identified, in national advertisements for cars and liquor. The volcano is especially photogenic. Jefferson rises above the Cascade divide to an elevation of 10,495 feet. Its sharp, pointed summit is as picturesquely horn-like as those of Thielsen or Washington. Abundant glacial ice keeps the peak white and glittering throughout the year. The Whitewater Glacier covers much of the east side, extending even below timberline. Jefferson Park, with its clear rivulets and profusion of wild flowers, provides an exceptionally attractive setting, which photographers have used to frame their pictures of this erosion-scarred volcano.

Jefferson first entered the literature of the West when Lewis and Clark sighted it from near the mouth of the Willamette River. Named to honor the great president who sponsored their expedition, Jefferson was the only High Cascade peak the explorers named.

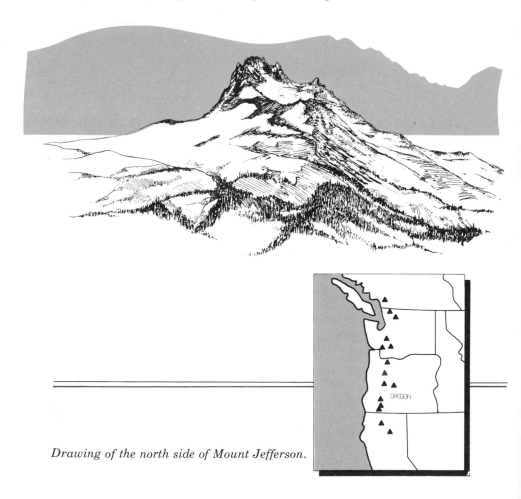

Drawing of the north side of Mount Jefferson.

164

Perhaps because neighboring Hood is higher and more visible from Oregon's population centers, very little has been published about Jefferson. The only lengthy printed report devoted exclusively to the volcano was written 60 years ago by Edwin T. Hodge, a colorful personality and former professor of geology at the University of Oregon. Other technical studies focused on the chemical composition of lavas or the mineral resources.

Fortunately, in 1974 Kenneth G. Sutton conducted a broad study of Jefferson's volcanic and glacial history, which provides the basis for much of this discussion. More recently, James G. Begét identified two extensive tephra layers ejected by Jefferson during one of its last eruptions, indicating that the volcano produced violently explosive activity late in its history.

The Geology of Jefferson

Jefferson stands atop a rugged terrain averaging 5500 to 6500 feet in altitude; its cone is thus no more than a mile high, although some andesite flows extend several miles into adjacent canyon bottoms to elevations as low as 3000 feet. Condie and Swenson estimated that Jefferson is composed about equally of pyroclastic material and lava flows, but Sutton's investigations showed that pyroclastic deposits make up only 20 to 25 percent of the volcano. Much of what appears to be pyroclastic debris is instead broken rock derived from the shattering and fragmenting of lava flows.

Jefferson is probably the most deeply eroded high stratovolcano in Oregon, so the sequence of events that built its cone is clearly evident. Like McLoughlin, Jefferson began as a highly explosive vent and erected a large cone composed of tephra and other pyroclastic material. This fragmental cone, which forms the core of the volcano, has been partly exhumed by glacial cutting. The most extensive exposure of Jefferson's pyroclastic interior occurs in the West Milk Creek Cirque, where a Pleistocene glacier cut deeply into the west flank of the volcano.

The original pyroclastic cone was later buried under sheets of basaltic andesite, which attained an aggregate thickness of at least 2000 feet. These lava flows, designated the Main Cone lavas, are also exposed in the West Milk Creek Cirque, where they overlie the older pyroclastic deposits. Glaciated remnants of the Main Cone flows also appear along the west and northwest flanks of the mountain, in the east wall of the Russell Glacier valley, beneath the terminus of Jefferson Glacier, in the ridge between the main summit and the North Complex, and above Waldo Glacier.

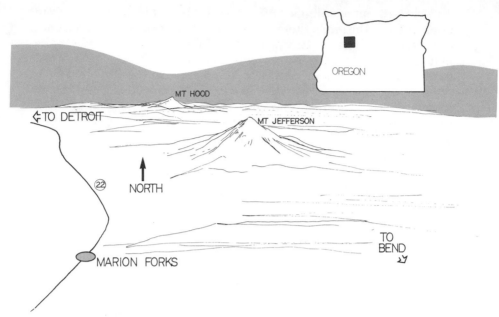

Relief map of the Mount Jefferson area.

These Main Cone lavas, some of which traveled seven miles or more into the nearby valleys, are relatively thin high on the mountain. On the northwest flank and in the east wall of the Russell Glacier valley, the lavas slant toward the north at about 20 to 30 degrees and range from five to 40 feet in thickness. Some lavas apparently poured east down steeper slopes of 30 to 45 degrees; most of these flows are about 15 to 20 feet thick.

Lava flows that form the north ridge between Jefferson Park Glacier and the Whitewater Glacier slant towards the north and west, indicating that there must have been an eruptive center east of the present main ridge near the head of Whitewater Glacier. The glacier apparently obliterated all traces of the subsidiary vent.

Sutton was unable to find either mudflow or avalanche deposits among the early pyroclastic material and main cone lavas, suggesting that these erupted during an interval of Pleistocene time when the volcano was mostly free of ice. After the flows of basaltic andesite composing the main cone had been emplaced, however, the climate turned colder and glaciers eroded parts of the volcano. Evidence of this erosion appears in the east wall of the Russell Glacier valley.

Following a dormant interval of unknown length, during which Jefferson's glaciers melted, the volcano began a new cycle of activity.

This episode produced andesite flows with a much higher silica content than the more fluid lavas that form the bulk of Jefferson's cone. These second stage flows were notably thicker and must have moved more sluggishly than the basaltic andesites. One 150-foot-thick andesite flow covered a 25 to 35 degree slope that had been cut in the main cone andesites by a glacier in the Russell Glacier valley. In one sequence, exposed in the wall behind Russell Glacier, a silicic andesite flow overlies glacial deposits that in turn lie atop the main cone lavas.

Renewed glaciation eroded much of the second stage cone, leaving only remnants of its flows capping the volcano's main peaks and ridges. A mass of this silicic andesite, approximately 100 feet thick, now forms the present summit of Jefferson. This flow erupted from a now vanished source west of the present summit.

Despite its resemblance to the spires of Thielsen and Washington, Jefferson's sharp peak is not, as is sometimes supposed, an eroded plug, a conduit-filling exhumed by glacial cutting. Instead, this flow remnant, dipping away from a crater that once existed above the present Milk Creek Glacier, indicates that Jefferson was once much larger and higher than it is now. Sutton estimated that the cone once reached an elevation as high as 12,000 feet. Glacial erosion after

Mt. Jefferson from the northwest, Jefferson Park in the foreground. —Photo by Bob and Ira Spring

eruption of the second stage lava destroyed at least a third of Jefferson's original cone, the summit and much of the western flank. Part of the volcano's original eastern slope became the present crest.

Intensely altered rock in the West Milk Creek Cirque walls may indicate the approximate location of the vent that produced the second stage lavas. North of this area, which was chemically transformed by heat and volcanic acids, is a mass of light gray lava that intruded the main cone lavas. Other second stage lavas flowed north and east from the summit area and now cap the cliffs north of the Whitewater Glacier. The crumbling banks of talus between the north and south summit horns, the Red Saddle, are also shattered remains of second stage lava flows.

Today the Whitewater Glacier on the east and the Milk Creek Glacier on the west continue to carve the mountain, digging ever deeper into the already narrow summit ridge. The glaciers will soon undermine the relatively thin barrier separating them, replacing the present low saddle between the north and south pinnacles with a deep cleft, perhaps allowing the two ice streams to merge.

While the second stage lavas were erupting, two large parasitic cones formed on the flanks of Jefferson. One of these secondary cones, the North Complex, is near Jefferson Park. Erosion stripped its original surface, revealing lava dikes that once fed flows. Since the last ice age, the creek issuing from the northwest part of the Whitewater Glacier cut further into the exposed lava flows, nearly obliterating their original form.

Another secondary vent opened two miles south of Jefferson's summit, producing a dome of reddish-brown dacite, Goat's Peak.

More fluid eruptions of silicic andesite erupted from fissures along the south flank of Jefferson. One poured from the top of the eastern fissure toward Goat's Peak before it cascaded over a ridge and turned southeast. Remains of this flow form the east ridge of the valley north of Shale Lake at an elevation of 7600 feet. More of the same flow underlies the cliffs southwest of Waldo Glacier.

Jefferson's Last Blast

In contrast to the relatively quiet outpouring of lava that built the volcano during its main stages of growth, Jefferson's final activity was violently explosive. Late in its history, but before the last two ice advances, Jefferson ejected large volumes of pumiceous tephra. In the early 1980s, Jim Begét discovered two previously unrecognized layers of airfall pumice mantling parts of the Metolius and Deschutes

River valleys in central Oregon. An older and thinner deposit, called layer U, is generally covered by a younger, thicker and more extensive tephra bed, layer E, which forms a blanket six to seven feet thick in areas about 12.5 miles downwind from the volcano.

The dating is uncertain, but perhaps at the same time that the tephra falls occurred, Jefferson also erupted high-temperature pyroclastic flows, containing blocks of dacite, that filled valleys west of the volcano. The tephra eruptions took place between about 40,000 and 140,000 years ago, when most of the Oregon Cascades were buried beneath a thick icecap. In the Jefferson region, this next-to-last advance of Ice Age glaciers is called the "Jack Creek Glaciation."

Since the tephra eruptions, which apparently represent Jefferson's final outburst, the volcano has been dormant or extinct for many tens of thousands of years. Like Washington and Three Fingered Jack, it has erupted no new lava to repair the ravages of the last two glaciations, which so deeply dissected its cone.

Other Formations in the Jefferson Area

An extensive lava formation in the Jefferson vicinity, known as the Santiam basalts, formed some time before the last major ice age. Like Jefferson, the Santiam basalts have normal magnetic polarity, which means that they erupted since the last significant reversal of the earth's magnetic poles approximately 700,000 years ago. Santiam basalts fill the North Santiam River valley to a depth of 1800 feet. There is no contact between these basalts and the Jefferson lavas, so it is impossible to determine their relative ages.

After the ice age glaciers melted, several pyroclastic cones and related canyon filling basalt flows erupted south of Jefferson. Prominent examples include the Forked Butte Cinder Cone with its associated flow in the valley of Cabot Creek, and the Bear Butte lava flow, which descended the valley of Jefferson Creek. Another young

Basalt vesicles one to two inches long.
—photo courtesy D.W. Hyndman

169

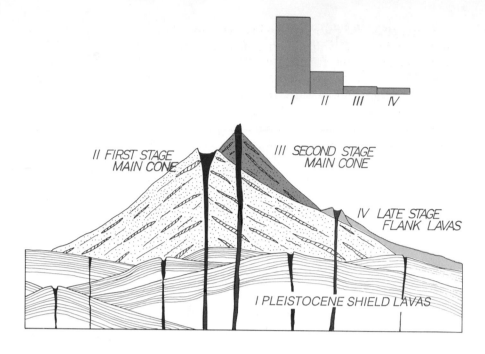

Simplified cross-section of Mt. Jefferson, showing relative volumes of lava erupted during the four eruptive cycles in the Jefferson area.
—After Sutton, 1973 and Williams and McBirney, 1979

pumice cone, not recognized until recently, is about a quarter mile northeast of Goat's Peak. Although partly eroded, it retains a conical form; a basalt lava flow extends east from its southeast flank.

Age and Volume of Eruptions in the Jefferson Area

All the lavas near Jefferson are younger than the last magnetic reversal, and therefore erupted during the last 700,000 years. The earliest of these, the Minto lavas, formed a broad plateau of coalescing shield volcanoes that were thoroughly glaciated before Jefferson was born. Both Jefferson's main cone and second stage lavas are younger than the Minto group, but older than the Jack Creek glaciation. Part of the cone may be older than the Abbot Creek glaciation of 120,000 to 200,000 years ago. The Forked Butte lava flows erupted after the last ice age which ended 10,000 to 12,000 years ago. The latest cinder cone activity has been tentatively dated at 6400 to 6500 years old.

170

In general, the younger the lava the less there is of it. The Minto lavas are the most voluminous, approximating 23.4 cubic miles; the main cone lavas total about five cubic miles, while those of the second stage are less than one cubic mile. The Santiam basalts represent about two cubic miles and the Forked Butte lavas only 0.8 of a cubic mile.

Thus, although Jefferson is the most imposing volcano in this part of the High Cascades, its volume is minor compared to the older flows from shield volcanoes and fissure eruptions. This is typical of the central Oregon Cascades, where some individual basaltic flows with little topographical relief but large areas, have volumes two or three times greater than those of a high stratovolcano.

Judging by its glacially denuded cone and its lack of eruptions since before the last two ice ages, Jefferson may well be extinct. Any future activity may be confined to the eruption of a few cinder cones and lava flows. On the other hand, the progressively higher silica contents of Jefferson's eruptive cycle may indicate that a silicic magma chamber is developing beneath its cone. Many of the world's most cataclysmic eruptions have involved volcanoes that had been dormant for thousands or tens of thousands of years. Reports of Jefferson's demise may be exaggerated.

Columnar jointing in plateau basalts. —photo courtesy D.W. Hyndman

Hood from the north, Eliot Glacier in the center. The smooth ridge at the left center of the photo is Cooper Spur. —U.S. Geological Survey photo by Austin Post

XIV
Mt. Hood:
Oregon's Most Popular Volcano

When the pioneers' covered wagons lumbered over the dry sage-brush plains of eastern Oregon on their way to the Willamette Valley, the travelers' first sight of Hood's glittering white spire promised a land of fresh, snow-fed waters, cool evergreen forests, and a long-hoped-for respite from the arid desert that stretched between the Rockies and the Cascades. The Cascade barrier had yet to be crossed, either via the Columbia River or Barlow Pass, which skirts Hood's southern base, but the end was in sight.

Hood is still a delightful sight to the descendants of those hardy people. Viewed from Portland, Hood seems a classically symmetrical cone terminating in a sharp peak. Seen from other directions, the volcano is highly asymmetrical. Eroded by glaciers that stripped away much of its original bulk and reduced its height by 1000 feet, Hood nonetheless is one of the most beautiful mountains in the world. It appears much loftier than its 11,245 feet.

As the second most often climbed snowpeak in the world—the first being Japan's sacred volcano, Fujiyama—Hood attracts thousands of alpinists, hikers, and other sportsmen. Because some slopes remain snow covered throughout the year, it is the site of an international summer ski school and competition. Best known of Hood's man-made attractions is Timberline Lodge, at the 6000-foot level on the mountain's south side. Constructed of native materials handicrafted in Indian and wildlife motifs, the luxurious and picturesque lodge makes a good point of departure for exploring the mountain.

Hikers pushing upslope from Timberline Lodge find Crater Rock a natural goal. A steep lava dome located about 1000 feet below the

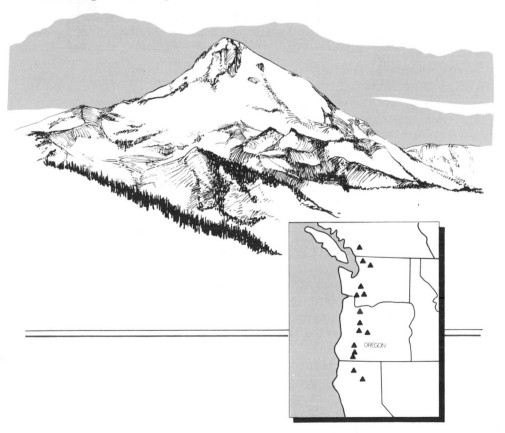

Drawing of Mount Hood from the north, showing its location in the Cascade Range.

summit ridge, it was emplaced only about 200 years ago during Hood's last major eruptions. Crater Rock, the nearby Hot Rocks, the base of Steel Cliff, and other adjacent areas comprise the largest solfataric field in Oregon. Here the corrosive effects of heat and acidic gases have altered the lavas to soft, crumbling masses, colored pink, white, orange, and yellow. Some vents leak vapor quietly; others on the north flanks of Crater Rock spew steam columns tens of feet high. Hood's fumaroles are the source of a powerful sulphur odor that climbers can smell more than a mile away. The temperature of many vents is approximately that of boiling water.

When escaping clouds of gas condense to form a visible plume curling about the mountain top, the Timberline Lodge staff occasionally receives phone calls from anxious Portlanders to whom the steam appears to portend a new eruption. Except for a brief period in 1907, the gas plumes do not seem to reflect increased heat emission. But they are reminders that Hood's volcanic fires still smolder. Hood is the only volcano in Oregon with a well-authenticated record of observed eruptions. It may also be the next Northwest volcano to erupt.

During the last century Hood and St. Helens erupted almost simultaneously. Both staged major outbursts about 1800 and both remained intermittently active through the mid-nineteenth century. Although Hood was quiet in 1845 when Samuel Barlow and Joel Palmer led a wagon train across its south flank, it produced a series of mildly explosive eruptions between 1853 and 1869, one or more of which may have deposited the pumice fragments now found scattered across its southern and eastern slopes.

A. Perrey, who kept month-by-month reports of earthquakes and volcanic eruptions world-wide, noted that Hood was active in 1853, August, 1854, August 15-17, 1859, and September 21 to October 8, 1865. Although Perrey seems accurate, his notations are brief and his sources generally unknown. It is not until 1859 that we have a detailed account of an eruption on Hood. Writing in the *Everett* (Wash.) *Record* of May 17, 1902, W.F. Courtney, an Oregon pioneer, recalled that:

> The eruption took place during the later part of September, 1859 . . . We were camped on Tie Ridge about thirty-five miles from Mt. Hood . . . It was about 1:30 o'clock in the morning . . . when suddenly the heavens lit up and from the dark there shot up a column of fire. With a flash that illuminated the whole mountainside with a pinkish glare, the flame danced from the crater . . . For two hours, as we watched, the mountain continued to blaze at irregular intervals, and when morning came Mt. Hood presented a peculiar sight. His sides, where the day before there was snow, were blackened as if cinders and ashes had been thrown out.

Aerial view of Mt. Hood from the southeast. Most of the visible cone is thickly mantled in pyroclastic flow and mud flow debris from Hood's last three eruptive cycles. The White River Glacier is in the foreground. Crater Rock, the thumb-shaped dome just below the summit, was erupted between 200 to 300 years ago and is still steaming. Newton Clark Glacier is on the extreme right.
—Photo by Austin Post, U.S. Geological Survey

Courtney adds "That was the only time that I ever saw flames issue from the crater, but I was a member of a party at one time when we encountered hot cinders on the mountainside." The sighting of actual flame and finding hot cinders indicates that fresh, possibly molten magma was discharged in 1859.

The Portland *Weekly Oregonian* of August 20, 1859 reported that "The Pride of Portland" was then in eruption. Although the *Oregonian* gives an earlier date for the activity than did Courtney, it seems likely that both accounts refer to the same event. After 43 years Courtney may have forgotten the exact month. Under the heading "Eruptions of Mt. Hood," the *Oregonian* stated:

> On Wednesday last, the atmosphere suddenly became exceedingly hot about midday. In the afternoon the heavens presented a singular appearance; dark, silvery, condensed clouds hung over the top of Mt. Hood. The next day several persons watched the appearance of Mt. Hood until evening. An occasional flash of fire could be seen by every one whose attention could be drawn to the subject. Yesterday the mountain was closely examined by those who have recently returned from a visit to its summit, when, by the naked eye or a glass, it seemed that a large mass of the northwest side had disappeared, and that the immense quantity of snow which, two weeks since, covered the south

side, had also disappeared. The dense cloud of steam and smoke constantly rising over and far above the summit, together with the entire change in its appearance heretofore, convinces us that Mt. Hood is now in a state of eruption, which has broken out within a few days. The curious will examine it and see for themselves.

The *Oregonian*'s claim that "a large mass of the northwest side had disappeared" and that the wholesale melting of snowpacks occurred on the south flank must be an exaggeration. No evidence of massive rockfalls or mudflows dating from this period has been recognized. The *Oregonian*'s emphasis on heat, humidity, and other atmospheric phenomena sounds suspiciously as if some of the "eruption" clouds could be accounted for by an unusually heavy condensation of the steam and gas effusions that normally rise from Crater Rock. Courtney's observation, from a high ridge much nearer the mountain than Portland, that Hood's snowcap was "blackened as if cinders and ashes had been thrown out" rather than extensively melted is certainly the preferred explanation of any sudden change in the mountain's appearance.

Six years later, on September 26, 1865, the *Oregonian* again reported an eruption. "It is some time since we have had an excitement about old Hood belching forth, but on Saturday last the active puffs, in dense black smoke, were witnessed by hundreds of people in this city. The fumes were so thick as to literally obscure the view of the summit at times." Since the source of the "dense black smoke" was a "deep gorge," apparently low down on the southwestern flank of the mountain, the phenomenon described could have been a forest fire. However, the reference to "puffs" rather than billows or columns of smoke hints at a mildly explosive ejection of vapor, similar to the way St. Helens expelled "puffs" of steam and ash in 1854.

The same issue of the *Oregonian* also carried a plausible eyewitness report that tends to rule out the theory that a forest fire was mistaken for an eruption. John Dever, of Company E. First Regiment, Washington Territory Volunteers, was stationed at Vancouver, during the early morning of September 21. He wrote the Portland newspaper that:

> Between the hours of 5 and 7 o'clock, and as the morning was particularly bright for this season of the year, my attention was naturally drawn towards the east . . . Judge, then, of my surprise to see the top of Mount Hood enveloped in smoke and flame. Yes, sir, real jets of flame shot upwards seemingly a distance of fifteen or twenty feet above the mountain's height, accompanied by discharge of what appeared to be fragments of rock, cast up a considerable distance, which I could perceive fall immediately after with a rumbling noise not unlike distant thunder. This phenomenon was witnessed by other members of the guard.

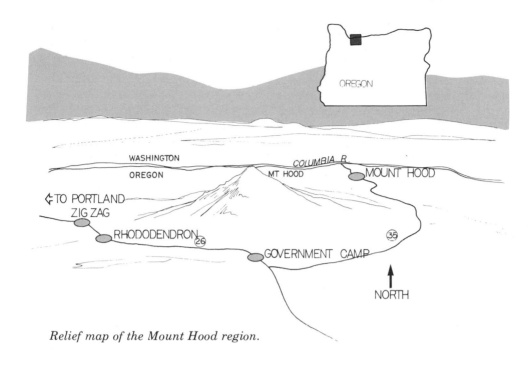

Relief map of the Mount Hood region.

Like Courtney in 1859, Dever saw clouds of "smoke" and "flame" issue from the volcano's summit rather than from a "deep gorge" as the *Oregonian* had reported. Dever's clear and precise description of flying shards of rock thundering downslope could apply only to a genuine eruption. On November 17 the same paper noted that Hood was again "steaming voluminously," repeatedly "jetting forth" ". . . puffs of black smoke."

An unpublished letter by Franklin A. Hinds, who witnessed the 1865 eruptions from Portland, reveals that Mount Hood remained sporadically active into early 1866. Writing to his family on January 28 of that year, Hinds advises his recipients not to believe press-exaggerated versions of the event, but that he can personally vouch for the reality of Hood's eruptions. He remarks that the volcano had been "smoking" for the previous three months and that "[he] saw it last eve just before sundown give unmistakable evidences [sic.] of emit[ting] smoke . . ." Laconically, Hinds adds that Oregonians are not overly impressed by an active volcano located fifty miles distant. Hood's flare-ups do not "create as much excitement here as you would naturally suppose, the morning paper speaks of it as an item of news and it is soon again forgotten in the hum of business." Then as now, the public is characteristically indifferent toward a potential volcanic threat.

Plummer cites another eruption in 1869, but no other written account of it seems to have survived. Except for the wreaths of smoke that frequently rise from Crater Rock to encircle the peak, the only 20th century incident involving a volcanic flareup was seen by a camping party in August, 1907. A member of the U.S. Geological Survey, A.H. Sylvester, was bivouacked at Government Camp, five miles south of the summit, when he noticed increased clouds of steam rising from Steel Cliff, slightly east of Crater Rock. Later that day (the 28th) his companions saw a dense column of smoke or steam rise from Crater Rock and extend high above the mountain's summit. This emission of vapor lasted throughout the day.

When night fell a member of the party, with the aid of field glasses, observed a glow from behind Crater Rock which he described as "looking like a chimney burning out." The next morning the White River, which originates from a glacier between Crater Rock and Steel Cliff, suddenly swelled from a gentle stream to a rushing watercourse triple its previous volume. Since weather conditions could not account for the abrupt increase in the river's flow, Slyvester concluded that volcanic heat had partly melted the White Glacier.

The Old Maid Eruptions

If Lewis and Clark had timed their trip down the Columbia River a few years earlier or later, they might have witnessed one of Hood's more destructive eruptions. According to a new study by Kenneth Cameron and Patrick Pringle of the Cascade Volcano Observatory, Hood erupted intermittently between about 1760 and 1810, producing numerous thin ashfalls, large-volume mudflows, extrusions of viscous lava at Crater Rock, and erupting at least one large pyroclastic flow into the upper White River valley. This eruptive cycle is known as the Old Maid eruptive period, named for the extensive mudflow deposits underlying Old Maids Flat along the Sandy River on Hood's west flank.

The Old Maid cycle probably began with a series of vent-clearing explosions that blanketed the mountain with thin layers of lithic ash. The erupting vents were apparently located between Crater Rock and the summit ridge so that little material avalanched directly southward onto the older Timberline debris fan. The presence of the Crater Rock lava dome deflected most of the new ejecta southeastward into the White River drainage. Lesser amounts were directed westward into the Sandy River and Zigzag River drainage.

During the first eruptive sequence, masses of dacite oozing from Crater Rock vents sent avalanches of hot rock onto the White River

Glacier. Glacier meltwater generated a series of large mudflows that formed the deposit known as Mesa Terrace in the White River canyon. Ash layers preserved in these mudflow deposits suggest that explosive activity and/or rockfalls cascading from growing lobes of the Crater Rock dome took place intermittently.

Between about 1770 and 1780 a second pulse of the eruptive cycle sent a mudflow traveling down Mount Hood's west slope into the Zigzag River valley at least as far as the town of Brightwood. This mudflow buried Old Maids Flat, dammed Muddy Fork and Lost Creek, tributaries of the Sandy River, and caused lakes to form in the impounded waters of these streams. Runout from the mudflow traveled as far as the Columbia River, significantly transforming the Sandy delta area.

Effects of this mudflow were still evident when Lewis and Clark explored the lower Sandy River in 1805 and 1806. Clark's journal describes the Sandy as a shallow river choked with sediment and manifesting a quicksand bottom. By contrast, the present Sandy River runs through a deep narrow boulder-lined channel. In 1792, Lieutenant Broughton of Captain Vancouver's expedition reported that he had observed logs stranded twelve feet above the river level and that a shallow bar of sediment extended from the mouth of the Sandy across the Columbia River. The presence of this bar, which the Columbia's mighty current would soon wash away, suggests that the Sandy River mudflow had occurred only a short time before Broughton's visit.

The climactic event of the Old Maid cycle occurred about 1800 when a massive pyroclastic flow was erupted into the White River Basin. Perhaps generated by the collapse of an elevating dome, the avalanche of hot rock traveled downslope as far as the point where Highway 35 crosses the White River. Large ashclouds accompanying the pyroclastic flow were probably responsible for killing the trees that form the modern "Ghost Forest," a stand of branchless snags high on Mount Hood's southeast shoulder.

The pyroclastic flow, which left deposits up to eighty-five feet thick, triggered two extensive mudflows. One swept the entire length of the White River, inundating Tygh Valley before emptying into the Deschutes River. Another mudflow, which took place between 1800 and 1810, extended downvalley only as far as the bridge below White River Station. Smaller mudflows were confined to the upper White River canyon.

A layer of light gray ash that covers most of Hood's lower slopes was also deposited during the Old Maid episode. As much as about seven inches deep near the Tilly Jane Guard Station on the northeast side, it has a thickness of about 5.5 inches on Yokum Ridge 2.5 miles west of the volcano's summit, about 2.5 inches on Cooper Spur and near Polallie Creek. Presence of the ash on all sides of Hood suggests that it was deposited over a fairly long time, while the winds repeatedly shifted direction. The ash was not fresh magma, but older rock, perhaps pulverized in steam explosions that cleared the volcano's throat during the first stage of the eruptions, or fallout from pyroclastic flows.

Hood's Geologic History

Thanks to the work of W.S. Wise, Craig White, and D.R. Crandell, upon whose reports the following summary is largely based, we now have a comprehensive picture both of Hood's early growth and post-glacial activity.

When Hood began to erupt, perhaps relatively late in Pleistocene time, the surrounding Cascade topography was apparently very much as it is now. The average elevation in this area 20 miles south of the Columbia River is and was about 4500 feet, but individual peaks stood at least 2000 feet higher. The streams, which now originate from Hood's glaciers—the Zigzag, Sandy, White, and Hood rivers—already flowed, although they then occupied narrower canyons. The Pleistocene glaciers were yet to remodel them into broad valleys.

As Hood grew, it gradually buried at least one ancestral volcano that had built a sizable cone during the late Pliocene or early Pleistocene. Erosion has exhumed remnants of this structure beneath Sandy Glacier on Hood's west flank. According to Terry Keith of the U.S. Geological Survey, thin flows interbedded with layers of tuff and cut by nearly vertical dikes are visible at the head of Muddy Fork.

During its first decade, the new volcano may have grown several thousand feet. Early andesite lavas flowed as far as eight miles down ancient river canyons. A few individual flows were 500 feet thick, half a cubic mile in volume. These flows, the oldest yet recognized on the mountain, now form several long ridges extending from the base of Hood. The streams they displaced hundreds of thousands of years ago have since cut new canyons on either side of each flow, leaving them standing as high divides between valleys. Such reversal of topography appears on nearly every Cascade volcano that has been subjected to long-term erosion.

After the first voluminous eruptions of lava ceased, Hood's eruptions became less frequent and intermittently explosive. Much of the upper 4000 feet of the cone is built of pyroclastics—many of which were probably derived from shattered flows—interbedded with relatively thin lava flows. Although at least 70 percent of the edifice

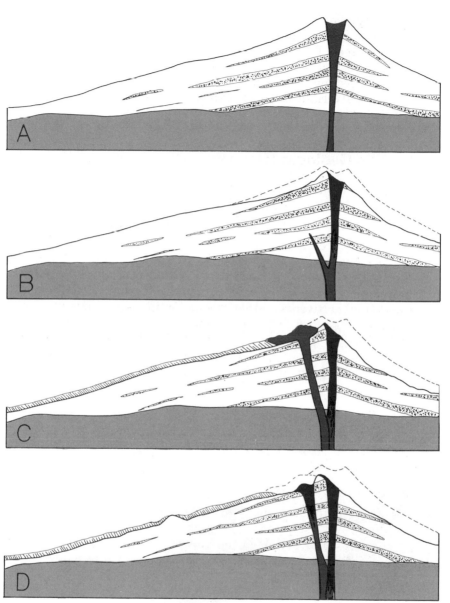

Four of the principal stages of development of Mt. Hood: A. At the time of its maximum height. Dark lines indicate lava flows; dotted areas represent deposits of pyroclastic material. B. After extensive glaciation. The dotted line above the cone indicates the former profile of the volcano. C. Eruption of a large hornblende andesite-dacite plug dome and formation of the debris fan on the south flank during the Timberline eruptive period about 1700 years ago. D. Mt. Hood today, after another dacite dome, Crater Rock, was erupted just south of the summit during the Old Maid Flat eruptive period of about 200 years ago. —Scheme based drawings by W.S. Wise (1966) and used here with his permission

181

consists of lava streams, Hood's cone contains a larger proportion of loose fragmental rock than most of the other principal Cascade volcanoes. Perhaps only St. Helens contains a higher percentage of pyroclastic material.

A few thick flows erupted from vents at 8000 feet or higher. Remnants of these form Barrett Spurr on the north flank and Steel Cliff. The latter, a massive outcrop east of Crater Rock, is so thick that it may have filled a high glacial basin. Andesite flows typically exceed a depth of 50 feet only on slopes of less than eight degrees, or when they are contained in canyons.

Expanding glaciers repeatedly sheathed the rising volcano, so that as hot rock was emitted onto icefields, meltwater generated large mudflows that poured into valleys heading on the mountain. Some old mudflows left thick deposits near the volcano's base, while others streamed many tens of miles west down the Sandy River valley and north down the upper Hood River Valley, perhaps as far as the Columbia River. Portland's east side is partly underlain by a succession of early and middle Pleistocene mudflows from Hood.

At its zenith, Hood stood 12,000 feet high, a full 8000 feet above its Cascade foundation. With a base spreading over 92 square miles, the cone attained a volume of 45 cubic miles. No other volcano in Oregon, except Mazama, could match it in bulk or stature. To the north, only Adams and Rainier were larger.

Summit of Mt. Hood from the southeast. Bare patches in the snow are sites of fumaroles and hot rocks. Steel Cliff, a massive flow or dome, appears on the extreme right. —Photo courtesy of the U.S. Forest Service

After Hood reached its maximum size, eruptions apparently ceased at the summit crater. Two new vents opened on the north and northeast flanks to erupt the largest lava flows in the volcano's later history. Some poured 15 miles into the upper Hood River Valley. The plug of one satellite vent forms The Pinnacle, a sharp spire to which three flows can be traced. The second vent, near Cloud Cap Inn, lies beneath moraines from Eliot Glacier.

When the Fraser Ice Age began, about 29,000 years ago, Hood may have been as symmetrical as St. Helens was before the 1980 eruptions. During the last glacial advance, the mountain was the center of an icecap that almost covered the volcano and extended far down adjacent valleys, particularly on the north and east sides. The grinding ice removed 1000 feet from the sides and top of Hood, transforming it from a smooth cone to the four-faceted horn it is today.

Glaciers scoured the peak so thoroughly that they exposed the original eruptive conduit, now a column of solidified lava. Recent investigations suggest that Hood's primary crater was *north* of the present summit ridge, in what is now thin air, rather than at the site of Crater Rock as had been supposed. The conduit plug is now virtually inaccessible among the precipices and ice-falls of the north face.

Hood's Latest Eruptive Cycles

Hood has experienced four principal eruptive episodes since the climax of the Fraser Glaciation about 18,000 years ago. All four produced dacite domes, pyroclastic flows, and mudflows. None are known to have produced lava flows or significant quantities of pumice. The earliest, the Polallie eruptive period, occurred between about 12,000 and 15,000 years ago, when Hood's glaciers had shrunk, but were still much larger than they are today. The glacier cover at least partly determined where freshly erupted material was deposited.

During this late Fraser activity, a series of silicic domes erupted at or near the summit of Hood. As the domes grew, avalanches of rock debris from their crumbling surfaces or from explosions along their bases repeatedly moved down all sides of the volcano, particularly the north and east flanks. Today Polallie deposits—pyroclastic flows interbedded with mudflows—are typically found on ridge-tops, valley walls, and valley floors. Where glacier ice filled canyons or cirques, pyroclastic debris accumulated on adjacent ridges or rocky divides between glaciers. As glaciers withdrew from valley sides, debris filled in between ice margins and valley walls. As valley glaciers retreated upslope, volcanic rocks accumulated on valley floors beyond the shrinking glaciers.

According to Crandell's study of Hood's late-glacial and post-glacial eruptions, typical Polallie age deposits are exposed on the mountain's northeast side, at the head of Polallie Creek, about half a mile southwest of Tilly Jane Guard Station. Perhaps the most impressive accumulation of Polallie debris underlies a broad triangular apron whose apex forms Cooper Spur (8500 feet) on Hood's northeast slope. A huge wedge-shaped mass of loosely compacted rock fragments rising above the east margin of Eliot Glacier; Cooper Spur points upslope toward its source in a now-vanished high-altitude dome or domes. The Polallie eruptions apparently continued intermittently for a long time; intervals of erosion occurred between successive volcanic deposits.

The Parkdale Soils, a yellowish-brown ash layer as much as 14 feet thick lie east and north of Hood, particularly in the upper Hood River Valley. They apparently accumulated during the Polallie eruptive period as fallout from ash clouds that billowed above pyroclastic flows, and drifted northeast on the prevailing winds.

By the close of the Polallie eruptive episode, Hood was probably a gray and grim spectacle. Shrouded in dark ash and virtually buried in dacite rubble, accumulations of pyroclastic rock formed broad aprons extending miles downslope and merged into mudflows that radiated many tens of miles downstream from the volcano. On the north side, mudflows reached as far as the upper Hood River Valley, and may have extended to the Columbia River.

The Timberline Phase

After a long dormant interval that apparently lasted more than 10,000 years, Hood reawakened, blasting a new crater high on the south slope, at about the 10,000 foot level, to begin its Timberline eruptive episode. Again, the dacite magma was too viscous to create lava flows, and another dome extruded, perhaps at or near the site of Crater Rock. Pushing through snow and ice, the rising dome shattered and crumbled, initiating hot avalanches of angular blocks as much as 20 feet in diameter. As it mixed with melted snow and newly erupted ash, the hot debris turned into mudflows that swept down the mountainside to engulf entire forests. Radiocarbon dates on charred wood indicate that these eruptions occurred between about 1400 and 1800 years ago.

Although the mode of eruption—domes, pyroclastic flows, and mudflows—duplicated that of the Polallie eruptive cycle, the areas affected were much smaller, restricted primarily to the south and southwest flanks of the volcano.

Pyroclastic flows and mudflows repeatedly descended the Zigzag and Little Zigzag canyons and the Salmon River valley on the south and southwest sides of Hood. They followed the Sandy River as far as the Columbia, where you can see them near Troutdale and Interstate Highway 80N, 50 miles downstream from the volcano.

Timberline ash lies about 6 to 24 inches deep on the south, east, and northeast sides of the volcano as far as six miles from the summit.

Recent and Future Activity

Since the Timberline episode, Hood has erupted frequently. The Zigzag eruptive period, which sent a pyroclastic flow into the Sandy River basin and mudflows down the Zigzag River, took place between about 400 and 600 years ago. Only a couple of centuries elapsed between this event and the Old Maid eruptive cycle, about 1760-1810, with the observed historic eruptions following a few decades later. Hood's latest activity scattered light gray pumice lapilli over the south, east, and northeast sides of the mountain, perhaps in 1859 or 1865.

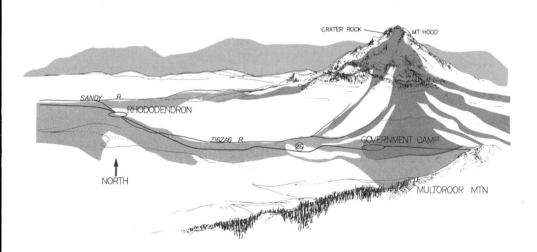

Volcanic hazards map of the Mt. Hood region, showing areas likely to be endangered from future eruptions. Pyroclastic flows derived from dome emplacement and collapse near Crater Rock and accompanying ash clouds and lateral blasts may devastate the south and southwest flanks of the cone, and the upper reaches of valleys heading on the volcano. Floods and mudflows may reach tens of miles down valley, some to the Columbia River. —After Crandell, 1980

It may be significant that Hood's latest eruptions have roughly coincided with eruptive cycles at St. Helens. Since both volcanoes erupted about 1500, 1800, and in the mid-nineteenth century, Hood may soon awaken to keep its Columbia neighbor company. As Crandell suggests, "Mount Hood will almost surely erupt again, and *possibly in the very near future . . .*"

When Hood becomes the second currently active volcano on the Portland skyline, it will probably behave as it did during the Timberline, Zigzag, and Old Maid eruptive periods. The present summit configuration may largely determine the effects of future eruptions. If the next eruptions are of about the same size and volume of the Old Maid activity and occur at similarly placed vents south of the summit ridge, Crater Rock will probably again deflect most of the erupted material southeastward into the White River valley canyon or westward into the Sandy and Zigzag River basins.

Unless Hood radically alters the behavior pattern set during the last 15,000 years, it will not erupt either lava flows or large quantities of pumice. Initial vent-clearing explosions will mantle the surrounding countryside in gray ash derived from pulverizing of old rock inside the cone. Thick, sticky tongues of dacite lava will ooze from vents near Crater Rock, collapsing to create avalanches of incandescent rock that will sweep down the White, Zigzag, or Sandy River valleys. Meltwater will again send floods and mudflows many tens of miles downstream, imperiling towns and other settlements all the way to the Columbia. As it did in the 1790s, debris swept into the Columbia may block the river, temporarily closing it to shipping.

Indian Lore

A discussion of Mt. Hood would be incomplete without some notice of the Indian legends it inspired. The most famous of these, the "Bridge of the Gods" myth, belongs to the chapter on St. Helens, but the one accounting for the supposed "Chief's Face" amid the crags and shadows on Hood's north side is equally significant and more poignant. Both legends illustrate that the peak, which now seems benign in its airy lightness, was in former times the incarnation of divine anger and the realm of spirits hostile to man.

According to a popular version of the Chief's story, Hood once towered so high that when the sun shone on its south side a shadow stretched north for a day's journey. It was majestic but also fearsome, for inside dwelt demonic spirits who sent forth streams of liquid fire to destroy the Indians' homes. One night a particularly brave chief had a vision in which a voice told him that he must conquer the evil forces that inhabited the mountain, lest they annihilate his people. The courageous warrior climbed the mountain, where he found a gaping hole from which issued fire and smoke. Hurling boulders into the opening in the hopes of killing the evil spirits who lived below, the chief was horrified when they were thrown back, heated red-hot. For many days the chief and the fire-monsters kept up the stone-throwing contest. Then, during a respite in the battle, the Chief looked down to the land he had left. His heart cracked in grief: the once-fertile valleys were desolate, choked with lava and ash. The rivers had dried up, the forest burned, and the people and animals fled. The chief's courage forsook him and he sank to the earth, to be buried by fresh streams of liquid rock. Today his sorrowful profile, with its distinctive scalp lock, can be seen halfway down the north side of the mountain.

Although some of the chief's people had taken refuge on hills above the devastated valleys and thus survived, the stricken land offered so poor a living that their children never attained the parents' former stature. Thus the Indians must remain small people until another brave leader can overpower the evil spirits hidden within the volcano.

How to See Mt. Hood

Driving the Mt. Hood loop provides some excellent views of the peak and its fellow Guardians of the Columbia, St. Helens and Adams. Highway 26, through Sandy, enters the Hood National Forest at Zigzag. Much of the west side of this highway parallels the old Barlow Road, the first to cross the Oregon Cascades. Near Government Camp, named for the contingent of U.S. Army Rifles who wintered here in 1849, take the road to Timberline Lodge, which affords a direct view of Crater Rock, Steel Cliff, and the recent debris fan. Only experienced climbers should attempt the summit.

To see the north side of the mountain, return to Highway 26, turn off on Highway 35 toward Hood River, and take a side trip to Cloud Cap Saddle. A short hike up from this 6000 foot elevation are viewpoints of Eliot Glacier, the second largest in Oregon. North across the Columbia River bluffs toward the Washington Cascades there is an incomparable panorama that includes St. Helens, Adams and Rainier.

Return to 35 and Hood River, where you take the Columbia River Highway back to Portland and Interstate 5. Include a stop at Cascade Locks where the legendary Bridge of the Gods once spanned the Columbia rapids. Multnomah Falls and Crown Point State Park are worth a special stop.

The two-armed Lyman Glacier, center, flowing from the summit icecap, bites deeply into Mt. Adams' northeast face. Red Butte, a partly eroded cinder cone, sits atop the extensive lava field in the foreground.

—U.S. Geological Survey photo by Austin Post

XV
Mt. Adams:
The Forgotten Giant of Washington

Adams is the second highest peak in the Pacific Northwest, its 12,276-foot elevation exceeded only by that of Rainier. Despite its height, massive bulk, extensive glaciers and superb alpine meadows, Adams remains relatively little known.

Towering a mile and a half above the Cascade crest about 35 miles due east of St. Helens, Adams lies within a rugged wilderness of jagged ridges, deep glacier-cut valleys and barren fields of lava.

Although visible from Portland, numerous viewpoints along the Columbia River Gorge, and several towns in southeast Washington, Adams is far enough removed from the Northwest's larger population centers and main highways to require a special effort to reach its natural flower gardens and spectacular icefalls.

Despite its vast size and gleaming summit icecap, Adams' peculiar shape and profile do not please everyone. Instead of tapering gracefully to a pinnacle like Jefferson or sweeping grandly to a truncated summit like Rainier, Adams is broad, asymmetrical and squat. Its wide, almost flat summit robs the mountain of the soaring quality that characterizes Hood, its legendary rival across the Columbia.

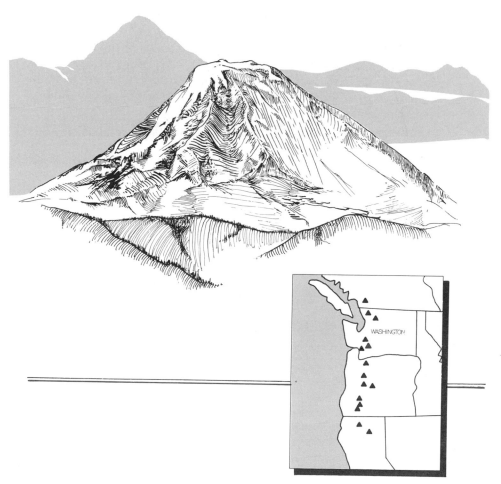

Mount Adams as seen from the south and showing its location in the Cascade Range.

A complex of several overlapping cones rather than a single symmetrical edifice, Adams is elongated along a north-south axis, giving it a broad, humpbacked appearance when viewed from east or west. The volcano consists of a steep main cone surrounded by a wide apron of gently-sloping lava fields that extends outward seven or eight miles. Adams' base, 18 miles in diameter, covers an area of 250 square miles. The estimated volume of the total structure—more than 85 cubic miles—makes Adams second after Shasta among the High Cascade stratovolcanoes! In contrast to the gentle lower slopes, the main cone above about 7000 feet rises steeply, with slopes between 15 and 35 degrees. Even steeper cliffs are common in the 2000-foot high headwalls of the Klickitat and Rusk glaciers on Adams' eroded east side.

Many small tributaries of the Klickitat, Cispus, Lewis and White Salmon rivers drain radially from Adams. The summit icefield and several glaciers, covering about 6.2 square miles of the upper cone, depend on annual rain and snowfall that average about 140 inches. Water drains through the highly porous, fragmented lavas of the main cone to emerge as springs and small streams along Adams' lower flanks.

The Mt. Adams Wilderness includes only the upper cone, a primitive terrain of glaciers, barren moraines and blocky lava flow surfaces, flowering meadows, huckleberry thickets and, below the 6000-foot level, a narrow circle of dense evergreen forest.

Relief map of the Mount Adams region.

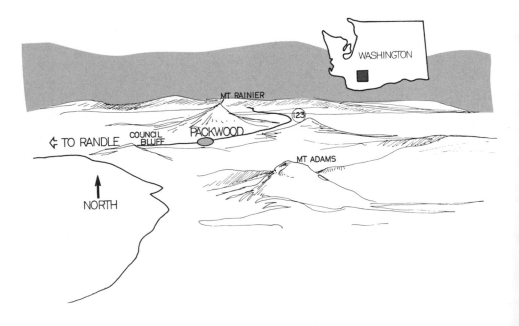

The wilderness area, 32,365 acres, borders the Yakima Indian Reservation, which in 1972 was awarded a grant of over 10,000 acres to be kept in a pristine state. The U.S. Forest Service manages land north, south, and west of Adams, creating a checkerboard of green timber and a steadily increasing number of clearcuts.

Adams' Glaciers

During the last major glaciation, about 29,000 to 10,000 years ago, ice covered approximately 90 percent of Adams. Today only 2.5 percent of the mountain's surface is mantled in glacial ice. Of Adams' ten principal glaciers, the Klickitat, Wilson, Lyman, Adams, and White Salmon glaciers descend from the summit icecap. Others originate in broad trenches or cirques excavated lower on the mountain. The Adams Glacier, a massive ice stream on Adams' northwest face, flows from the summit icefields in a series of spectacular icefalls down a steep channel, then spreads in a large sheet down to an elevation of about 7000 feet.

The Klickitat Glacier has carved an enormous amphitheater deep into the volcano's eastern flank. Two large ice streams plunging from the summit icecap feed this vigorous glacier, whose mile-wide cirque is the second largest of any active Cascade glacier, after that of the Carbon Glacier on Rainier's north slope. Klickitat Glacier terminates at an elevation of about 6000 feet. Down valley from its snout lie a succession of well-preserved moraines dating from the 14th to 20th centuries. They record a recent Neoglaciation and a subsequent series of ice retreats.

Hiking along Adams' well-named Ridge of Wonders, a thousand feet above the floor of Hellroaring Canyon, affords impressive views of the Klickitat and Mazama glaciers. No maintained trails lead to this rugged highland, but hikers venturing there are rewarded by an incomparable panorama of Adams' eastern cirques, cleavers, lava dikes, icefalls and crumbling piles of multicolored lava. Avalanches of ice, snow, and rock frequently thunder down the volcano's steep eastern face.

The Summit Sulphur Fields

In the 1930s Adams acquired the dubious distinction of becoming the only High Cascade volcano to have its crater invaded by commercial speculators. In 1929 and 1931, Wade Dean, a promoter from the Columbia River village of White Salmon, filed sulphur claims covering the 210-acre summit plateau. A horse and mule trail was built on the comparatively easy south slope and a diamond core drilling machine was laboriously hauled to the crater. Dean's employees dug several pits into the heat-decayed rocks of a 70-acre tract lying due north of the true summit. In 1934, the would-be miners drilled through 305 feet of ice and up to 38 feet of altered rocks, finding sulphates and sulphur in sludge from the drill holes. Despite Dean's advertising campaign, it seems unlikely that any significant amount of sulphur was ever mined or sold. By 1959, when the last assessment work was done, the estimated cost and trouble of attempting to remove the sulphur became prohibitive and the project was abandoned.

The 1930s prospectors apparently overestimated both the quantity and the quality of the sulphur obtainable. In the early 1980s Hildreth and Fierstein evaluated 60 samples of summit rock, and found only small quantities of sulphur. They concluded that while the potential sulphur resource in Adams' crater had been grossly exaggerated, "the difficulty of recovering it cannot be."

Indian Legends and Early History

According to the Columbia River Indians, Adams was one of three mountains that smoked continuously, the others being Hood and St. Helens. Like his brother mountain, Wyeast (Hood) across the Great River, Adams (Pahto or Paddo) was the son of the Great Spirit. Like Wyeast he also courted the fair St. Helens (La-wa-la-clough), the damsel of the trio. When St. Helens preferred Adams to Hood, the latter struck his northern brother a mighty blow that flattened his head. The legend attempted to explain Adams' ungainly, almost bulbous appearance, but it does not represent a geologic fact. Unlike Rainier and St. Helens, Adams' did not lose a formerly higher summit through volcanic action. In fact, the volcano probably still stands within 100 feet of its maximum height.

The story of how Adams received its rechristening by white men is only slightly less fantastic than the Indian legends about its romantic past. Although Lewis and Clark sighted the peak as early as 1805, they mistook it for St. Helens and consequently made no attempt to bestow an appropriate name upon what they took to be "the highest pinnacle in America." Since Captain Vancouver did not see the mountain, it did not already bear the name of yet another 18th century British diplomat. In the period 1830-1834, Hall J. Kelley, an enthusiastic American patriot, led a movement to call the Cascades the "Presidents' Range." All the major peaks were to bear the last names of former U.S. presidents. Oregon's Three Sisters and California's Shasta were to become Madison and Jackson, while Hood and St. Helens were to be renamed Adams and Washington. Adams, apparently unknown to Kelley, was not included in his plan.

Implementing Kelley's project, Thomas J. Farnham, working from indadequate maps, inadvertently interchanged the Kelley names for Hood and St. Helens. Farnham made a major error in latitude and placed "Adams" on his map about 40 miles east of St. Helens and north of Hood. As mountaineer Ray Smutek commented, "In what has to be one of geography's greatest coincidences, there was a mountain there to accept the name." Ironically, Adams was the only name in the Kelley-Farnham scheme that took, and it was applied to a mountain whose existence they never suspected.

Because of its relatively isolated position, Adams has not developed into a tourist attraction as have St. Helens and Hood. The volcano was likewise isolated from scientific attention. It was not until 1901, when a local settler and mountaineer, C.E. Rusk, guided the celebrated glaciologist Harry Reid on a circuit around Adams, that it was systematically explored and its largest glaciers named.

Another eighty years passed before the U.S. Geological Survey investigated Adams, including its potential as a geothermal resource. Most of our present knowledge of the volcano derives from the field work of two survey geologists, Wes Hildreth and Judy Fierstein.

Perhaps the most surprising finding about Adams is the relative youth of the main cone. The part of the volcano above 7000 feet was built during latest Pleistocene time, between about 25,000 and 10,000 years ago. That modern cone stands on the deeply eroded remains of older volcanoes. The oldest rocks date from about half a million years ago.

Like several other High Cascade volcanoes, Adams began to form in comparatively late Pleistocene time and grew sporadically in several distinct eruptive stages separated by long periods of inactivity. Voluminous cone-building episodes that lasted for decades or intermittently for a few centuries were separated by intervals of quiet that spanned tens of thousands of years. Potassium-argon dates on Adams' lava suggest that the volcano erupted between about 275,000 and 200,000 years ago and again between about 150,000 and 100,000 years ago. During each long dormant interval, glaciers destroyed much of the cone, each time reducing it to an elevation below 9000 feet.

Unlike its explosive neighbor St. Helens, Adams consists almost exclusively of lava flows. Hildreth and Fierstein were unable to find any widespread tephra deposits associated with the volcano. The rubbly, fragmental material that composes much of the volcano's core, partly exposed by glacial erosion, derives from brecciated lava flows. These deposits of angular, broken rock probably formed as lava flows shattered when they ploughed through icefields, or disintegrated on steep slopes. Similar breccias formed on Hood and Rainier.

During its phases of vigorous cone-building, Adams erupted copious streams of lava from central and peripheral vents. Many flows spread out into relatively thin sheets over the flanks of the cone, some apparently fed by fountains of molten rock like those that accompany modern eruptions of the Hawaiian volcanoes. Showers of molten fragments accumulated to form streams of lava. The wide, gently-sloping base that encircles the volcano's central cone consists largely of andesite and basalt lava flows that range from abut 20 to 200 feet in thickness. The thicker flows apparently ponded in canyons or other topographical depressions. The heavily brecciated flows above about 7000 feet are generally less than 20 feet thick.

A few dacite flows and occasional pyroclastic flows erupted early in the volcano's history, but the bulk of the mountain consists of andesite. Thus there was no progression in the chemical composition of Adams' lavas toward a more silicic component, as at some other Cascade volcanoes, such as Mazama and Hood.

Because the present main cone developed during the last Pleistocene glaciation, most of the exposed lavas were at various times covered by glacial ice. Magma erupted at Adams' central vent probably emerged beneath an icecap similar to, but significantly larger and thicker, than the one that now mantles the volcano's summit. Hot rock shattered as it met glacial ice, creating a chaotic jumble of lava fragments, some of which avalanched down the cone. Repeated eruptions of this type erected a central cone whose interior is composed largely of andesite rubble, friable rock that was further weakened by prolonged exposure to heat and gas emissions.

Where glaciers have cut deeply into Adams' interior, the volcano reveals a core of chemically altered breccia derived from shattered lava flows. Pervasive leaching by acidic hot water and steam converted much of the original rock into a mixture of kaolinite clay, quartz, sulphur, and iron oxides. The oxidized minerals, deposits of yellow sulphur, and white, green, and red-stained gypsum contrast vividly with the dark grays and blacks of Adams' summit lavas.

Holocene Activity

During the last 10,000 years, Adams has erupted at least seven times, mostly from vents below the summit, but at elevations above 6500 feet. One of the younger flows is the A.G. Aiken Lava Bed, a

Mt. Adams from the north, with the Potato Hill cinder cone in the foreground. Adams Glacier spills from the summit icecap to the right, Lava Glacier occupies a cirque in the center and Lyman Glacier extends icy arms to the left.
—Photo by Bob and Ira Spring

rubbly lava stream a half mile across at its widest point, which emerged from a fissure at the foot of South Butte and traveled about 4.5 miles. The flow surface supports only a few shrubs and trees, but must be more than 3500 years old because St. Helens ash of that age lies on it. The only recent lava not overlain by the St. Helens ash is the extensive Muddy Fork series of flows on Adams' north side.

The Summit

Adams has several summits, the result of cone-building eruptions from a cluster of vents. The long south slope rises to a false summit at about 11,500 feet. From this shelf a plain nearly half a mile across slopes gently up to the highest peak. From the south summit to about the 8000-foot level is a long buttress, Suksdorf Ridge, the only un-glaciated high part of the volcano. It was the site of repeated erup-tions during late Pleistocene time. The northwest summit, The Pin-nacle, is part of the main summit that was carved and steepened by the Adams and White Salmon glaciers. Adams' highest point is a small lava cone atop the summit plateau, approximately 800 feet above the southern false summit. The snow-filled crater is open along the west rim, as if part of the wall slid down the steep western face of the volcano. The summit icecap has cut into the east base of this cone, exposing thin flows of dark gray to black andesite.

Secondary Structures Along Adams' Flanks

Adams' central pile is surrounded by a least 30 secondary cones, shields, and lava-flow complexes. One of the most prominent is Little Mt. Adams, a symmetrical cinder cone about 450 feet high perched atop the Ridge of Wonders on Adams' southeast side. From a distance the cone seems freshly built, its steep slopes of reddish scoria resembling the walls of a Roman stadium. But it is older than the last glaciation because moving ice trimmed its lower slopes and deposited till against it.

South Butte, on Adams' south flank at an elevation of about 7500 feet, consists largely of lava flows, with some scoria. Its east slope has been truncated by the Gotchen Glacier. South Butte's lavas are younger than most of those composing Suksdorf Ridge, but are all glaciated, indicating that the structure formed about 15,000 to 20,000 years ago, when most of the mountain was shrouded in ice. The vent from which the young Aiken flow issued is at the foot of South Butte, but is not related to it, and is at least 10,000 years younger.

Goat Butte, a basaltic shield at Adams' east base, is the largest flank structure. It is probably older than 150,000 years. A recent lava flow, erupted from a vent on Goat Butte, is andesite and not related to the older shield.

The Potato Hill cinder cone, standing about 800 feet above the surrounding lava plain on Adams' north side, formed during late Pleistocene time. Lava flows erupted from its base extend into the Cispus River valley, where their surfaces are glaciated. Its location saved Potato Hill from bulldozing by Pleistocene glaciers. The cone's logged over surface wears a blanket of gray dacite ash, in places several inches thick, erupted from St. Helens in 1980.

The Trout Lake Mudflow

Adams has produced only one large debris flow since the last ice age, the Trout Lake Mudflow. It blocked the former Trout Creek drainage, forming Trout Lake, and buried the floor of the White Salmon River valley for 25 miles. The Trout Lake Mudflow followed an avalanche of altered rock high on Adams' western face. The avalanche removed a large section of the summit plateau's western rim, leaving The Pinnacle as a prominent outcrop. It also created the steep cliffs in the headwall of the White Salmon Glacier, which formed in the basin the debris flow opened.

The largest avalanche on Adams in historic time, the Great Slide of 1921, also started near the headwall of the White Salmon Glacier. A large quantity of rock and ice fell a mile to inundate an area of about one square mile in the upper Salt Creek region. Active steam vents were observed near the rockfall source three years after the event, suggesting that the landslide was precipitated by a small steam explosion. Hildreth noted that "debris avalanches of small to moderate volume" occur almost every year on Adams when masses of rock and ice tumble from the glacially undercut headwalls of the Adams, Lava, Lyman, Wilson, Rusk, Klickitat, White Salmon and Avalanche glaciers.

Present Thermal Activity

American Indian legends suggest that Adams was at least mildly active within tribal memory but the youngest recognized deposits— the Muddy Fork lava flows—were emplaced between about 2500 and 3500 years ago. Although Adams has no record of historic eruption, the presence of thermal anomalies, "hot spots," on the mountain and continuing gas emission from the summit area suggest that a heat source exists beneath the mountain. During exploration of the summit for sulphur deposits in the 1930s, Fowler discovered steam issuing only from a fumarole on the crater's south wall, but he found hydrogen sulfide issuing from numerous crevasses around the crater.

In 1924 Judge Fred W. Stadler discovered "a group of steam vents west of the summit and less than 1000 feet below it, in a canyon south of the west summit and above White Salmon Glacier." When Phillips explored the summit more than a decade later, he could find no steam

Mt. Adams' smooth south slope (left) contrasts with the eroded east (right) side. Icefalls spilling from the summit icecap carve the canyons into the volcano's east flank. Mazama Glacier is in the left center foreground; right center is the Klickitat Glacier and beyond it the Rusk. Mt. Rainier stands about 50 miles to the north. —Photo by Delano Photographics

197

vents, but he did discover a small fumarole about one quarter mile north of the summit in the center of a large snowfield that had a maximum temperature of 100.4 degrees Fahrenheit, the lowest his thermometer could register.

Gas emission on Adams' summit is highly diffuse, originating at many unmarked places, instead of from distinct fumaroles. Although high winds commonly rake the icy summit, climbers typically smell hydrogen sulphide. During a 1982 climb, the author smelled a strong odor of rotten eggs, and despite subfreezing temperatures, saw water running beneath the ice along the southwestern crater rim.

Future Volcanic Hazards

All of Adams' recent eruptions have quietly produced andesite lava flows. Future activity will probably resemble that of the recent geological past, producing flank emissions of andesite or basalt. Lava fountaining at or near the vent site may erect small spatter or cinder cones similar to those dotting the volcano's flanks.

Although Adams has produced only one large debris flow, the Trout Lake Mudflow, during recent time, small to moderate debris flows may occur in the future. The volcano's interior consists of a rubbly pile of hydrothermally altered andesite fragments. The same corrosive gases that deposit sulphur and sulphates in the porous summit rock combine with oxygen to produce sulphuric acid that continually seeps downward into Adams' pyroclastic core, leaching out minerals originally present in the andesite lava and leaving behind a residue of highly unstable altered rock. The upper part of Adams' central cone thus consists of a soft permeable core encased in a thin shell of solid andesite. In many locations glaciers have cut through the outer layer to expose the weak material within.

The Goat Rocks Wilderness

A few miles north of Adams, along the serrated crest of the Washington Cascades, lie the deeply eroded remains of a much older stratovolcano. This cluster of ridges and sheer cliffs, named Goat Rocks after the bands of mountain goats that gambol about its precipices, offers a stunning preview of what all the lofty Cascade volcanoes will one day become long after they are extinct. The Goat Rocks volcano may once have been a towering peak resembling today's large composite cones.

The Goat Rocks Wilderness, 82,680 acres, and the adjacent Goat Rocks Roadless Areas, 25,240 acres, lie in the Gifford Pinchot National Forest west of the Cascade divide, in the Snoqualmie National Forest east of the crest. Elevations range from 2930 feet along Upper Lake Creek to 8184 feet at Gilbert Peak, the highest remnant of the Goat Rocks volcano.

The area's volcanic history is complex, involving a number of vents that erupted during different periods of time. Some eruptions were extremely explosive; those left thick layers of rhyolitic ash and pyroclastic flow deposits. Ash flow tuffs up to 2100 feet thick are now exposed beneath the ridge from Tieton Peak to Bear Creek Mountain. These rhyolitic tuffs, about 2.5 million years old, may record an eruption large enough to open a caldera now partly buried under the Goat Rocks volcano. Eruptions continued into late Pleistocene time, but glaciers destroyed the original form of the Goat Rocks complex.

The view across the ruined foundations of a mountain that may once have rivaled Hood conjures an image of vanished majesty. This prehistoric colossus was brought to ruin by the inexorable forces of erosion that now work to level the younger Cascade volcanoes.

Exploring Mt. Adams

No paved roads reach Adams, but narrow, rutted dirt roads take one to Cold Springs, nearly at timberline on the south side. Drive to the village of Trout Lake from Randle on the recently improved Trout Lake-Randle Road or from Vancouver, Washington, on Highway 14 along the north bank of the Columbia River. From Trout Lake drive north, turning off the surfaced highway in about 1.5 miles onto Road 80, which becomes 8040 after a few miles. Cold Springs is 12.8 miles, from which one can hike to timberline and then ascend Suksdorf Ridge to the summit. Ice axes and crampons are recommended for a safe and successful climb.

A round-the-mountain trail encircles Adams at about the 6000 foot level, near timberline. Bird Creek Meadows, a particularly scenic alpine field on the southeast flank, is no longer open to vehicles, but you can walk there on the trail from the road's end. Trails also lead from the Mirror Lake-Bench Lake road to Hellroaring Viewpoint. Another trail from Bench Lake along the Ridge of Wonders is not well defined or maintained. To discover which trails are open, inquire at the Forest Service ranger station in Trout Lake or Packwood. For permission to hike or camp in the Yakima Indian Reservation on Adams' east flank, call or write the Yakima Agency in Toppenish, Washington, 509-865-2373.

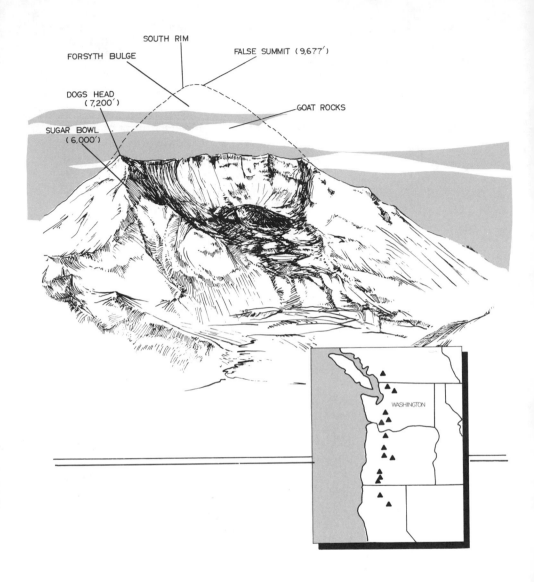

Drawing of post-1980 Mount St. Helens as viewed from the northeast, showing profile of the former summit and the volcano's location in the Cascade Range.

XVI
Mt. St. Helens:
A Living "Fire Mountain"

It was once known as the "Fujiyama of America" for the almost ideal perfection of its 9,677-foot cone. Before 1980, St. Helens rivaled Hood as the most graceful of the three guardians of the Columbia. One of the youngest and smallest of the principal Cascade volcanoes, St. Helens stood amid wooded hills and long steep ridges about 35 miles west of Adams. Twelfth in stature among Cascade peaks, it rose nearly 6500 feet above the crystal waters of Spirit Lake at its northern base. Although almost untouched by erosion, it supported 11 glaciers, the largest of which, the Forsyth and Wishbone, mantled its north face in thick sheets of crevassed ice.

St. Helens was first sighted by the British explorer George Vancouver in October, 1792 and named after his friend, Alleyne Fitzherbert, whom King George III had recently dubbed Lord St. Helens for his service as ambassador to the Spanish court. Northwest tribes, such as the Klickitats, who knew this youthful volcano more intimately, called it Tah-one-lat-clah—"Fire Mountain." That name was more appropriate. During the last 4500 years, St. Helens has been the most frequently and violently active volcano in the 48 adjacent states.

Mt. St. Helens before 1980. This north-side view, with Spirit Lake, shows the dark mass of the Dog's Head dome on the left, the Forsyth Glacier center, and the Goat Rock dome (erupted in the 1840s) on the right. —Photo by Bob and Ira Spring

A sharp earthquake at 3:45 p.m., March 20, 1980 marked St. Helens' return to life after 123 years of repose. That magnitude four earthquake ushered in a swarm of tremors that peaked on March 25, when seismographs were so thoroughly saturated that they could no longer distinguish individual shocks. Although the total number of earthquakes declined after March 25, larger jolts of magnitude 3.2 or above continued at a sightly increasing rate during April and May. Shocks greater than a magnitude of four occurred at an average rate of five per day in early April and eight per day during the week preceeding May 18. Like most quakes heralding a volcanic eruption, these centered at shallow depth. Most came from beneath the north flank of St. Helens, apparently as rising magma fractured crustal rock.

Heat and gas rising from the underground magma caused ground water within the porous volcanic cone to flash into steam, producing an eruption at 12:36 p.m. on March 27, the first since 1857. The initial burst of steam, which carried ash pulverized from old rock inside the volcano, opened a small oval vent about 250 feet across within the ice-filled summit crater. As earthquake swarms continued to focus beneath the mountain, a series of steam explosions sent ashclouds billowing 10,000 to 11,000 feet above the summit. Most of the ash fell within three to a dozen miles of the peak, although variable winds bore small amounts as far away as Bend, Oregon, 150 miles south, and Spokane, 285 miles to the east.

Some of the early eruptions were essentially single bursts, while others were sustained events lasting several hours. By March 29 a second crater appeared and observers noted a blue flame, probably caused by burning volcanic gases, that flickered and arched from one crater to the other. Ashclouds rolling downslope generated static electricity from which flashed lightning bolts, some nearly two miles long. Eruptions were frequent during this period and produced towering columns of ash. On March 30, no fewer than 93 separate outbursts were recorded. By April 8, the two new vents inside the old summit crater merged into one, which eventually measured 1700 feet across and 850 feet deep.

Spring snow flurries repeatedly restored St. Helens' customary white cloak, only to have new ashfalls blacken first one side of the mountain and then another as winds shifted direction. The volcano rapidly lost its ethereal appearance and began to take on a dark and sinister aspect.

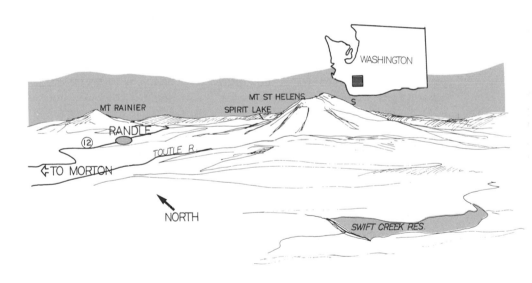

Relief map of the Mount St. Helens area.

The Bulge

St. Helens symmetrical form was also distorted by a series of fractures and the displacement of massive blocks of the summit dome, deformations that apparently began about the time of the initial eruption. By March 27, a nearly continuous east-trending fracture system some 16,000 feet long had formed across the summit, nearly bisecting the old summit crater. Another system of fractures that paralleled the major one along the old north crater rim bounded a rising block of the volcano's north flank.

The extent of deformation on St. Helens' north slope was not fully recognized until the last week of April, when a U.S. Geological Survey team determined that a section of the north face approximately a mile and a half in diameter had moved up or out at least 270 feet. Despite two weeks without eruptions in late April and early May, earthquakes continued, and the bulge steadily grew five to six feet per day. By mid-May the bulge had expanded more than 400 feet north, fracturing and buckling the Forsyth Glacier. On April 30, geologists announced that potential sliding of this unstable rock mass was the greatest immediate danger, and that might initate an eruption.

May 18—The Largest Cascade Eruption in Historic Time

When the author flew around St. Helens' summit on Sunday afternoon, May 11, the crater was in continual activity, with explosions vigorously blowing ash clouds to a height of several thousand feet.

Deceptively peaceful less than a day before the catastrophic eruption of May 18, 1980, Mt. St. Helens reveals the notorious bulge in the north flank (left-center). From left to right: Dog's Head dome; ash-blanketed Forsyth Glacier; Goat Rocks dome and avalanche fan below; the Floating Island lava flow, which cuts a swath over two miles long through the forest encircling St. Helens' base.
—U.S Geological Survey Photo, May 17, 1980

These eruptions, which had resumed on May 7 after nearly two weeks of relative quiet, seemed identical to the dozens he had seen in March and April. It first appeared that the mountain was successfully venting whatever pressure had built up within. Close flights over the ash-draped north side, however, revealed that the bulge had grown to enormous proportions, an impressive indication that whatever forces were causing the deformation remained active. Fellow passengers agreed that it seemed impossible for the oversteepened north flank to swell much farther without collapsing.

Disaster struck early on Sunday, May 18, with no warning increase in either seismic activity or the rate at which the bulge was expanding. At 8:32 a.m. the strongest earthquake to date, with a magnitude of 5.1 centered directly beneath the north side, shook the already weakened cone, triggering one of the largest landslides of historic time. Keith and Dorothy Stoffel, two geologists who were then flying almost directly over the mountain, recalled seeing "the whole north side of the summit crater began to move instantaneously as one gigantic mass." For a few seconds the north summit "began to ripple and churn" in place, then peeled away from the main edifice and moved northward along a "deep-seated plane."

Catastrophic events followed in quick succession. Moving at a velocity of about 155 to 180 miles per hour, the disintegrating slope avalanched down St. Helens' north flank, swept across the west arm of Spirit Lake, and slammed against a ridge about six miles to the north. Momentum carried one lobe of the avalanche over the nearly 1200 foot ridge, but the main body of the slide—lubricated by steam, ground water contained in the porous volcanic rock, and melting blocks of glacier ice—poured 17 miles down the North Fork Toutle River, filling the valley wall-to-wall with hummocky debris up to a mile or more wide and as much as 600 feet deep.

As the avalanche began to move, a small plume of inky ash shot upward from the bisected summit crater, while lighter ashclouds burst through fractures in the slide itself. Within seconds, these searing clouds overtook the slide, expanding in an arc that nearly encompassed the northern half of the volcano. Hugging the ground, the racing cloud accelerated from an initial speed of about 220 miles per hour to about 670 miles per hour. The moving front of the ashcloud may have surpassed the speed of sound, 730 m.p.h. In a hurricane of hot ash and rock fragments, the cloud rolled nearly 19 miles north from the crater, devastating a fan-shaped area 23 miles across from east to west. Approximately 230 square miles of forested mountainscape, tens of thousands of trees, including Douglas Firs six feet in diameter and 200 feet tall, were laid flat. Trees growing within about eight miles from the crater were felled, shredded, and carried away. In this region, the earth's surface looked as if it had been swept clean by a giant sandblaster, the topsoil scoured down to bedrock. Along the margins of the blow-down area, trees were left standing but thoroughly seared and dead. Altogether about a billion board feet of timber toppled, although much of it was later salvaged.

The avalanche and surging ashcloud struck its victims with cyclonic force and temperatures as high as about 680 degrees, killing about 57 people within a few minutes. Several people died of burns, but autopsies showed that most died of suffocation, from inhaling the ash. The few people who survived in the devastated area recounted unforgettable horrors, of darkness, heat, and inability to breathe.

A study of the deposits left by the surging ashcloud showed that it contained angular fragments of old rock from St. Helens' cone as well as blocks and bombs of fresh lava. Of the approximately 0.045 cubic mile of material carried in the pyroclastic surge, one-third was new magma.

The avalanche included magma that had invaded St. Helens' cone. When it plunged into the west arm of Spirit Lake and the headwaters of the North Fork Toutle River, the hot magma boiled the water to create a second explosion far greater than that opening the crater. This secondary explosion was heard as far away as Vancouver, B.C., and it sent waves of hot rock debris sweeping over Coldwater Ridge and beyond. Photographs showing enormous ash columns rising from the volcano's northern foot reveal the intensity of its blast.

Before the lateral eruption reached its maximum extent, the small jet plume from the summit crater began to expand into a titanic mushroom cloud that rose to an altitude of 12 miles. For the next ten hours this vertical column swept millions of tons of ash into the stratosphere, where strong winds bore it northeastward in a narrow plume. The ash canopy cast an eerie pall across the ground beneath. At 10:10 a.m. ash fell in Yakima, 90 miles away, by 2:00 p.m. the plume hung over Spokane, nearly 300 miles distant. By 10:15 p.m. it reached West Yellowstone, Montana. Ash fell in quantity as far east as central Montana. On May 19 a light dusting settled over Denver; later ash fell visibly in Minnesota and Oklahoma.

Yakima was blanketed with four to five inches of ash, although this compacted to a considerably thinner layer. Yakima officials reported that 600,000 tons of ash that weighed 95 pounds per cubic foot covered 12 square miles of their city. Because of peculiar wind conditions, the small town of Ritzville, Washington, received an unusually heavy deposit of ash the consistency of talcum powder. Near the northern edge of the cloud, Spokane got about half an inch, which reduced visibility to ten feet. Airports closed and traffic crawled as street lights switched on in mid-afternoon.

The approximate distribution in the United States of ash fall from the May 18 eruption of Mt. St. Helens. Most areas beyond western Montana received only trace amounts. —After Foxworthy and Hill, 1982.

Pyroclastic Flows

As the great cauliflower cloud broiled upward, a series of pyroclastic flows swept down St. Helens' flanks. The first were thin ash flows that bathed the upper surface of the cone. They fell from the initial blast cloud as it expanded over the summit and from the rising column of ash. Later pyroclastic flows consisted of incandescent pumice bombs and lapilli, as well as pumiceous ash. They poured north through the vast gap opened in the crater as the volcano's north side collapsed. Throughout the day, these flows spread over the earlier avalanche deposits, racing past Spirit Lake at more than 60 miles per hour, further damming its western outlet, then down the north Toutle valley. The frothy dacite composing these flows makes a pale contrast to the dark debris deposited from the avalanche and subsequent mudflows.

Where the hot debris flows covered ice or water, steam exploded through the flow to open craters as much as 65 feet across, and sent columns of steam and ash as high as 6500 feet. Some of these steam explosion craters along the southern shore of Spirit Lake were later filled in by pyroclastic flows from later eruptions.

Mudflows

As in many earlier eruptions of St. Helens, floods and mudflows were part of the May 18 activity. They affected three of the four river systems draining the mountain. Water from melting snow and glacial ice mixed with ash and other debris on the volcano's southern and eastern slopes, to form debris flows that resembled wet concrete as they rapidly streamed down Smith Creek, Muddy River, and Pine Creek to their confluence with the Lewis River. The mudflows scoured the stream valleys, leaving only a few trees standing along the lower reaches of the Muddy River, and destroyed bridges at the mouth of Pine Creek and head of Swift Reservoir. Between 9:00 a.m. and noon on May 18, they dumped enough water, mud, and other debris, including thousands of logs, into the reservoir to raise its level about 2.6 feet.

A Plinian eruption cloud soars miles above Mt. St. Helens' truncated summit, May 18, 1980. Falling ash blackens the sky to the east, while steam explosions (left of center) near Spirit Lake send white plumes thousands of feet into the air. Hot ash flows mantle most of the cone, partly covering blocky lava flows erupted earlier (right bottom). —U.S. Geological Survey Photo

Much larger mudflows poured down the north and south forks of the Toutle River, which drains the northwest slopes of the volcano and flows west to join the Cowlitz River near the town of Castle Rock. An observer, trapped about 8:50 a.m. in the valley of the South Fork about 4.5 miles west of the mountain watched "a huge mass of water, mud, and trees" crash past him, "snapping off trees." Shortly after 10:00 a.m. a 12-foot wall of water and logs passed Weyerhauser's Camp 12, 27 river miles from the volcano. The mudflow had covered that distance in 90 minutes. The Toutle River gauging station at Silver Lake, near the South Fork's confluence with the North Fork, recorded a flood stage of 23.5 feet, the highest ever observed there. By 1:00 p.m. the mudflow, carrying its burden of logs, a shattered railroad trestle, and other debris, had entered the Cowlitz River.

The North Fork of the Toutle, which originates at the west end of Spirit Lake, received the greatest damage. By 9:00 a.m. on the 18th most of the former north side of St. Helens lay as a 17-mile fill, 600 feet thick a mile below Spirit Lake and 150 feet thick at its western terminus. Part of the avalanching north slope temporarily displaced the water from Spirit Lake, generating waves that splashed about 600 feet up the ridge to the north. As it flowed back into the lake basin, the water carried with it thousands of trees uprooted by the lateral blast. Avalanche debris had filled the lake bottom, raising the lake surface about 200 feet above its previous level, but still about 200 feet below the avalanche deposit that now blocked its western outlet.

Meltwater from the snow and glaciers, perhaps with some displaced from Spirit Lake, mobilized the biggest mudflows. Unlike others on the east and west slopes, the North Fork mudflow did not begin until early afternoon, and moved more slowly than its predecessors. At 2:30 p.m. Weyerhauser's Camp Baker was obliterated by a massive onslaught of rock and mud pouring down the North Fork valley. High water preceding the mudflow picked up logs stacked there, which then served as battering rams to destroy many of the seven bridges swept away downstream. The Highway 504 bridge, which had survived the South Fork mudflow earlier that day, was destroyed shortly after 6:00 p.m. A nearby gauge showed that the later mudflow rose about 30 feet higher than its predecessor, probably as the result of temporary ponding.

Possessing a "mortar-like consistency," the mudflow filled the lower Toutle where it emptied into the Cowlitz River. Although most of the flow went downstream, part backed up the Cowlitz for two and one-half miles. The Cowlitz transported the now-diluted mud, logs, and other debris 17 miles farther south, where about 3,900,000 cubic yards of sediment dumped into the Columbia, reducing the river's depth from over 38 feet to about 13 feet for a distance of four miles.

Deposits from the mightiest landslide in historic time choke the North Fork Toutle River. Collapse of the entire north side of Mt. St. Helens, 8:32 a.m., May 18, 1980, sent a debris flow up to 600 feet thick 17 miles down the valley. Looking eastward toward St. Helens; part of the devastated area appears at the left. —U.S. Geological Survey Photo, June 30, 1980

Dredging soon reopened the Columbia to ocean-going freighters, but the city of Portland reportedly lost $5,000,000 during the temporary closure.

At the Castle Rock Bridge over the Cowlitz, more than 45 miles below the volcano, the temperature of the mixed mudflow and river water was 85 degrees at 1:15 a.m. on May 19, and remained that warm at least five hours. Temperatures in the lower Toutle at the Highway 99 Bridge registered 91 degrees at 9:45 on the 19th.

Scores of buildings, bridges, roads, and river levees, were damaged or destroyed by the North Fork mudflow and flooding. Near Castle Rock, the Cowlitz overflowed its banks, depositing as much as ten feet of sediment on farms and grazing land. The fair grounds across the river from Castle Rock were buried under three to five feet of muck, while farm buildings one to two miles south of town were swamped with mud 15 feet thick.

Aftermath

After about 5:30 p.m. on May 18, the eruption column markedly declined, although milder eruptions continued throughout the night

210

and for the next several days. The amount of energy released during the climactic explosion was staggering: a force equivalent to 27,000 Hiroshima-size bombs. Most of the slightly more than one cubic mile of material erupted consisted of old rock blown from the summit and north side of the cone. Approximately one-fourth of that volume was fresh magma ejected in the form of pyroclastic flows and ash. Removal of an estimated 13 percent of the former cone reduced St. Helens' height by about 1300 feet, leaving a crater between one and two miles in diameter and half a mile deep open to the north.

During the summer and fall of 1980, St. Helens produced five more explosive epidsodes, as well as several quieter periods of dome-building. At 2:30 a.m. May 25, during a sudden increase in seismic activity, an ashcloud rose to an altitude of nine miles. It was a stormy night and winds above the mountain were erratic, bearing fine ash to the south and west where it fell over wide areas of western Washington and Oregon. Pyroclastic flows again raced through the northern breach in the crater wall to spread over earlier deposits.

Map showing major features of the paroxysmal May 18, 1980 eruption of Mt. St. Helens. Gray stippled area indicates effects of the pyroclastic surge (blow down area). Mudflows extended down nearly every valley heading on the volcano. —After Lipman and Mullineaux, 1981.

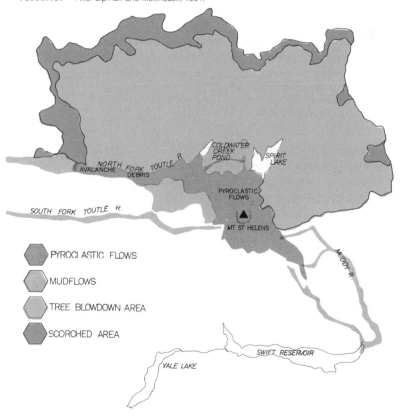

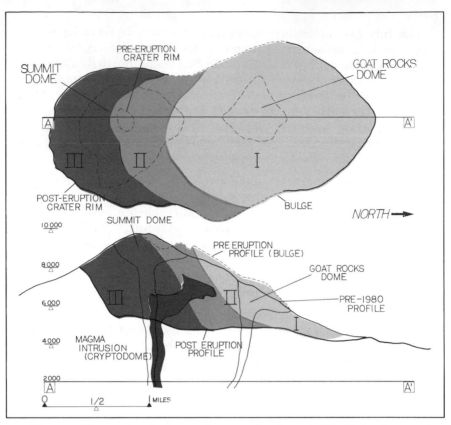

Diagram illustrating the body of magma that intruded St. Helens' cone prior to May 18, producing the bulge on the north flank. The three major blocks (I, II, and III) shown here collapsed in quick succession to form the massive landslide that swept across the west arm of Spirit Lake and poured 17 miles down the North Fork Toutle River valley. —After Tilling, 1982

Two explosive bursts occurred on June 12. The first, at 7:05 p.m., sent an ash plume about two and a half miles into the air. A stronger pulse began at 9:09 and soared to a height of about ten miles. Boiling ash flows, six to 12 feet thick, swept from the crater toward Spirit Lake. Then a viscous dacite dome about 650 feet wide rose above the crater floor; it gained a maximum height of about 200 feet within a week.

A month of repose ended on July 22 in a series of spectacular explosions, which were seen throughout western Washington and Oregon. Suddenly at 5:14 p.m., an ashcloud shot to an altitude of ten miles; another erupted with even greater velocity at 6:25 p.m., rising over 55,000 feet within seven and a half minutes. The final eruption began at 7:01 p.m. and lasted for more than two hours. Despite their impressive size, the clouds dropped only a light sprinkling of ash over eastern Washington.

The July 22 explosions completely demolished the dome formed in June, hurling dense rock fragments at least four miles. A member of the U.S. Geological Survey flying over St. Helens' north flank saw ". . . an ash fountain . . . ejected to about 1500 feet above the vent. As the projections of the fountain arched over and reached the surface in the vicinity of the vent, they gave rise to a pyroclastic flow and began to rapidly flow northward out of the amphitheatre." He escaped to tell the tale, but few geologists would care to risk watching the origin of a pyroclastic flow at such close range.

The next explosive phase, like that of July 22, followed increases in earthquake frequency and gas emission. On August 7, a slowly expanding ash cloud appeared at 4:26 and reached a height of about eight miles. Weaker pulses of ash continued throughout the evening, sending small pyroclastic flows down the north flank. Activity culminated in a second large eruption at 10:32 p.m.; in the days following a second dacite dome filled the vent.

After a two-month interval, renewed eruptions on October 16-18 destroyed the dome and produced ashclouds ten miles high, as well as small but red hot pyroclastic flows. Southwest winds again dusted Portland with ash.

Although St. Helens produced numerous bursts of gas and tephra after 1980, most of the eruptions were quiet dome-building events. By 1987 the dome measured more than 3000 feet in diameter and rose nearly 800 feet above the crater floor. At its present rate of growth, geologists estimate that it will take 150 to 200 years to replace the volcano's former summit. Before that happens, however, St. Helens is likely to erupt in other ways. A review of the volcano's past behavior offers some guide to the kinds of eruptions that we can expect in the future.

The Geologic History of St. Helens

The abrupt transformation of St. Helens from a classically symmetrical cone glistening with glacial ice into a blackened, shattered amphitheatre opened to the north vividly illustrates how swiftly a single catastrophic event can transform a familiar landscape. Geologic processes, usually noted for their painful slowness, did all that in a few minutes.

Most of St. Helens' white-clad, elegantly tapering profile had been built within the last thousand years; the smoothly rounded summit dome formed only about 330 years ago. Since the time of Christ, St. Helens had repeatedly changed its form and appearance; it can be expected to continue its protean behavior in the years to come.

Summary of the Eruptive History of St. Helens

Erupitve Period	Approx. Age	Eruptive Products
Present		Pyroclastic flows, dome growth, tephra
Goat Rocks	150-100 years	Dome growth, lava flows, tephra
Kalama	470 - 300 years	Pyroclastic flows; dome growth; numerous lava flows; tephra
Sugar Bowl	1150	Dome growth; pyroclastic flows, lateral blast
Castle Creek	2,200 - 1,700	Lava flows; pyroclastic flows
Pine Creek	3,000 - 2,500	Pyroclastic flows, dome growth
Smith Creek	4,000 - 3,000	Pyrclastic flows; voluminous tephra
Swift Creek	13,000 - 8,000	Pyroclastic flows, dome growth, tephra
Cougar	20,000 - 18,000	Pyroclastic flows, dome growth
Ape Canyon	40,000 - 35,000	Pyroclastic flows

Pleistocene glaciers had whittled Hood to its present sharply pyramidal form and gouged cirques and canyons into broad-shouldered Adams long before St. Helens was born. The present cone stands atop and partly conceals an older volcano which began to erupt less than 40,000 years ago. The oldest recognized products of this ancestral volcano are a pumice deposit dated at 37,600 years and a weathered mudflow deposit about 36,000 years old. Glacial sediments that contain fragments of the earlier mountain are between 14,000 and 18,000 years old, and reveal that the ancestral St. Helens was high enough to support an icecap during the last ice age.

The chief difference between the ancestral and modern versions of St. Helens is the chemical composition of their lavas. The older vent erupted a characteristic variety of dacite and andesite until about 2500 years ago. Then the lavas became strikingly more diverse, ranging from olivine basalt to andesite and dacite, of which the modern St. Helens is mostly built.

Although now almost buried beneath the recent cone, ancestral St. Helens left impressive evidence of its past energy. Conspicuous blankets of ash extend for hundreds of miles; valleys heading on the cone contain deep fills of explosion rubble. A recent study of the Pine

Creek valley on the southeast side of St. Helens revealed that during the past 18,000 years the volcano has repeatedly erupted glowing avalanches of hot gas and pyroclastic debris, shattered rock fragments, and showers of air-borne pumice.

The known eruptions of about the last 4500 years can be roughly grouped into four principal cycles: 2500 to 1600 B.C., 1200 to 800 B.C., 400 B.C. to about 300 A.D., and 1480 A.D. through the present. Radiocarbon dating of deposits laid down during these eruptive periods shows that each phase included activity that continued intermittently over several centuries. The longer periods of repose were sporadically broken by isolated eruptions of apparently short duration.

The first of these four cycles, about 2500 to 1600 B.C., followed a dormant interval that may have lasted as long as 4000 years. It produced large volumes of ash, which covered thousands of square miles. One ash plume blanketed the site of Mt. Rainier National Park, 50 miles distant, with yellowish-brown pumice up to 18 inches in depth. Traces of this ash have been identified as far northeast as Banff Park in Alberta, Canada. Another ash fall extends southeast from St. Helens at least as far as eastern Oregon. The total volume of material blown out during these eruptions may approximate 2.5 cubic miles.

The next eruptive period, 1200 B.C. to 800 B.C., differed from the first in that individual outbursts produced relatively smaller quantities of material. This period, like the previous cycle, also produced numerous glowing avalanches. These repeatedly swept down the flanks of the volcano and left thick fills in the adjacent valleys. The debris flows, which apparently erupted at temperatures approaching a red heat, consisted of both frothy pumice and solid rock fragments. The dense rock debris may have come from viscous domes and spines which were shattered by explosion or collapse as they rose.

Until about 2400 years ago lava flows played a minor role in St. Helens' construction, but mudflows seem to have occurred throughout the volcano's development. During all eruptive periods mudflows streamed many miles down valleys west and south of the cone. Between 2500 and 3000 years ago one large mudflow inundated at least 40 miles of the Lewis River valley. About 2000 years ago similar mudflows traveled as far as 30 miles down the Toutle and Kalama River valleys. Some may have reached the Columbia River.

Building the "New" St. Helens

The present cone of St. Helens thus rises above a thick accumulation of avalanche debris, explosion rubble, and mudflow deposits generated by its predecessor. Beginning about 400 B.C., the volcano abruptly modified its behavior, emitting—over short periods of time—a wide variety of lava types. Although continuing to produce the highly silicic domes and tephra characteristic of its earlier history, St. Helens now produced large flows of andesite and basalt. About 1900 years ago, fluid streams of lava flowed down the south flank of the mountain and cascaded into the Lewis and Kalama River valleys. Some moved as far as eight or nine miles from their source. The best known of these is the Cave Basalt, which filled a stream valley cut into earlier pyroclastic deposits. Named for the elaborate system of passageways, chambers, and tunnels that honeycomb its interior, the Cave Basalt is a favorite haunt of Washington state spelunkers.

The last eruptive episodes before 1980, also notable for their diversity, may offer a clue about what the present activity may bring. As in 1980, the two previous episodes began with the violent explosion of dacite tephra. A decline in both gas and silica content then produced andesite lava flows, which were in turn followed by the extrusion of volatile-poor dacite domes, which ended the cycle. The first of these phases, about 1500 years ago, produced small volumes of ash and lava flows. It consisted mainly of the Sugar Bowl dacite dome and relatively small pyroclastic flows low on the north flank of the volcano.

Approximately 700 years after the Sugar Bowl dome was emplaced the Kalama eruptive period saw St. Helens produce large quantities of dacite pumice and ash, the most voluminous since the eruptions 3000 years earlier. These pale gray deposits consist of at least seven different beds. The lowermost layer, erupted about 500 years ago, forms a thick lobe northeast of the volcano. It is at least three feet thick six miles northeast of the summit, two inches deep about 50 miles away. During or between the successive ash eruptions, large pyroclastic flows and mudflows raced down the west flanks of the volcano into the Kalama River drainage. Some of the debris flows may have followed the destruction of a dacite dome at or near the mountain's summit.

St. Helens then began to erupt less silicic magma, producing andesitic ash. A basal unit of lapilli is overlain by at least eight alternating light- and dark-gray ash beds and topped by a dark-gray to black ash blanket. After this explosive phase, quieter eruptions of andesite lava issued from the summit crater. Thick tongues of blocky lava oozed down the mountainside, piling up in particularly conspicuous ramparts along the southeast flank. Hot pyroclastic flows then swept downslope to cover the andesite flows in the Kalama River valley.

The culminating event in this century-and-a-half-long eruptive sequence was the emplacement of a large dacite dome at the summit of the volcano, accompanied by more pyroclastic flows and mudflows. The stiff, pasty dacite filled and overflowed a large explosion crater. Crumbling fragments from the growing dome, which towered hundreds of feet above the old crater rim, repeatedly avalanched downslope, mantling much of the cone with talus. A hot mudflow derived from the summit dome partly filled the upper course of Swift Creek. Explosions during elevation of the dome scattered fine ash. Meanwhile, lateral blasts cut a notch in the southeast wall of the crater. These explosions may have aided in excavating the narrow ravine which extends down the east slope of the cone and ends in a chaotic fan of broken rock torn from a crater rim. Before 1980, the Shoestring Glacier descended from the crater through the narrow gorge; partly buried remnants of the beheaded Shoestring Glacier, longest and narrowest icesheet on the volcano, still fill the lower part of the ravine.

By the end of the Kalama eruptive period, about 1647 A.D., St. Helens had reached her maximum height and attained her graceful pre-1980 form. A dormant interval of about 150 years ensued, during which glaciers mantled the peak, and narrow stream valleys were cut into the cone.

The Goat Rocks Eruptive Period

The 57-year sequence of eruptions that began in 1800 with an explosion of dacite tephra is the only one for which we have accounts from native American oral traditions, as well as written reports by Caucasian observers. This 19th century activity duplicated the Kalama sequence: explosions of dacite pumice were followed by a flow of andesite lava, which in turn was succeeded by the emplacement of a dacite dome. At least a dozen or more small eruptions of ash, only one of which left a recognizable layer, were logged by eyewitnesses between 1831 and 1857.

The dacite apparently erupted from a vent at or near the site of Goat Rocks, and drifted hundreds of miles northeast over central and eastern Washington, northern Idaho, and western Montana. This ash fall; although smaller than those of many earlier eruptions, alarmed the Indians. A Nespelim account from northeastern Washington tells of their "dry snow" with prayer and dancing to the detriment of their usual food gathering and storing. Many starved and some died during the desperate winter that followed.

In his 1845 narrative of an expedition to Oregon territory, Charles Wilkes quoted a 60-year-old chief of the Spokane tribe, named Cornelius, telling of an eruption that had occurred "some fifty years ago, before [the Indians] knew anything of white people, or had heard of them." When Cornelius was about ten years old, he was

> suddenly awakened by his mother, who called out to him that the world was falling to pieces. He then heard a great noise of thunder overhead, and all the people crying out in great terror. Something was falling very thick, which they at first took for snow, but on going out they found it to be dirt: it proved to be ashes, which fell to the depth of six inches, and increased their fears, by causing them to suppose that the end of the world was actually at hand.

The geologist James D. Dana was with Wilkes' expedition. He probably referred to Cornelius' story when, in 1849, he stated that "an account is on record of ashes falling fifty years since." Two Presbyterian missionaries, Elkanah and Mary Walker, colleagues of Marcus and Narcissa Whitman in eastern Washington, also heard tales of the ashfall from the Spokane. In an 1839 report to his mission board, the Rev. Walker summarized the tribal accounts:

> They say some forty or fifty years ago there was a great fall of ashes, the truth of which is very plain now to be seen in turning up the ground. They say that it was a very long night with heavy thunder. They feared the world would fall to pieces and their hearts were very small. The quantity which fell was about six inches.

In spite of the obviously exaggerated depth of ash blanket, the Spokane traditions are remarkably consistent in stressing the darkness, confusion, and terror inspired by the 1800 eruption, experiences that several hundred thousand Northwesterners relived in 1980.

Older memories of St. Helens' eruptions were orally transmitted through many generations and eventually recorded by 19th century missionaries. The most celebrated is the Bridge of the Gods legend, in which St. Helens' fellow guardians of the Columbia—Adams and Hood—play leading roles. The Klickitat Indians, who called St. Helens' Tah-one-lat-clah (Fire Mountain), told of a mighty bridge of stone that once panned the Columbia River near the present town of Cascade Locks. Territory north and south of the bridge was ruled respectively, by Pahto and Wyeast, powerful sons of the Great Spirit. Both young chiefs fell in love with a beautiful maiden who could not decide which she preferred. Pahto and Wyeast waged war by hurling fire and hot rocks at each other across the river, devastating the countryside and making the earth tremble. Earthquakes toppled the

Bridge of the Gods into the river, forming the rock-strewn cascades of the Columbia River Gorge, now submerged beneath the waters impounded by Bonneville Dam. Angered, the Great Spirit changed his querulous sons into the two "masculine" peaks overlooking the river. Wyeast became Hood, Pahto became Adams, while the feminine cause of the battle became lovely St. Helens.

The legend's reference to St. Helens' metamorphosis from a dark and ugly form into one of snowy beauty may reflect tribal memories of the radical changes that took place on the volcano during the Kalama eruptive period about 350 to 500 years ago. St. Helens may then have resembled a black slag heap with a large crater in its truncated summit. Following the eruption of the andesite lava flows and the massive dome that formed its pre-1980 summit, the mountain assumed the pleasing configuration admired by both Indians and white settlers.

By 1835 legend and oral history were supplanted by authenticated eyewitness reports of St. Helens' activity. In March of that year Dr. Meredith Gairdner, official physician for the Hudson's Bay Company at Fort Vancouver, observed an eruption of the volcano. Unfortunately, this brilliant 26-year-old doctor from Edinburgh had to forego his proposed climb of the volcano. He did send an undated letter to the *Edinburgh New Philosophical Journal*, which appeared in January, 1836.

> We have recently had an eruption of Mount Saint Helens, one of the snowy peaks of the Marine Chain of the north-west coast, about 40 miles to the north of this place [Fort Vancouver]. There was no earthquake or preliminary noise here: the first thing which excited my notice was a dense haze for two or three days, accompanied with a fall of minute flocculi of ashes, which, on clearing off, disclosed the mountain destitute of its cover of everlasting snow, and furrowed deeply by what through the glass appeared to be lava streams . . . I believe this is the first well ascertained proof of the existence of a volcano on the west coast of America, to the north of California on the mainland. At the same season of the year 1831, a much denser darkness occurred here, which doubtless arose from the same cause, although at that time no one thought of examining the appearance of this mountain.

Shortly after Gairdner's account, the American missionary Samuel Parker arrived at Fort Vancouver. Although he noted that the volcano was still active, "down to the present . . . [sending] forth smoke and fine cinders to a considerable distance," he did not himself see the eruptions of 1831 or 1835. Parker's secondhand version of the 1831 activity corresponds almost verbatim to that of J. Quinn Thornton, a judge of the Oregon Supreme Court, who claimed a "Dr. Gassner, a distinguished naturalist of England" as well as "gentlemen connected with the Hudson Bay Company" as authorities for his report. Gairdner's researches no doubt stand behind both accounts: In the 1846 edition of his *Journal* Parker wrote,

> . . . there was in August 1931, an uncommonly dark day, which was thought to have been caused by an eruption of a volcano. The whole day was nearly as dark as night, except a slight red, lurid in appearance, which was perceptible until near night. Lighted candles were necessary during the day. The atmosphere was filled with ashes of wood, all having the appearance of having been produced by great fires, and yet none were known to have been in the whole region. The day was perfectly calm, without any wind. For a few days after, the fires out of doors were noticed to burn as though mixed with sulphur. There were no earthquakes.
>
> By observations which were made after the atmosphere became clear, it was thought the pure white perpetual snow of Mount St. Helens was discolored, presenting a brown appearance, and therefore it was concluded that there had been a slight eruption.

In a footnote, Parker adds "I have been creditably informed that lava was ejected at that time from St. Helens." Because they viewed the activity from a considerable distance, both Gairdner's and Parker's references to lava flows must be taken with caution. None of St. Helens' recent flows has been correlated with specific historic eruptions, but at least one blocky andesite stream, the Floating Island lava flow, erupted early in the 19th century from beneath the Toutle Glacier and extended below timberline into the surrounding forest. The age of trees rooted in patches of soil on the flow give a minimum date of 1838, but Lawrence concluded that it may have erupted shortly after the 1800 pumicefall.

The Goat Rocks dome probably erupted after the lava flow, perhaps as late as the observed eruptions in the 1840s. This large dacite mass was extruded high on St. Helens' northwest slope. Virtually all eyewitnesses agree that the craters active during the mid-19th century were below the summit on the northwest, northeast, or south sides of the cone. Angular fragments cascading from the dome built a prominent fan-shaped deposit along the northwest flank.

The "Great Eruption" of 1842-43

The most notable eruption observed by missionaries and other settlers occurred in the late fall or early winter of 1842, when ash drifted south and southeast of the volcano for several tens of miles. Although small in volume, this eruption produced impressive ashclouds remarkably similar to those generated by the steam explosion of March, April, and early May of 1980.

The 1842 outburst, which introduced a mildly explosive phase that continued intermittently for 15 years, apparently began without warning. According to J.L. Parrish, a Methodist missionary, "no earthquake was felt, no noise was heard." From Parrish's vantage "ten miles below Salem" in the Willamette Valley it was a silent if awe-inspiring spectacle: "He saw vast columns of lurid smoke and fire shoot up; which spread out in a line parallel to the plane of the horizon, and presented the appearance of a vast table, supported by immense pillars of convolving flame and smoke." Fifty years later, in the letter to the editor of a short-lived mountaineering publication, *Steel Points*, Parrish added further details:

> upon looking at the mountain we saw arising from its summit immense and beautiful scrolls of what seemed to be pure white steam, which rose many degrees into the heavens. Then came a stratum just below those fine huge scrolls of steam, which was an indefinite shade of gray. Then down next the mountain's top the substance emitted was black as ink.

Parrish also described how the ash-fall changed the volcano's appearance:

> The next day after the eruption I was out on French Prairie where I had a good view of the mountain, and I noticed that she had changed her snowy dress of pure white for a sombre black mantle, which she wore until the snows of the ensuing winter fell upon her.

This darkening of St. Helen's snowfields was a familiar sight to those who watched the early 1980 activity, when competing snow and ashfalls gave the mountain a black and white mottled appearance.

Another, less well-known account, preserved in F.G. Plummer's remarkable article on Northwest volcanism, is that of Rev. Gustavis Hines, an early missionary to the Columbia River country. He emphasizes the vast eruption cloud that suddenly burst from St. Helen's snowy cone.

> ... in the month of October [sic] 1842, St. Helens was discovered all at once to be covered with a dense cloud of smoke, which continued to enlarge and move off in dense masses to the eastward, and filling the heavens in that direction, presented an appearance like that occasioned by a tremendous conflagration viewed at a vast distance. When the first volume of smoke had cleared away it could be distinctly seen from different parts of the country that an eruption had taken place on the north side of St. Helens, a little below the summit, and from the smoke that continued to rise from the chasm or crater, it was pronounced to be a volcano in active operation. When the explosion took place the wind was north/northwest and on the same day and

extending from thirty to fifty miles to the southeast there fell showers of ashes or dust, which covered the ground in some places so as to admit of its being gathered in quantities.

St. Helens' 1842 ash fall was extensive. Captain J.C. Frémont's journal for November 13, 1843, recounts that "on the 23rd of the preceding November [Nov. 23, 1842], St. Helens had scattered its ashes, like a light fall of snow, over the Dalles of the Columbia, 50 miles distant [actually about 65 air-line miles]. A specimen of these ashes was given to me by Mr. Brewer, one of the clergymen at The Dalles."

One of these missionaries later published an anonymous memoir in which he described the eruption's effect on The Dalles community:

> On a pleasant evening, in the month of November 1843 [sic] the missionaries at The Dalles were favored with a visit . . . About the time these friends arrived at the mission, a dark, heavy cloud was seen rising in the direction of Mount Saint Helens. No special remark was excited by this fact, but, on going to the door the next morning, the missionaries were surprised to see the ground, the trees, the grass—everything—sprinkled with ashes. A dark cloud shrouded the sky. It seemed to rain; but the clouds were not dropping water. Something descended gently to the earth, in form like fine sand—in color, it appeared like ashes. Its odor was that of sulphur. The Indians said it had descended in larger quantities toward Mount Saint Helens. Soon the mystery was solved: that mountain was broken forth in a spendid eruption, and the winds had wafted its ashes to the door of the missionaries.

Although the date given for this occurrence is a year later than the accepted one, it undoubtedly recalls the outburst of November 22, 1842. Climbing a hill near the Columbia River from which they commanded a sweeping panorama of St. Helens, Adams, and Rainier, The Dalles missionaries noted that—striking contrast to the snowy repose of the other peaks—St. Helens tumultously sent up dense masses of steam and ash: "Amid this group of lofty mountains, Helens threw out its dark cloud of smoke. Its fires seem smothered, but the issuing volumes of smoke and ashes contrasted impressively with the sparkling snow of the surrounding peaks."

After the initial explosions, St. Helens remained in almost continuous eruption for many weeks or even months throughout the winter of 1842-43. On December 13, 1842, another Methodist missionary, John H. Frost, observed "a column of smoke to ascend from the N.W. side of Mt. St. Helens, toward the top . . . It has been ascertained since that it was an actual volcanic eruption. I know not that it has as yet emitted anything but smoke. Have learned since that ashes have been thrown out in great abundance, even as far as The Dalles."

Earlier in December another spectacular outburst was observed nearer at hand from the Cowlitz Mission, in what is now Washington State. On December 5, 1842, Father J.B.Z. Bolduc, wrote to his superiors in Quebec:

> To the northeast and southeast are two mountains whose height I still do not know, but which are at least 4,000 [a 10,000-foot underestimate for Rainier!] They are snowcovered, even in the greatest heat of summer. One of them—the one toward the southeast—is in the shape of a cone, and is opposite my dwelling. On the 5th of December, toward three o'clock in the afternoon, one of its sides opened and there was an eruption of smoke such that all our old voyageurs have never seen anything equal to it. These eruptions of smoke took place for several days at intervals not far apart, after which eruptions of flame began. They take place almost continually, but with an intensity that varies greatly from time to time. I am led to believe that there are three craters at least, for I have observed several times three eruptions at once and at different places, although close to each other. Especially in the evening are these phenomena well observed, and they offer a magnificent sight to the spectator. There is at the foot of this mountain a little river whose waters empty into the Cowlitz. After the volcano manifested itself, almost all the fish that it used to feed died—which is attributed to the quantity of cinders with which the waters were affected.

Father Bolduc does not mention the eruption of November 22, probably because he did not arrive at the Cowlitz Mission until the 30th of that month.

Three years later the Rt. Rev. Modeste Demers, also stationed at the Cowlitz Mission, wrote that St. Helens continued active and that "since the month of December 1842, the time when the mountain opened its sides from the drive of subterranean fires, the waters of this [the Toutle] river have carted cinders and scoria. After the first eruption the natives assured us that they had found dead fish."

In October 1843, Overton Johnson and William H. Winter noted in a journal published in 1846 as *Route Across the Rocky Mountains with a Description of Oregon and California 1843*, that St. Helens was then erupting: "Mt. St. Helens, a lofty snowcapped Volcano rises from the plain, and is now burning. Frequently the huge columns of black smoke may be seen, suddenly bursting from its crater, at the distance of thirty or forty miles. The crater is on the south side, below the summit." The following February 16 brought a particularly spectacular display. Peter H. Burnett, a lawyer who later became governor of

California, observed it from a point near the confluence of the Willamette and Columbia rivers: "being a beautiful and clear day, the mountain burned magnificently. The dense masses of smoke rose up in one immense column, covering the whole crest of the mountain in clouds." In what seems to be a slightly altered version of Burnett's description, Plummer adds that "in the evenings its fires lit up the mountainside in a flood of soft yet brilliant radiance." Burnett also indicated that lava, or at least hot mud, perhaps generated by hot ash falling on the melting snow fields, was erupting:

> On the side of the mountain, near its top, is a large black object, admist the pure white snow around it. This is supposed to be the mouth of a large cavern. From Indian accounts this mountain emitted a volume of burning lava about the time it first commenced burning [November, 1842?]. An Indian came to Vancouver with his foot and leg badly burnt, who stated that he was on the side of the mountain hunting deer, and he came to a stream of something running down the mountain, and when he attempted to jump across it, he fell with one foot into it; and that was the way in which he got his foot and leg burned. This Indian came to the fort to get Doctor Barclay to administer some remedy to cure his foot.

The story of the Indian with the burned foot—one of the two pre-1980 casualties reportedly caused by a volcanic eruption in the United States—with various embellishments, eventually became an established part of early Northwest lore. It may, however, be apocryphal. Napolean McGilvery, who represented himself as having been in charge of the Fort Vancouver commissary at the time, disclaimed any knowledge of the incident. In a late reminiscence published in 1899, McGilvery recalled his impression of the eruption as follows:

> The mountain was not visible from Vancouver at any time. The eruption probably occurred on one day, and was not discovered by us until the next, when, upon going out early in the morning, gray white ashes were found to cover the ground as a light fall of snow. Both days were beautiful and clear. There was not traveling at that time away from the water course, except by Indians, and very little by them. It has been published that during this eruption an Indian was caught in the hot lava, was badly burned and taken to Vancouver, where he was treated by Dr. McLoughlin. I had charge of the commissary, so that such an incident could not have happened without my knowledge, and I never heard of it until recently.

In support of the original tale, it should be noted that Burnett's version states that the Indian was treated by Dr. Barclay, not the more famous Dr. McLoughlin, and that he is vague about the time, "about the time it [St. Helens] first commenced burning."

Whatever the fate of Indians rash enough to approach the "fire-mountains," St. Helens "continued to burn" fitfully throughout the mid 1840s. On May 30, 1844, the Rev. George Gary, then on a ship off the Oregon coast, had "a very distant view of a volcano in action, throwing up clouds of smoke." At first Gary could not determine if the clouds were arising from the snowy peak of St. Helens or from a vent near it, but:

> on further inquiry I have learned that this volcano is in Mount Helen [sic] itself, and that either the snow is diminishing or the soot settling upon the white covering of the mountain presents the appearance of wasting snow. It is so cold near these snowy mountains and the snow is so deep I believe there has been no very thorough examination of them, and this volcano [active crater] is as high up the mountain as that the temperature at its base is but little, if any, affected by it. The falling ashes or soot have been seen and gathered from boards or anything of a smooth surface, fifty miles from the crater.

The mountain was still in action on December 28, 1844, when Burnett noted that it continued "a burning volcano." On February 15, 1845, Samuel B. Crockett wrote that "St. Helens, which is the highest peak that stands nearest the Columbia on the north side, sends forth columns of smoke from its frozen top." These eruptions of the later 1840s do not appear to have been as violent as those of 1842-43 when large areas south and east of the peak were coated with sulphurous ash. They did, however, continue to attract attention.

Two intinerant artists, Henry J. Warre and Paul Kane, were among those attracted by reports of St. Helens' fireworks. In the summer of 1845, Warre, a British lieutenant, sketched the mountain from the Columbia and the Cowlitz River settlement, in each case depicting a moderate ashcloud rising from the volcano's west or

This romantic painting of Mt. St. Helens during the night eruption in the 1840s, by Canadian artist Paul Kane, correctly shows the active crater located considerably below the volcano's summit, probably at the site of the Goat Rocks dome. —Reprinted by permission of the Royal Ontario Museum, Toronto, Canada

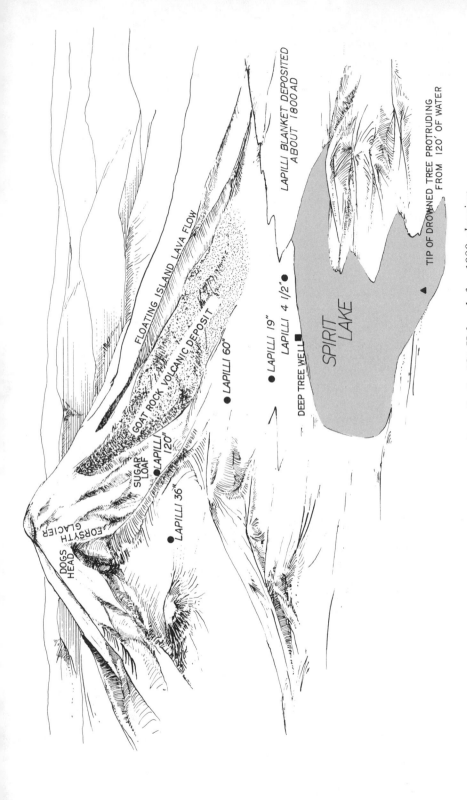

FLOATING ISLAND LAVA FLOW

LAPILLI BLANKET DEPOSITED
ABOUT 1800 AD

TIP OF DROWNED TREE PROTRUDING
FROM 120' OF WATER

GOAT ROCK VOLCANIC DEPOSIT

SUGAR
LOAF

• LAPILLI 19"

• LAPILLI 4 1/2"

• LAPILLI 60"

SPIRIT
LAKE

DEEP TREE WELL ■

LAPILLI
120"

• LAPILLI 36"

FORSYTH GLACIER

DOGS
HEAD

*Diagrammatic map of the northeast flank of St. Helens before 1980, showing
deposits of recent eruptions.* —After Lawrence. 1954

northwest flank. Kane, a Canadian, was on a long tour from Toronto to collect material for paintings. On March 26, 1847, from a point near the mouth of the Lewis River that commanded an unobstructed view of the peak, Kane made a preliminary drawing. As he noted in his journal, "There was not a cloud visible in the sky at the time I commenced my sketch, and not a breath of air was perceptible; suddenly a stream of white smoke shot up from the crater of the mountain, and hovered a short time over its summit; it then settled down like a cap. This shape it retained for about an hour and a half, and then gradually disappeared." A watercolor apparently based on this sketch is in the Stark Museum of Art in Orange, Texas.

Kane's best known painting of the volcano is now in the Royal Ontario Museum of Archaeology in Toronto. It is a romantically dramatic work that shows a group of awed Indians watching a night eruption, the glare reflected on their upturned faces and on the river where their canoe is harbored. If it seems unlikely that the volcano would so obligingly put on a display for the artist, remember that Kane saw the mountain erupt more than once. He depicted the position of the erupting vent with great precision. Four days after making his first sketch, he had another "fine view of Mt. St. Helens' [sic.] throwing up a long column of dark smoke into the clear blue sky." This experience may have inspired a painting, now in the Stark Museum, that shows the volcano in daylight, ejecting a gray steam plume into azure skies. Kane painted the eruption clouds emerging from a vent about a third of the way down the west or northwest side, perhaps at the site of Goat Rocks—which, in contrast to the surrounding snowy slopes, is depicted as bare rock. Only by having witnessed a genuine eruption is Kane likely to have placed the crater and rendered the new dome so accurately.

That St. Helens was still lighting up the night skies a full year after Kane had so painted her is shown by Robert Caufield of Oregon City in a letter dated April 1, 1848:

> St. Helens which is still a volcano and continually covered with snow stands . . . about 70 miles north of this place. There has been two emptyings of this mountain since we came here. The report we could hear distinctly and the *reflections seen in the sky at night.*

Although references to "flame" and luminosity are frequent in descriptions of the volcano, Caufield is one of the very few who mentioned hearing "reports" or explosive noises.

On February 25 of 1854 the *Oregonian* reported that:

> The Crater of Mt. St. Helens has been unusually active for several days past. Those who have been in a position so as to obtain a view of the

mountain, represent clouds of smoke and ashes constantly rising from it. The smoke appears to come up in puffs, which was the case at the time we visited in August last. There is now more smoke issuing from it than there was then, which indicates that the volcanic fires are rapidly increasing within the bowels of this majestic mountain.

The *Oregon Weekly Times* for the same date published a report from W.H.H. Halls, pilot of the Whitcomb, a Columbia River steamer, to the effect that "volumes of smoke . . . were thrown out at intervals." Stevens also noted the activity in his diary, but added that he was not himself in a position to see it.

Several of St Helens' historic eruptions may have gone unnoticed. Low-temperature night eruptions that produced no glow or flare, as well as outbursts during winter months when the mountain was veiled in thick clouds could not have been seen. How many of the milder expulsions of steam or ash mingled imperceptibly with the murk of night or storm will never be known.

After the activity of 1854, the mountain was quiet until April, 1857, St. Helens' last recorded outburst before 1980. This eruption was publicized by the Steilacoom, Washington *Republican*, which carried the following item on April 17:

> Mount St. Helens, or some other mount to the southward, is seen from the Nisqually plains in this county, to be in a state of eruption. It has for the last few days been emitting huge volumes of dense smoke and fire, presenting a grand and sublime spectacle.

Again, the reference to "fire" may indicate either a high-temperature explosion of molten material or the electrical flashes common during such a disturbance. Since this final outburst seems to have produced no significant ash deposit nor to have been hot enough to initiate a recognized mudflow, it is probable that no more than billowing clouds of steam and dust were ejected.

Position of the Crater

From the testimony of Frost, Parrish, Kane, Stevens, Hines, Gibbs, and others, it seems certain that most of St. Helens' historic eruptions centered on the north or northwest side below the summit. The Goat Rocks dome, destroyed in 1980, probably emerged from the same northwest-side vent that had produced the 1800 tephra. These eruptions repeatedly fractured the north slope, while internal heat concentrated there decayed St. Helens' core, preparing for the catastrophic collapse of the north flank in 1980.

A light autumn snowfall blankets Mt. St. Helens' mile-wide crater, contrasting with the dark furrowed surface of the dacite dome rising in the central vent. At upper left the channel of the Shoestring Glacier indents the decapitated summit rim. Mt. Hood appears 50 miles to the south.
—U.S Geological Survey Photo, October 4, 1981

The Future

If St. Helens repeats its recent eruptive behavior, the volcano may remain sporadically active for decades. The Kalama eruptive period lasted for more than a century and a half; the Goat Rocks cycle spanned nearly six decades. Following the pattern of those two episodes, St. Helens will eventually produce flows of andesite lava and another large dacite dome. The volcano is notoriously unpredictable, however, and the present lava dome may be blasted skyward by explosive outbursts far more violent than those of 1980.

In the past St. Helens has erupted with many times the energy and volume of ejecta than those observed in historic time, and may do so again. Rainier, fifty miles to the north, received only a light dusting of ash on May 18; about 3400 years ago it was buried under two feet of tephra. Future lava flows and pyroclastic flows will probably be directed through the northern breach in the elliptical crater and may travel several miles down the North Fork Toutle River. If eruptions occur when the snowpack is heavy, mudflows may rush down any valley heading on the mountain, some perhaps as far as the Columbia River. Although geologists do not now expect it to happen, if the

229

debris damming the west end of Spirit Lake were to fail, floods and mudflows of much greater dimension than those of May 1980 could decimate communities far down the Toutle and Cowlitz rivers.

Although St. Helens destroyed its former summit, it is a young volcano still in the cone-building stage. It may erupt copious streams of andesite or basalt lava and construct a new cone perhaps as symmetrical as that existing before 1980. Before it rises, Phoenix-like, to become again the "Fujiyama of America," however, St. Helens is likely to be known to our children and grandchildren as the "Lighthouse of the Columbia."

How To See Mt. St. Helens

The Mount St. Helens National Monument was opened to summit climbers in 1987, although week-end alpinists must obtain a permit from the U.S. Forest Service and use the south-side route up this still-smouldering peak. To view the Devastated Area and Spirit Lake on the north side, turn south off Washington state highway 12 at Randle and follow Forest Service road No. 25.

Rainier at the end of a dry summer, the Tahoma Glacier cascades from the summit down the volcano's west face. Above the Puyallap Glacier, center, is Sunset Amphitheatre, a huge scar left by a massive rockfall that generated the Round Pass Mudflow about 2800 years ago. —U.S. Geological Survey photo by Austin Post

XVII
Mt. Rainier:
The Mountain That Was "God"

Built astride the Cascades of central Washington, Rainier towers 14,410 feet above the shores of Puget Sound, the inland sea to the west. Rainier's bulky mass is the most conspicuous landmark on the skyline of almost every major city in the area.

Rainier supports the single largest glacier system in the 48 contingent states. Its 26 officially named glaciers are the sources of several major Northwest rivers—the Nisqually, Puyallup, Carbon, Cowlitz, and White rivers—invaluable producers of hydroelectric power.

Thanks to its status as a national park, much of the mountain's base remains covered by stands of Douglas fir, cedar, alder and hemlock. The park-like areas between timberline and the permanent snowfields are carpeted in summer with wildflowers, mosses and heather. Rising from virgin forest, garlanded with alpine blossoms,

231

and crowned with its great icecap, Rainier presents a superlative study in contrasting colors. Whether veiled in clouds or looming immense on a clear day, the volcano stands apart, an arctic island in a temperate zone.

Because of its exposed northern position and heavy load of glaciers, Rainier is not a smooth-sided symmetrical cone as St. Helens was before 1980. Each face of the mountain has been glacier-carved into a distinctive shape; the mountain reveals a totally different profile from various directions.

The most familiar view is from Paradise Valley, a mile high on the south slope. From there, Rainier appears as a broad ice-covered dome with long rocky ribs standing black against surrounding snowfields. High on its east shoulder, the reddish-brown mass of Gibraltar Rock gives the mountain a bulky, elongated look. Farther east, Little Tahoma Peak, points a jagged triangle toward the sky. From the west or southwest, Rainier shows something of the classic volcanic cone, albeit with irregular slopes and a truncated crest. From Olympia, it resembles Matthes' description of "an enormous tree stump with spreading base and broken top." The western view also reveals the three separate peaks of Rainier's broad summit.

From the north, Rainier hardly seems the same mountain familiar to tourists at the Paradise Visitor Center. The Carbon Glacier has dug deeply into the northwest flank of the cone, gouging out the largest natural amphitheatre in the Cascade Range. Curtis Ridge on

Relief map of the Mount Rainier as viewed from the northeast.

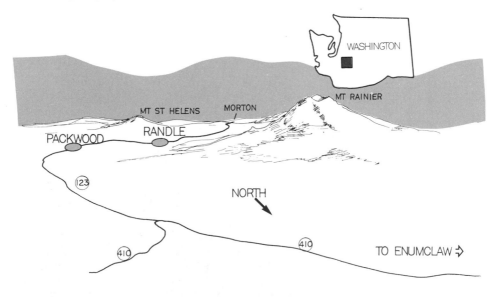

the east and Ptarmigan Ridge on the west half enclose an enormous cirque with headwalls rising almost vertically 3600 feet above the glacier to the summit icecap. These slopes form the Willis Wall, in which countless lava flows and breccia layers are exposed in cross-section. Avalanches of rock and ice often tumble from the overhanging edge of the 200- to 500-foot-thick summit glacier.

Rainier is most impressive from the east. The volcano looks grandest from the Sunrise Visitor Center on the grassy tableland of Yakima Park. The Emmons Glacier descends five miles from the summit to the depths of White River Canyon 10,000 feet below. To the south rears the irregular pyramid of Little Tahoma, the third highest peak in Washington state. It is a remnant of a once-continuous eastern slope, surviving to remind one how much Rainier's glacial and other erosional processes have cost it in size and volume. Emmons Glacier is bordered on the north by another conspicuous escarpment, the Wedge or Steamboat Prow, another indicator of what the volcano has lost through glacial and volcanic action. Beyond it lie the Winthrop Glacier, Russell Cliff and the shadowed cirque of the Carbon Glacier.

This brief survey shows that Rainier, imposing as it is, is much diminished. Not a scrap of the volcano's original constructional surface remains; Gibraltar Rock and Little Tahoma do not represent the mountain's former contours. An unknown number of lava flows once moved down slopes that lay *above* the present surface of these prominent outcrops.

Rainier has undergone tremendous changes during its long history. Many of these, particularly those of the last 10,000 years, can be described and even approximately dated by a careful examination of

A rocky island in a sea of ice, Little Tahoma Peak is the most conspicuous remnant of Rainier's once continuous eastern slope. Little Tahoma is rapidly being leveled by the Emmons Glacier (foreground), the Ingraham Glacier (right), and the Fryingpan Glacier (top center). Photo by Bob and Ira Spring

233

the rocks. Rainier has left abundant evidence that it was once larger and higher than now, and that it experienced outbursts of cataclysmic proportions. Layers of ash, congealed streams of lava, and remnants of mudflows that once filled adjacent valleys to depths of 1000 feet record some of the events that shaped the modern cone.

In the reenactment of the mountain's early history that follows, I have liberally interpreted the work of Dwight R. Crandell, Richard S. Fiske, Clifford A. Hopson, and Aaron C. Waters. Although details of prehistoric eruptions are necessarily invented, descriptions are based on the observed activity of volcanoes such as Paricutin, which first appeared in 1943. Rainier's later history, that of the last 10,000 years, relies principally on the work of Crandell and Donald R. Mullineaux, as cited below. Freely drawing on the discoveries of these, and other geologists, we can enjoy a kind of "time-machine" adventure into the volcano's remote past.

The Setting

If we were transported nearly 1,000,000 years back in time to just before the birth of Rainier, we would find the neighboring Cascade terrain surprisingly recognizable. Then, as now, saw-toothed peaks and sharp ridges were intersected by deep canyons containing swift rivers. Some of the lower canyons were probably relatively narrow, while valleys higher in the mountains were broader, with rounded floors, the result of glacial action. The Tatoosh Range, which now borders Rainier's southern base, probably was nearly as craggy as today.

Rainier probably appeared during an interglacial period as warm as today. The few glaciers were then small and confined to higher elevations. Dense forests, composed at least partly of familiar fir and hemlock, extended to a timberline at about 6000 feet. Animal life was relatively sparse but included such exotic species as the wooly mammoth, which occasionally strayed up from the neighboring lowlands.

The immediate scene of Rainier's birth was a rugged plateau more than a mile above sea level, surrounded by peaks from 2000 to 3500 feet higher. Several rivers, originating from alpine snow fields, flowed from this mountainous upland. The ancestral Carbon, White, Puyallup, and Mowich rivers had already excavated large ravines extending north, northwest, and west to the Puget Sound lowlands. In some places, the loftiest peaks stood 4000 feet above the valley floors.

The Carbon Glacier, mostly covered with rocks and dirt (center), bites deeply into Mt. Rainier's north face, and descends to the lowest altitude (about 3000 feet) of any glacier in the conterminous United States. Winthrop Glacier (left center) flows from the youthful summit cone. Little Tahoma Peak appears at the extreme left. The dark outcrops on the right center are Echo and Observation rocks, remnants of two ice-age satellite cones. —Photo by Austin Post, U.S. Geological Survey

Rainier Is Born

The first warning that a new volcano was about to appear came in a series of earthquakes that gradually increased in frequency. During the most severe of these shocks, the earth cracked in long fissures to emit streamers of white vapor. Tremors continued until the ground vibrated almost incessantly and the wisps of vapor became surging columns of steam. Then, with an ear-splitting roar, an explosion ripped through one of the central fissures, tearing loose huge fragments of rock and hurling them high into the air. As each explosion was succeeded by another more violent, dust and volcanic ash billowed upward and spread out to darken the sky.

After nightfall, a glow vividly lit the new vent. Pasty globs of molten rock soared through the air and showered down around the opening, piling up in sizzling heaps. Throughout the night, the eruptions continued unabated.

Daybreak revealed desolation. The panorama of rock, snowfields, and clear alpine streams looked as if dirty gray snow had fallen over the entire landscape. Nearby creeks and streams carried heavy loads of ash, cinders, and floating pumice. To the northeast, trees were defoliated and bent beneath a load of gritty volcanic dust.

The new volcano, which had overnight built a cinder cone 150 feet high, produced enormous quantities of pyroclastics, chiefly andesite pumice. At night, when the eruption cloud was not too dense, blazing sprays of red-hot particles shot skyward, while fountains of incandescent rock played against an inky curtain of falling ash. For years the magma from Rainier's subterranean reservoir was charged with too much gas to allow the lava to escape quietly; it was blown out with pulverizing force.

A. Thick streams of andesite from a broad lava cone inundate the surrounding valleys.

After Rainier's pyroclastic cone reached a height of about 1000 feet, explosive activity temporarily declined. Following several years of quiet, the volcano introduced a variation in its eruptive pattern. Liquid rock rose into the summit crater, which for a moment appeared brim-full and ready to disgorge its first real lava flow. But the loosely consolidated fragmental structure could not bear the weight of so much lava. From the foot of the cone to the crater rim, the volcano split into a network of fissures from which issued streams of fluid lava. The whole side of the cone was swept away by this irresistible torrent of molten rock.

Seeking the lowest channels in which to flow, the lava poured into a vacated riverbed near the base of the cone. Full of seething gases and registering temperatures of perhaps 1800 degrees Fahrenheit, the liquid andesite moved swiftly downslope. A few miles from its source, the flow lost momentum. As its crust hardened and thickened, the lava advanced only a few hundred yards per day. Where the old riverbed widened and deepened, the slowly-congealing lava accumulated to a depth of several hundred feet.

Eventually, gas pressure within the magma chamber was sufficiently relieved to allow the magma to sink back into its underground reservoir. Rainier's initial lava flow, the first of thousands, had ceased, but its slag-like surface remained hot for nearly a year, melting snows of winter.

Because the Rainier known to us is built almost exclusively of andesite flows, we might suppose that the new volcano grew chiefly by quiet emission of lava. This was not the case. For an unknown number of centuries, the pyroclastic cone ancestral to modern Rainier, and now concealed beneath its later lavas, erupted violently. The earliest evidence of this activity is in the Puget Sound lowland where volcanic mudflows derived from an active volcano at the site of Rainier are interbedded with glacial drift deposited by the vast ice sheet that moved south from Canada.

Some time about 700,000 years ago, after the ancestral cone had been partly removed by erosion, Rainier began to erupt voluminous lava flows. At this stage in its growth, the volcano probably resembled a broad lava dome, with tentacles of black, red, and gray andesite extending from a large central crater.

These prodigious flows probably issued from vents along the flanks of the volcano. Some traveled many miles. One flow entered the canyon of the Grand Park River, since vanished, which trended northeast from Rainier. This flow filled the valley for a distance of more than 12 miles to a depth of 2000 feet. The remnants of the flow,

Aerial view of Mt. Rainier's twin summit craters, the rims of which are kept snow-free by heat and steam emission. The North summit, Liberty Cap, appears on the left. —Photo by Bob and Ira Spring

since transformed into a free-standing ridge by erosion on either side, are now a mountain called Old Desolate and the flat bench underlying Grand Park.

Deep gorges west of Rainier met a similar fate. Since the growing volcano sat atop a complex of ridges and ravines, its lavas did not pile up around the base of the cone. Instead, they drained into nearby valleys, such as those of the ancestral Puyallup and Mowich rivers. The rivers were crowded from their ancient beds and forced to cut new channels along the flow margins. The old valleys of the Puyallup and Mowich rivers were buried beneath more than 2000 feet of andesite. On the east side of the volcano, a flow, 900 feet thick and several miles long, deflected the ancestral White River from its course. Another moved along the west face of Goat Mountain. Now, the displaced rivers have eroded most early intracanyon flows into isolated ridges that radiate like spokes of a wheel from the base of Rainier.

The War of Fire and Ice

As Rainier grew, the climate turned cooler. The glaciers, which had retreated to cirques high on neighboring Cascade peaks shortly before Rainier was born, began a new advance. Canyons formerly occupied by streams were now filled to capacity with rivers of ice, broadened as glaciers undermined canyon walls and scraped out

238

valley floors. Rainier then lapsed into its longest period of inactivity as the ice scoured its flanks, hollowing out cirques and canyons. The mountain, so newly built, seemed doomed to piecemeal destruction. Only a column of vapor drifting lazily from the summit crater indicated that the volcano merely slept.

When Rainier finally awoke, the effects were catastrophic. Explosions tore through the solidified plug in the volcano's conduit and sent fragments of old rock, mixed with fresh pumice and ash, thousands of feet into the air. Day became night as millions of tons of ash descended in a blinding curtain, sealing off light from the sun and blackening Rainier's snowy mantle. Incandescent lava erupted onto glacial ice produced enormous mud flows. Steam, meltwater, chunks of molten rock, and shattered lava blocks swept downslope in a chaotic mixture that moved at speeds up to 50 miles per hour. One lava flow poured down Rainier's north slope, where it consumed and replaced a glacier at least 1200 feet thick. It survives as Ptarmigan Ridge.

B. At its maximum, about 75,000 years ago, Mt. Rainier towered 16,000 feet above the Puget Lowland.

During the recurrent episodes of Pleistocene glaciation, fire and ice continued to shape the mountain. When the volcano was dormant, ice carved new trenches in its slopes. When it erupted, these gullies were filled by fresh streams of lava. At the height of glacial expansion, Rainier's entire cone was encased in ice, which extended far west to join the great ice sheet that smothered the Puget lowland.

Rainier further changed its eruptive habits. It no longer emitted canyon-filling lava flows, but produced thinner tongues of lava that piled up around the central crater. Earlier flows had constructed a broad foundation covering more than 100 square miles; the later lavas, rarely exceeding a thickness of 50 to 200 feet, erected a towering superstructure. The mountain began to assume the graceful shape characteristic of the world's great stratovolcanoes.

The Last 75,000 Years

About 75,000 years ago Rainier attained its maximum size. Although it then stood about 16,000 feet above the sea level and was far more symmetrical than it is today, Rainier was never a perfect cone. Because of its great height, it undoubtedly supported glaciers even during the warmest Pleistocene inter-glacial periods.

New vents then opened on Rainier's northern flank and built two satellite cones, the eroded remains of which are Echo Rock and Observation Rock. Fluid streams of olivine andesite issued from these parasitic volcanoes and flooded the valleys and depressions at Rainier's northern foot to a maximum depth of 500 feet. Their northeastward extent was blocked by Old Desolate mountain, a remnant of the gigantic flow that filled the Grand Park River canyon more than half a million years earlier.

After the north side eruptions, Rainier's volcanic energies declined. Lava flows occasionally broke through the sides of the cone, including one that filled a meltwater stream valley along the margin of a glacier in the Stevens River canyon. Activity at the summit was mainly restricted to the emission of corrosive gases.

At least three times during the last 65,000 years Rainier underwent intense glacial erosion. During the first two glaciations, ice completely smothered the mountain as well as the surrounding peaks, except for the very highest elevations. One glacier flowed 65 miles down the Cowlitz River. During the last glacial episode, sheets of ice stripped what remained of Rainier's original surface. Between approximately 25,000 and 10,000 years ago, about 2000 to 3000 feet of material were removed from all sides of the cone. By the time the

Mt. Rainier from the east. From left to right are: Little Tahoma Peak; wedge-shaped Gibraltar Rock; descending from the summit is Emmons Glacier; dividing the Emmons from the Winthrop Glacier is the inverted-V of Steamboat Prow; above the Winthrop Glacier stands Russell Cliff. The road leads to Yakima Park and Sunrise Visitors Center.
—Photo courtesy of the Washington State Dept. of Commerce and Economic Development

glaciers melted, one third of the mountain had disappeared. Although Rainier erupted little fresh lava during the last ice age, steam and other gases rising from magma deep inside the volcano gradually decayed the summit rocks and converted much of them into clay. This hydrothermal activity also prevented glaciers from forming inside the crater and significantly lowering the volcano's crest. At the end of the last ice age, Rainier still stood nearly 16,000 feet high.

Rainier in Post-Glacial Time

Most of Rainier's post-glacial eruptions have further diminished the peak. Between 6600 and 5700 years ago, a particularly destructive cycle of activity took place. At the outset of this cycle, Rainier was nearly 2000 feet higher than now. Little Tahoma Peak and Steamboat Prow extended much higher up the mountain, standing little more than a few hundred feet above the general level of the eastern slope. The cirque that now separates them was only a broad and relatively shallow basin, containing the ancestral Emmons Glacier. Loosened by blasts of steam, the glacier-worn eastern slope of the mountain collapsed, and an enormous mass of shattered rock plunged into the canyon of the White River. Water from condensed steam within the rockfall transformed the mass into a mudflow hundreds of feet thick that streamed many miles down the valley. Carrying blocks of andesite as much as 30 feet in diameter, the mudflow also incorporated boulders of granodiorite that tumbled from the walls of the White River canyon.

C. After repeated glaciations, the cone was reduced to an irregular mass of steep cliffs and cirques. Seen from the east are Little Tahoma, Steamboat Prow and Russell Cliff.

The rockfall and mudflow stripped another layer from Rainier's east side, exposing altered lavas that lay beneath the volcano's outer shell. Some of the newly-exposed lavas and breccias of Rainier's interior were tinted green, white, dull orange, or sulphur yellow, showing that hot water and chemicals had decomposed once solid rock into soft permeable materials. The steam explosions that showered rock fragments over a wide area east of the volcano and triggered the east-side mudflow also started a massive avalanche that swept down the Nisqually Glacier on the south side of Rainier. Transformed into a mudflow, it rushed across Paradise Valley in a single wave 800 feet high.

The Destruction of the Summit

Rainier lost more than a fifth of a cubic mile in volume during the Paradise and Greenwater mudflows, but the loss was small compared to the destruction that followed. About 5700 years ago, after at least two more eruptive episodes, the most cataclysmic of all Rainier's recent eruptions occurred.

To Indians camped on a hill near the present site of Enumclaw, about 25 miles northwest of the volcano, this eruption at first must have seemed little different from others they or their ancestors watched. True, the mountain's summit was wrapped in an unusually large cloud of gray ash that prevented them from seeing exactly what was happening. It is also true that the roaring and thundering of Tahoma, the mountain that was "God," was more frightening than usual. But these cautious hunters were on high ground, 35 miles down the twisting valley that heads on the *east* side of the erupting volcano.

When they first noticed what was advancing on their camp, the Indians must have dropped their stone implements in terror. A wall of rock and mud nearly 100 feet high was rushing toward them. The Indians ran toward a nearby ridge, but the wave of volcanic debris was moving perhaps 40 miles per hour. It buried them and their settlement.

The unfortunate Indians never understood what hit them, but geologists have reconstructed the events that led to this disastrous mudflow, the Osceola, more than a half a cubic mile in volume. It was caused by the sudden collapse of Rainier's summit, which disintegrated like the dome of some great stone cathedral during an earthquake. Steam explosions directed a hurricane of shattered rock over

D. During the eruption that triggered the Osceola mudflow, about 5700 years ago, Mt. Rainier lost 2000 feet of its former summit. A broad caldera, dipping eastward, now occupies the summit area.

the northeast side of the volcano. Simultaneously, the undermined summit toppled eastward, forming an avalanche of hydrothermally altered rock hundreds of feet high that easily overrode the apex of Steamboat Prow, momentarily submerging the entire structure.

One wave descended the Emmons Glacier between the Prow and Little Tahoma to flood the White River canyon; another sped down the Winthrop Glacier into the West Fork of the White River. Converging beyond the base of the mountain they flowed another 65 miles to inundate 125 square miles of the Puget lowland west of the Cascade mountain front. One lobe of the flow reached an arm of Puget Sound, burying the sites of Kent, Auburn, Sumner, and Puyallup. Within hours, rocks that had stood nearly 16,000 feet above sea level lay beneath the waters of Puget Sound. Other parts of the former summit filled lowland valleys to create a level plain. Seldom has a volcanic mudflow buried an area so far from its source.

After this catastrophe, Rainier's summit housed a void almost two miles in diameter—a bowl tipped toward the east. The highest point on the crater walls were, on the north, Liberty Cap, 14,112 feet, and, on the southwest, Point Success, 14,150 feet. The western wall, relatively intact, stood somewhat higher until about 2800 years ago, when it collapsed during another series of rockslides.

Originating high on the west face of Rainier, at a point below Liberty Cap, an avalanche of chemically altered rock, transformed into a mudflow, temporarily filled the upper South Puyallup and Tahoma Creek valleys to a depth of at least 1000 feet. The western wall of the summit caldera also crumbled, leaving a wide gap through

Mount Rainier's heavily glaciated summit area with the youthful summit cone, center, within remnants of the caldera wall formed by the collapse of the volcano's former summit about 5700 years ago. The caldera's highest remaining elevations are Point Success, left, and Liberty Cap, right center. —U.S. Geological Survey photo by Austin Post

E. About 2000 years ago eruptions of lava built the present summit cone, largely filling in the summit depression.

which icefalls of the Tahoma Glacier now descend. These rockfalls made the mountain even more vulnerable to erosion by uncovering the volcano's central plug, a yellowish mass so decayed by chemical action that it offered little resistance to glacial cutting. Puyallup and Tahoma glaciers ate rapidly into the headwalls of this exposed area, creating the large western cirque known as the Sunset Amphitheatre.

Building of the Present Summit Cone

About 2500 years ago Rainier introduced a welcome variation in its eruptive behavior. For the first time in perhaps 25,000 years a cycle of activity actually repaired some of the damage done by earlier outbursts.

Renewed construction began with a discharge of hot ash and molten breadcrust bombs blown from vents within the summit caldera. These accumulated around the mountaintop until repeated explosions sent them sliding down the west flank of the mountain. This hot avalanche, at least 200 feet high, traveled down the South Puyallup River valley and ignited groves of timber on the valley floor. Engulfed in a flowing mass with a temperature at least 600 degrees Fahrenheit, many tree trunks were reduced to charcoal. Carbon samples from this deposit enabled scientists to date the eruption.

Rainier next erupted large volumes of pumice, which the winds carried northeast of the peak to blanket Yakima Park, where it lies a foot thick. Then, streams of liquid andesite poured from fissures in the caldera floor and spilled east through the rim. These thin tongues of black, glassy lava did not travel far, but they did cause sudden melting of the Emmons and Nisqually glaciers. More volcanic mudflows poured into the White River and Nisqually valleys, raising canyon floors at least 80 feet above their present levels.

The lava flows quickly built a small volcano atop the old, shattered summit. As the cone grew, it filled most of the summit depression left by the catastrophic avalanches of about 3700 years before. When complete, the beautifully symmetrical young mountain sitting atop the ruins of its predecessor had a base a mile across and a crest that stood at least 1000 feet above the now-buried eastern lip of the old caldera. A second, briefer, eruption of lava occurred some time after the first, forming a somewhat larger crater east of the older vent.

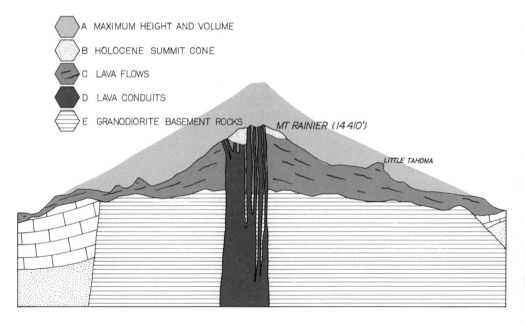

Simplified cross-section of Mt. Rainier. The present cone is marked in heavy outline. After Fiske, Hopson and Waters, 1963.
A. Maximum height and volume. B. Holocene summit cone. C. Lava Flows. D. Lava conduits. E. Granodiorite basement rocks.

Thirteen hundred feet across and perhaps 500 feet deep, this younger crater tilts noticeably toward the east. The point where the two recent craters overlap—now called Columbia Crest—marks the highest elevation of the volcano.

Despite its youth, the summit cone shows minor indications of glacial cutting—a few shallow depressions separated by low ridges preview future glacial trenches and basins. Were it not for constant heat and steam emission, glaciers might already have breached the crater rims. Occasional brief eruptions of steam and ash have come from the summit vents during the past 2000 years, the last within historic time. But, contrasted with the countless tons of material removed daily from the volcano by landslides, glaciers, and meltwater, the amount of new lava produced in recent centuries is negligible.

Our journey into Rainier's past has shown the volcano growing from a small cinder cone to an ice-covered giant; from a classic smooth-sided cone to its present craggy mass. Although its life story has already spanned more than a million years, Rainier will be with us and our descendants for a long time.

Rainier's Historic Eruptions

Historic time in the Pacific Northwest goes back scarcely a century and a half. Almost until the American Civil War, white inhabitants of western Washington included a mere handful of itinerant trappers, traders, or employees of the Hudson's Bay Company. Olympia, Tacoma, and Seattle did not really take root until the 1850s. Despite the scanty population, however, observers logged 16 apparent eruptions of Rainier between 1820 and 1894.

The earliest historic activity supposedly occurred "about 1820." Years after the alleged event, an aged Indian named John Hiaton reported having seen "fire," heard "noise," and felt "an earthquake" emanating from Rainier. He also said that his ancestors had witnessed many such disturbances in the past. The next reference to activity on Rainier appears in the bulletins of the noted Belgian seismologist, A. Perrey. For 1841 he recorded "violent eruptions of Mt. Raynier [sic.], Oregon." Perrey also cited eruptions for 1842 and November 23, 1843. Captain John Frémont wrote on November 13, 1843, that "at this time, two of the great snowy cones, Mount Regnier [sic] and St. Helens, were in action." Frémont specified "the 23rd of the preceeding November, 1842," as the date when St. Helens scattered ashes as far as The Dalles, Oregon. It seems suspiciously coincidental that Perrey cites November 23 as also marking an eruption of Rainier.

An early Northwest missionary, Father De Smet, alluded to "signs of activity" at Rainier in 1846. In his book on Mt. Rainier National Park, F.E. Matthes further asserts that "actual records exist of slight eruptions in 1843, 1854, 1858, and 1870. Neither De Smet nor Matthes stated his authority or gave any confirmatory details. It seems strange that Matthes' dates correspond precisely to times when Baker is known to have erupted.

Because so many of these early accounts are brief to the point of obscurity, most geologists dismissed them outright until recently. Many assumed that the historical records probably referred to unusual cloud formations, dust from rock-falls, and other non-volcanic phenomena. In 1967 Donald R. Mullineaux discoverd a previously unrecognized deposit of fresh, light brown pumice from Rainier. By dating young glacial moraines on which the thin sprinkling of pumice lay, he determined that at least once between 1820 and 1854 the volcano had indeed been active.

Indirect confirmation of unusual heat in the crater comes from the account of a Yakima Indian named Saluskin. According to his story, told when he was an old man, he had, in June, 1855, guided two "King George men," Indians' term for Caucasians, to the east flank of Rainier. After their safe return from the summit, they told Saluskin, who had waited at base camp, that there was "ice all over top, lake in center and smoke or steam coming out all around like sweat-house." One is inclined to believe this story because the men could not have observed these features without reaching the crater rim.

The eastern crater now contains snow and ice equivalent to one billion gallons of water, quite enough to form a small lake. If volcanic heat in 1854 was sufficient to melt the crater icepack and create a body of standing water, there was more thermal activity at Rainier then. The "eruption" of 1854 may have been a column of water vapor rising from fumaroles surrounding the crater lake.

When Hazard Stevens and Philemon Van Trump climbed Rainier in August, 1870, they saw quantities of steam and sulphur fumes in both summit craters, but no sign of a lake. Thermal activity at the summit had evidently decreased to the relatively mild but constant state it now maintains. The 1870 party took miserable refuge overnight in the steam caves of the west crater, where they were nauseated by hydrogen sulphide and alternately scalded by steam jets or frozen by icy gusts.

No evidence of newly erupted material was noticed in August, 1870, which casts doubt on the authenticity of an eruption reported for that year. However, a violent earthquake the following September triggered a large rockfall from Liberty Cap, the mountain's northern peak. About 80 acres of material cascaded downslope, leaving a nearly perpendicular face on the southern side of Liberty Cap. Very likely the dustcloud from this avalanche was mistaken for a volcanic explosion.

Several other signs of activity reported in the 1870s and 1880s were viewed from a considerable distance, so they may have been trailing streamers of fog, or dust kicked up by landslides. If any were actual expulsions of steam or ash, they left no recognizable signs.

Frederick G. Plummer, an early Northwest scientist, described activity that began on October 18, 1873, and lasted for a week. The *Washington Standard*, an Olympia newspaper, reported on October 25, 1873, that "clouds of smoke were seen pouring from the highest peak of Mount Rainier. The smoke was seen until nearly dark when clouds shut down upon the mountain hiding it from view." The following November 29 the same paper again noted that "smoke has been ascending from the highest peak of Mount Rainier, within the past few days."

Len Longmire, a pioneer settler, recalled sightings of "a series of brown, billowy clouds" issuing from the crater in 1879 and again in 1882. Plummer recorded a later disturbance: "On June 16, 1884, at about 7 p.m., jets of steam were seen shooting upward from the summit of Mount Tacoma to a considerable height. This phenomena [sic.] was repeated at short intervals until darkness cut off the view. There was no fire, and no earth tremors were reported."

The last 19th century report of an "eruption" was highly publicized by the Seattle *Post Intelligencer* and the *Tacoma Daily Ledger* in November-December, 1894. Although the stories are highly specific and detailed, it seems doubtful that anything more than journalistic imagination was active. In a preface to his compilation of historic eruptions in Oregon, Washington, and Alaska, Plummer wrote: "There can be no doubt that many eruptions were reported which might be contradicted if examination were possible. For example, the reports of the eruption and change in the summit of Mount Tacoma [Rainier] from November 21 to December 25, 1894, filled many columns of press dispatches, and possibly were intended for that purpose. December 25th was the most perfect day for observation, and, with my 6½-inch refractor, the crater-peak and its surroundings were carefully examined, and no change could be seen. No eruption was noted, other than the usual emission of steam, which varies with the barometer."

Recent Thermal Activity

Although Rainier has not erupted in more than a century, neither has it remained entirely at rest. Areas of hot rock and fumaroles exist on the summit cone and the volcano still produces occasional steam explosions. Beginning in the early 1960s summer climbers were sometimes startled by hearing loud reports and seeing columns of vapor rise from crevices. In 1961, steam blasted a hole near Gibraltar Rock, sending a column of pressurized vapor 200 feet into the air and scattering debris over the nearby Cowlitz Glacier. This vent remained active throughout the summer, though with diminishing energy. In March, 1965, skiers saw clouds of steam spouting from a ridge above the Kautz Glacier and setting off an "avalanche."

A much greater avalanche, the largest in historic time, may have been initiated by a steam explosion on December 14, 1963. About noon on that date, forest rangers about 12 miles northeast of the mountain heard "a very loud, sharp boom in the direction of Mount Rainier." When clouds and falling snow cleared enough for the rangers to observe the eastern slope, they saw rock debris covering the lower Emmons Glacier.

What the rangers could not see was that approximately 14 million cubic yards of lavas and breccias had fallen from the north face of Little Tahoma Peak. Plummeting straight downward for 1700 feet onto the glacier's surface, the avalanche struck with tremendous force. Because of its large mass and the steepness of its landing site, the avalanche shot across the glacier, at speeds up to 100 miles per hour. When it reached the glacier's snout the mass soared into space. A stream-gauging station, six feet high, was untouched as millions of tons of rock hurtled overhead. Where the upper White River valley curves or is constricted, the flowing mass of rock fragments and trapped air surged up the canyon walls as high as 300 feet. When it finally came to rest, a scant half mile from White River Campground, the avalanche had traveled about four miles from its source while dropping some 6200 feet in altitude. Some of the boulders are as large as buildings. One measures 60 by 130 by 160 feet and weighs approximately 50,000 tons!

Later studies revealed that at least seven separate rockfalls and avalanches had fallen in quick succession. The plywood gauge house, undisturbed by one avalanche, was later carried several hundred feet by a blast of air escaping from the margin of another avalanche that stopped a short distance away. At least two square miles of the Emmons Glacier and upper White River valley were covered with broken rock.

More recent steam explosions have happened on the west side of the mountain. In August and September, 1967, clouds of water vapor and steam were seen billowing from cliffs above the South Tahoma Glacier. During the same period, floods and mudflows repeatedly descended the Tahoma Creek valley, sweeping boulders downstream and raising waves of mud up to 15 feet high. These rockfalls and slurry floods may have been initiated by a steam vent beneath the South Tahoma Glacier.

A more threatening manifestation of Rainier's internal heat occurred high on Emmons Glacier during the summer of 1969. Between an elevation of 10,000 and 13,000 feet the ice surface broke into a network of potholes and crevasses. In some places, gaps of ice opened wide enough to reveal bare rock beneath the glacier. This melting by subsurface heating was brief, however, and by the summer of 1970 new ice and snow had filled most of the caved-in area. Nonetheless, such melt-depressions on glaciers, unrelated to weather conditions, are the kind of warning to be expected when a dormant volcano is preparing to erupt.

The "Volcano Watch"

Because of these—and other—signs of geologic restlessness, scientists at the University of Washington and the United States Geological Survey keep a "volcano watch" on Rainier.

In the 1960s, the U.S. Geological Survey began taking aerial photographs and infrared images of the summit craters, Little Tahoma Peak, and the Emmons and South Tahoma glaciers. Seismographs were placed at various locations on the mountain, one as high as Camp Muir at the 10,000-foot level to record swarms of microearthquakes centered beneath the peak. An increase in the number and intensity of earthquakes may mean that magma is rising through the volcano's conduit toward the surface. The tremors demonstrate that some activity continues in Rainier's subterranean magma chamber.

Infrared surveys have confirmed what climbers have always known—Rainier's summit craters are hot. The zones of most intense heat lie along the northwestern rim of the crater, the north side of the west crater, and along a pattern of concentric arcs on the western flank of the summit cone. A later survey indicated a possible increase of thermal anomalies along the southern rim of the west crater. In general, the hottest areas correspond to swaths of exposed rock, which stand out as black patches amid the summit icefields.

When geologist and mountaineer Dee Molenaar measured the heat generated at fumaroles in both summit craters, he found that the steam issues at temperatures of about 186 degrees Fahrenheit, the boiling point of water at 14,000 feet. In the summer of 1970, Robert Moxham of the U.S. Geological Survey installed inside the west crater an instrument programmed to take the volcano's daily temperature and relay it via space satellite to a ground station for interpretations. This device operated only five weeks before the weather destroyed it. Even so, the instruments functioned long enough to demonstrate the feasibility of using space satellites to monitor volcanic temperatures.

The Summit Steam Caves

Weary climbers who reach 14,410-foot Columbia Crest seldom have either the desire or time to explore one of Rainier's most interesting features. These are the steam caves, a 1.5 mile-long maze of twisting corridors and high-ceilinged caverns melted beneath the snow that almost fills the twin summit craters. So far as is known, Rainier, Baker, and Wrangell in Alaska have the only volcanic craters that contain such a labyrinth of ice tunnels.

Snow fills the east crater to an estimated depth of 360 feet, but steam jets and fumaroles along the inner crater walls have melted out a system of passageways with a cumulative length of at least a mile. These tunnels, actually space between the inside slopes of the crater and the overlying icecap, are extremely steep and littered with blocks of andesite, mud, and pumice. The lowest point to which scientists have thus far descended is about 540 feet below the highest elevation of the crater rim, 340 feet beneath the surface of the crater snowpack. The crater floor which slopes at an angle of 32 degrees, here touches the ice ceiling and prevents further exploration.

The western crater, only about 1000 feet across, has a smaller cave system. But it does contain a pool of meltwater which, in August, 1972, measured about 120 by 40 feet and was 18 feet deep. Rainier thus boasts another unique feature—a crater lake 14,000 feet above sea level and under an arching canopy of pure ice. This unnamed lake may someday vanish as the delicate balance between snowmelt and snowfall changes.

The Future

What will happen when Rainier erupts again? Members of the U.S. Geological Survey have recently studied the probable hazards of future eruptions. As with other high Cascade volcanoes, the chief

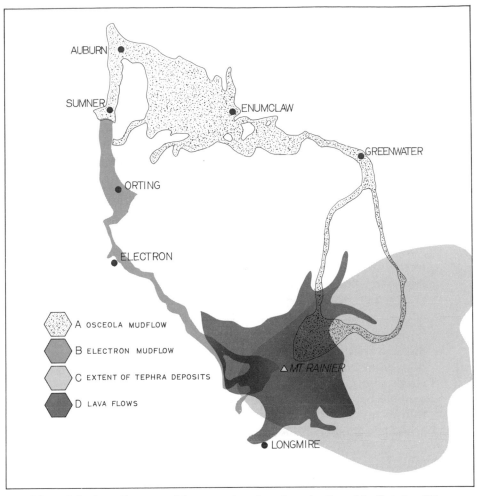

Map of the lava flows, mudflows, and tephra deposits from Mt. Rainier. The Osceola and the Electron mudflows overwhelmed the sites of several towns and cities in western Washington. —Based on Crandell, 1973
A. Osceola mudflow (5700 years ago). B. Electron mudflow (600 years ago). C. Extent of tephra deposits. D. Lava flows.

hazard stems from volcanic mudflows created when hot lava or ash suddenly melts large quantities of ice and snow. The resulting floods and debris flows are especially dangerous because they move downslope at high speeds and can reach into low areas many miles beyond their origin. Recall the immense impact of the Osceola Mudflow.

If renewed activity centers in the summit craters, the chemically altered rocks along the northern crater rim could easily give way, releasing perhaps a billion or more gallons of water derived from the existing crater snowfields. Should fresh lava rise into the eastern crater, steam explosions caused by hot magma in contact with melt-water might rupture the crater walls, perhaps causing another Osceola-type mudflow.

Judging by the effects of volcanic eruptions of the last 7000 years, the Puyallup, Tahoma, Nisqually, and White River valleys are among the zones of highest risk. The Cowlitz and Carbon River valleys seem to have seen few mudflows during the past several thousand years.

Prevailing winds from the Pacific would probably carry the eruption clouds east, so it seems unlikely that the Puget Sound cities would suffer a heavy ashfall. Towns downwind, such as Ellensburg or Yakima, might find their air polluted, water supplies muddied, and machinery clogged by ash. Only if Rainier exploded on a scale comparable to the catastrophic eruption of Mazama 6900 years ago would it pose a serious threat to population centers to the west.

With pyroclastic eruptions no larger than those that preceded the building of the present summit cone 2000 years ago, appreciable damage would probably be confined to a radius of 10 to 15 miles downwind from the crater. Curious persons standing on ridges at a closer range, even though they would be safe from mudflows in the canyons below, might easily be struck by flying chunks of rock or by falling pumice. Remember that previous eruptions have thrown rocks up to four feet long as far as eight miles.

Of course, Rainier might produce an eruption much larger than any of the past 7000 years. In that event, no one can predict what would happen. If such a paroxysm should seem imminent, speedy removal to high ground as far upwind as possible would be wise.

Visiting Rainier

As a major national park visited by hundreds of thousands of persons each year, Rainier Park is accessible from almost every side by paved roads. Hundreds of miles of dirt logging roads and hiking trails criss-cross the park from heavily forested valley floors to wind-swept meadows at timberline and above.

The most popular automobile route is that from Longmire at the mountain's southwest base, to Paradise Valley, a mile high, where an inn, restaurant, observation tower and museum, information center, and other tourist facilities cater to the public.

The Nisqually Glacier Vista trail provides memorable views of the three-mile-long river of ice. An undemanding one-mile round-trip through flowering meadows and across clear rivulets, the trail runs uphill from the Visitor Center.

For a panorama of the Cascades south of Rainier, including the snowy cones of Adams, St. Helens, and Hood, hike 2.5 miles from the Visitor Center to a meadow at 7000 feet. Early morning and after five in the evening, when summer haze is at a minimum, are the best times.

Sunrise Visitor Center at Yakima Park, on the volcano's east side, offers a close view of Emmons Glacier. Follow the Stevens Canyon road, with its spectacular bluffs and water falls, around Rainier's southern base, and follow the signs to the 6000-foot level. White River Campground, a short distance downriver from the snout of the Emmons Glacier, has excellent campsites with water.

Climbing Rainier, at any season, should be attempted only with experienced guides and proper equipment, including ice ax, rope, crampons, overnight supplies, warm clothing and other indispensables. Rainier is so high that it makes its own highly changeable weather; sudden storms may rage around the summit while valleys below bask in sunlight.

Heavily glaciated, Glacier Peak surmounts a sea of sharp ridges and twisting valleys in the rugged North Cascades. One of the most violently explosive volcanoes in the Pacific Northwest, it repeatedly produced large-volume pyroclastic flows and mudflows during the last 12,000 years. —Photo by Bob and Ira Spring

XVIII
Glacier Peak:
White Goddess of the North Cascades

According to Greek mythology, Mt. Olympus was the home of the 12 major Hellenic gods. If the dozen greatest Cascade peaks were to be identified with the ruling deities of classical antiquity, lonely Glacier Peak would undoubtedly typify Artemis. Known to the Romans as Diana, this virgin goddess shunned civilized life and withdrew into the wilderness to roam with the wild creatures. Given her choice of a Cascade dwelling place today, she would find Glacier Peak ideal. It is the most remote and inaccessible volcano in the range.

Although its 10,451-foot crest is only about 300 feet lower than Baker's, this solitary peak is unfamiliar even to many alpine enthusiasts. No roads approach the mountain, and one must hike many miles through extremely rough terrain to reach its base. Glacier Peak bestows familiarity only upon those willing to strive for it.

Glacier Peak and the surrounding mountains seem to be in the midst of a contemporary ice age. Although existing glaciers are smaller than those that buried the region during Pleistocene ice ages, aerial views resemble Arctic landscapes—a sea of ice through which loom isolated peaks and ridges.

Not only are the Cascades broadest in this northern stretch, summits are considerably higher—many exceed 8000 feet. Glacier Peak stands on a long ridge of this elevation, immediately west of the Cascade divide. Hikers can reach the volcano from the west via the White Chuck Valley, or the Suiattle River valley; from the east, approach from the western tip of Lake Chelan, which extends from the eastern border of the Cascades deep into their interior.

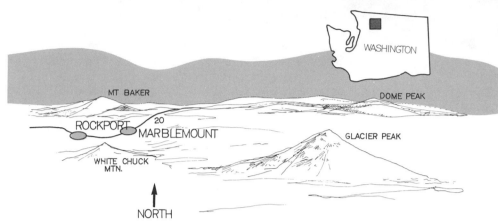

Relief map showing the location of Glacier Peak.

The Geology of Glacier Peak

Glacier Peak has long been noted for the enormous volumes of pumice it erupted about 12,000 years ago, one of the most extensive ash deposits in the Pacific Northwest. It was generally assumed that the volcano had been quiet ever since, and was probably extinct. In the early 1980s, however, James Begét showed that reports of Glacier Peak's demise had been greatly exaggerated. During the last 5500 years this relatively small cone has repeatedly demonstrated its undiminished vigor, producing additional tephra, domes, numerous pyroclastic flows, and mudflows that traveled many miles down valley. The latest eruption apparently occurred only about two centuries ago. Far from approaching extinction, Glacier Peak is one of the half dozen most active volcanoes in the Cascade Range.

Although best known for its great pumice eruptions, Glacier Peak was built principally by quiet effusions of dacite lava. The earliest flows, unusually fluid for dacites, issued from a vent on the east slope of Lime Ridge, a northwest-trending spur on the Cascade crest. The first lavas poured east into valleys tributary to the Suiattle River, which drains the north and east flanks of the volcano. The flows were blocked on the west by the high mass of Lime Ridge until after the cone had grown perhaps 2000 feet; then lavas spilled west into the ancestral White Chuck River and related stream valleys.

All Glacier Peak lavas exhibit normal magnetic polarity, indicating that they erupted since the last magnetic reversal about 700,000 years ago. It seems likely that the cone-building eruptions were over before the beginning of the last ice age. Since then, the original contours of the volcano have been strongly modified by glacial erosion.

The Late Pleistocene Eruptions

As the last Pleistocene ice sheets diminished and alpine glaciers retreated upvalley, Glacier Peak began a series of violently explosive eruptions. Between about 12,750 and 11,250 years ago, the volcano ejected at least 2.5 cubic miles of pyroclastic debris. A minimum of 1.25 cubic miles of pumice and ash were thrown high into the air and carried east as far as 620 miles to blanket most of eastern Washington, northeastern Oregon, the Idaho panhandle, western Montana, and parts of southern Canada. There are at least nine separate layers in this series, two of which account for most of the ash. One layer extends southward along the northern Cascade crest. Since the last ice age, only the ashfall from Mazama covered a larger territory. Because these two eruptions left distinctive ash layers, scientists use their deposits to help date other geological formations or archaeological remains. Glacier Peak produced an extremely useful time marker.

Although commonly referred to as Glacier Peak ash, in the North Cascades the tephra consists mainly of yellowish or light gray pumice bombs and lapilli, which have a thickness of up to 12 feet as far as 12 miles downwind from the volcano. Pumice and rock fragments progressively diminish in diameter with increasing distance from their source. Ash predominates only along the margins of the deposit and east of the Columbia River.

The same eruptions also produced large pyroclastic flows and mudflows that buried valley floors many miles from the volcano. The oldest recognized mudflow, which preceded the major tephra outburst, consisted of altered rock that presumably avalanched from Glacier Peak's summit. This clay-rich mudflow streamed down the volcano's west slope into Kennedy Creek, thence into the White Chuck River valley, where it traveled at least 18 miles.

Perhaps only a short time before the big pumice eruptions, Glacier Peak produced at least ten pyroclastic flows and mudflows that poured down the west side of the mountain into the White Chuck and Sauk valleys, where they are as much as 350 feet deep. They consist mostly of rock fragments, which suggests that they may have been derived largely from collapsing lava domes. One mudflow buried a charcoal deposit near Darrington that dates at about 11,670 years. The top of the sequence is a thick pyroclastic flow, the White Chuck Tuff, that was still hot enough when it settled partially to weld itself into solid rock, as far as ten miles from Glacier Peak.

Renewed eruptions poured more pyroclastic flows and mudflows into stream valleys west of Glacier Peak. Some of these later deposits consist of reworked pumice derived from earlier eruptions, but others were freshly erupted pumice. Large mudflows and floods went down the White Chuck, Sauk and Skagit rivers all the way to Puget Sound, so filling the stream beds that the rivers were forced to cut new channels.

Some rivers draining the Glacier Peak region radically changed their courses. Before these valley-filling eruptions, it appears that the Sauk and, perhaps, the Suiattle rivers flowed west through the Stilliguamish valley. After the Stilliguamish was chocked with pyroclastic debris that extended more than 30 miles to near Arlington, the Sauk and Suiattle rivers abandoned their original routes to flow north into the Skagit River at Concrete.

By the end of this late Ice Age eruptive period, Glacier Peak had transformed not only the North Cascade landscape but a broad fan-shaped area extending for hundreds of miles northeast, east, and south of the volcano. Glaciers, mountaintops, canyons, the great Columbia River Plateau, and the Snake River plain beyond were blanketed by gray drifts of pumiceous ash. Huge aprons of debris sloped outward from the cone, smothering both the Stilliguamish and Skagit drainages and perhaps extending tens of miles downstream to the shores of Puget Sound. Deposits from Glacier Peak's late Pleistocene pyroclastic eruptions can be traced as far west as Arlington, near the sound's ancient shoreline.

A long dormant interval followed during which streams eroded new canyons into the pyroclastic and mudflow deposits that filled their old valleys. The new ravines were to channel material ejected during future eruptions, which would deposit new debris terraces on freshly excavated valley bottoms.

After about 5700 to 6000 years of apparently total silence, Glacier Peak produced a mudflow composed of altered rock debris that is now exposed in the White Chuck River above Kennedy Hot Springs. This mudflow may correlate with a similar deposit in Bakos Creek on the southwest side of the mountain, and with a small layer of altered rock fragments east of the volcano. At this point, the volcano was probably blowing off steam and broken fragments of old rock. The eruptions that followed discharged large quantities of fresh magma. Dacite domes extruded at or near Glacier Peak's summit exploded and disintegrated, sending wave after wave of incandescent rock fragments avalanching into adjacent valleys. Dozens of individual block-and-ash flows, as well as pumice pyroclastic flows buried large forests under debris aprons extending many miles downvalley. Radiocarbon dates on charred wood entombed in the oldest flows shows that the trees burned about 5520 years ago.

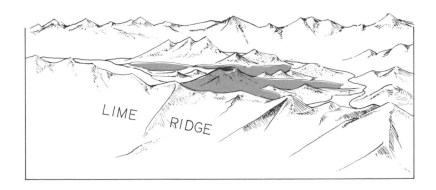

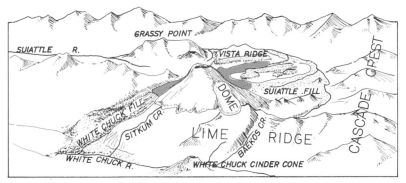

The growth of Glacier Peak volcano.

A. In Pleistocene time, the earliest Glacier Peak lavas erupted east of Lime Ridge and flowed into valleys tributary to the Suiattle River. B. Glacier Peak overtopped Lime Ridge and sent streams of dacite lava westward into tributaries of the White Chuck valley. On the east, erosional remnants of the oldest flows, which filled valleys, are now perched on top of ridges. Younger flows descended ancestral Vista Creek and Chocolate Dusty Creek. C. During ice advances Glacier Peak and the surrounding valleys were deeply glaciated. Remnants of the younger flows cling to valley sides. Valley bottom flows have descended Vista Creek and Kennedy Creek. Disappointment Peak Dome grew on the south flank of Glacier Peak and the Suiattle fill came from a dome on the east side. Explosions mantled the area with pumice. Late Pleistocene and Holocene pyroclastic flows and mudflows extended for many tens of miles down valleys. —After Tabor and Crowder, 1969; Beget, 1982a and 1982b

261

Dusty Creek, which issues from Dusty Glacier on the west flank of Glacier Peak and empties into the Suiattle River, received especially large amounts of pyroclastic material, as did Chocolate Creek just to the south. Hundreds of individual pyroclastic flows with an aggregate thickness of over 1000 feet fill Dusty Creek with at least one cubic mile of debris, probably about 5500 to 5100 years old. The Dusty Creek fill and adjacent ridges are blanketed with ash that probably fell from ash clouds that billowed above the pyroclastic flows. Altogether, the eruptions of about 5500 to 5100 years ago produced as much as three cubic miles of lava. Buried soils, lake sediments and peat layers interbedded with layers of volcanic rock indicate that debris flows dammed the White Chuck River several times. Water impounded behind these dams formed a succession of lakes.

Much of the Darrington area is underlain by deposits of this eruptive period. Flood sediments laid down at about the same time reach as far as Minkler Lake, about 70 miles downstream from Glacier Peak.

Between about 1700 and 1800 years ago more large pyroclastic flows poured down the east and west slopes of Glacier Peak, partially refilling the ravines cut into older volcanic deposits. Floods and mudflows transported debris more than 60 miles to the Skagit River delta. Sometime after 1800 years ago, another large mudflow streamed at least 19 miles west of the cone. It may have been caused by stream explosions that dislodged portions of the former summit. Shortly before 1000 A.D. numerous mudflows and some pyroclastic flows left deposits locally as much as 350 feet thick in valleys heading on Glacier Peak. Sometime before 300 years ago, a series of unusually large floods and mudflows poured into canyons near the volcano, carrying material 20 to 30 miles down the valleys, perhaps as the result of an eruption.

A widespread layer of fine gray ash covers much of the north and east slopes of Glacier Peak to depths of as much as a foot near the mouth of Dusty Creek, to over seven inches near Gamma Peak. The ash deposits on Gamma Ridge are in turn overlain by scattered lapilli up to about four inches in diameter. These youngest ash layers and pumice fragments lie on recent glacial moraine that supports trees up to 316 years old, suggesting that the latest eruptions occurred only two or three centuries ago. In the 1850s, Indians told George Gibbs that the mountain had been smoking within living memory.

Glacier Peak shows few obvious signs of its recent volcanic past. Lava ribs stand hundreds of feet above the icefields that surround them, indicating that much, if not all, of the mountain's original surface has been removed. Glacier Peak's earliest lava flows, which

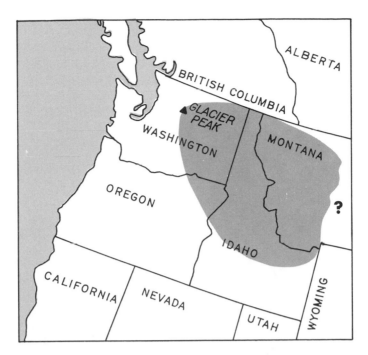

Map showing the area covered by ash falls from early Holocene eruptions of Glacier Peak.

filled narrow gorges to the east, now underlie ridge tops or stand as isolated features, their connection to the cone severed by ice and stream cutting. Nonetheless the outlines of a summit crater remain, breached at both the eastern and western walls, from which, respectively, issue the Chocolate and Scimitar glaciers. The north crater wall is highly oxidized, with some of its rock stained a blotchy yellow, suggesting that they have been attacked by steam and acid fumes.

Glacier Peak has no active fumaroles. Even so, three hot springs on the flanks of the cone—the Gamma, Kennedy, and Sulphur Hot Springs—suggest that magma still exists at depth. The number, intensity and voluminousness of Glacier Peak's late eruptions strongly argue that the volcano is almost certain to erupt again. Lonely and remote as it is at the center of the Glacier Peak Wilderness, the volcano has repeatedly affected areas many tens of miles downwind or downstream from its cone. Two factors make Glacier Peak potentially dangerous: First, it characteristically produces violent Peléan eruptions, extruding domes at or near the summit that typically collapse to form large pyroclastic flows. Second, its position high above stream valleys that slope steeply westward into inhabited regions directs pyroclastic flows, mudflows, and floods toward populated areas. Future eruptions are likely to eject moderate to catastrophically large quantities of pyroclastic material. Glacier Peak may again bury forests on the volcano's flanks and send floods and mudflows downvalley to farms and settlements where thousands of people now live.

The smooth surface on this rhyolite formed as it extruded. Compare it to the surface on freshly squeezed toothpaste. —Photo courtesy of D.W. Hyndman

Baker's relatively smooth cone contrasts with the Black Buttes, the eroded remnants of an older volcano. —U.S. Geological Survey photo by Austin Post

XIX
Mt. Baker:
"Fire" Under Ice

When Captain Vancouver sailed through the Strait of Juan de Fuca and anchored near the site of Dungeness, Washington, his third lieutenant, Joseph Baker, sighted "a very high, conspicuous, craggy mountain. . . towering above the clouds; as low down as it was visible, it was covered with snow; and south of it was a long ridge of very rugged snowy mountains, much less elevated, which seemed to stretch to a considerable distance." Vancouver named Baker after the junior officer who had first observed it. Two years earlier, the Spanish explorer Manuel Quimper had seen the peak and christened it *La Gran Montana del Carmelo* (The Great Mountain of the Carmelite), presumably observing the feastday of the Carmelite Order.

To many, Quimper's name may seem preferable. It evokes an appealing image of the mountain as a white-robed nun standing guard over virginal wilderness. Baker receives extremely heavy snowfall and supports 20 square miles of active glaciers. It is almost entirely sheathed in ice and, in clear weather, visible throughout northwestern Washington and southwestern British Columbia. The Nooksak name *Quck-Sam-ik* (White Rock Mountain) is apt.

To the Lummi Indians, Baker was known as *Komo Kulshan*, (Shot at the Point), referring to a legend in which the Great Spirit, Sochhalee Tyee, shot it with fire from heaven, inflicting a wound that bled, burned, and smoked. The Lummis' apparent reference to a

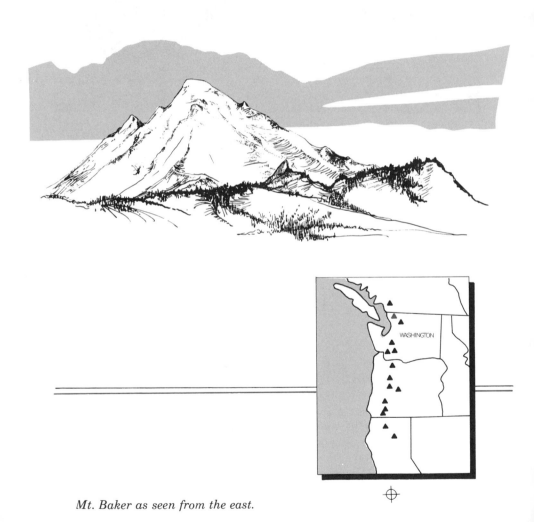

Mt. Baker as seen from the east.

Resembling "an ice floe on a frying pan," the Sherman Crater ice pack disintegrates as fumaroles with temperatures approaching 270 degrees Fahrenheit release columns of hot gas and deposit thick layers of sulphur on the Sherman Crater floor. —Photo courtesy of Fred Munich

prehistoric eruption that wounded the mountain is suggestive, for Baker's principal active vent, the Sherman Crater, is on the south flank about 1000 feet below the summit. It was blown open in early post-glacial time, radically altering the former contours of the summit. To the Indians, the new crater must have resembled a gash that bled fire.

Next to that of St. Helens, Baker's crater is the most active in the Cascades. On March 10, 1975, a dramatic increase in heat and steam emission took place as an unusually large plume of vapor rose above Sherman Crater. Aerial photographs taken the next day revealed striking changes in the snow and ice-filled crater: large new steam vents, melt pits, semicircular crevasses, and ponded meltwater had appeared; in addition, a thin dusting of gray ash extended approximately 300 to 1000 feet beyond the east rim.

Rapid changes continued throughout the spring and summer of 1975. Several more layers of dust or ash dusted most of the crater and drifted east down the surface of Boulder Glacier. Most of the ash came from a large new fumarole at the base of Lahar Lookout, a crumbling mass of decayed rock that borders a 500-foot deep gap in the crater's eastern rim. In April, the central ice depression developed into a 130-foot deep pit that held a shallow lake of warm and distinctly acidic water. On July 11, the lake drained through the eastern cleft in the crater wall into a stream that runs through the Boulder Glacier.

By August, the threefold increase in heat flow had caused a spectacular break-up of the crater glacier. Two years later, a large section of the glacier collapsed, producing an avalanche of snow and ice that buried half of the crater floor.

Steam emission has since varied in power and intensity. At their most active, the vents spew dark clouds an estimated 2000 feet above the crater rim. Climbers have reported that the vapor was released under pressure and that the jets pulsated violently. The amount of hydrogen sulphide gas being released increased from 2800 pounds of sulphur per hour in March, 1975 to about 21,000 pounds per hour the next July. Baker, along with St. Helens, is one of the worst air polluters in the Pacific Northwest.

A monitoring system set up by the U.S. Geological Survey, the University of Washington, and others includes making periodic aerial photographs of the craters, positioning seismic sensors to register earth movement that might indicate rising magma, and visits to the crater to detect changes in gas emission and temperature. As of August, 1987, the most significant changes were the gradual westward migration of thermal centers, which caused an increase in temperature, pressure, a number of fumaroles in the west and northwest parts of Sherman Crater, and corresponding decrease in activity

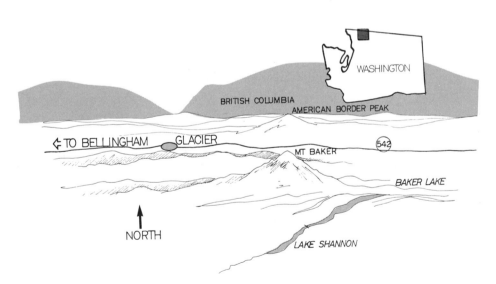

Relief map of the Mount Baker region.

at the central pit. Ice has refilled the central pit. Meanwhile, augmented heat emission along the north rim has, for the first time this century, melted the ice cover and exposed a band of rock from east to west across the north crater wall. Fumaroles near the eastern breach in the crater continue to produce steam. Despite the heat and gas emission, the almost complete absence of earthquakes suggests that an eruption is not imminent, but the situation could change at any time.

Baker's Geologic History

Baker is a relatively young stratovolcano that stands on the rugged and deeply eroded western edge of the North Cascades. The earliest lavas drained into spacious canyons, where they typically ponded to depths of several hundred feet. Like the early intracanyon flows from Rainier and Hood, some of Baker's first lavas have been eroded into high-standing ridges that now radiate from the base of the cone.

The Black Buttes

Baker's first eruptive center is marked by the Black Buttes, Lincoln Peak and Colfax Peak, which stands about two miles west of the main cone. This ancestral volcano erupted flows of basalt and andesite, at least one of which reached as far as seven and a half miles north of the vent. After those eruptions built a pile perhaps 10,000 feet or more above sea level, activity shifted eastward to Baker's main cone. Glaciers then carved the Black Buttes into extremely rugged and precipitous crags, their cliffs too steep to hold snow or ice. The Black Buttes stand out in bleak contrast against the smooth snowfields of the younger cone.

Eruptions at Baker's new central vent produced a series of relatively thin, fluid lavas, which spread out to form an 80 square mile base upon which the modern cone stands. Lava flows comprise 95 percent of Baker's volume, indicating that pyroclastic eruptions were few and intermittent. Some tuff and pumice are present, but the most common airborne deposit is a dark crystalline ash.

Several large cirques on Baker's north side indicate that the main cone was erected before the last Pleistocene glaciation. Baker is, however, much less eroded than many of its southern counterparts, such as Rainier, Adams, or Hood. Whereas Rainier has lost thousands of feet from its original surface through glacial erosion, Baker has lost only a few hundred. Existing glaciers may account for much of the moderate dissection of the cone that has thus far occurred.

Post-Glacial Eruptions

During the last 10,000 years Baker has erupted intermittently, producing small to moderate quantities of tephra at least four times, lava flows twice, one series of pyroclastic flows, and eight large avalanches and mudflows. Some of the avalanches may have accompanied eruptions. The thickest tephra layer covered Table Mountain, six or seven miles northeast of the volcano, with as much as 18 inches of black, sand-sized ash sometime between 500 and 6900 years ago. The youngest tephra deposit, composed of fragments of altered rock as much as four inches across, reaches depths of four inches about six miles northeast of the summit. It may have been erupted on one or more occasions during the period of historic activity between 1792 and 1880.

Baker's most significant post-glacial eruptions happened about 8700 years ago, when a succession of pyroclastic flows, mudflows, and lava flows formed a large fan-shaped fill in Boulder Creek valley on the volcano's east flank. The fill extends across the Baker River valley, now partly submerged by Baker Lake. The deposits that compose this fill probably formed as Sherman Crater was blasted out high on Baker's south flank. The assemblage includes two flows of andesite lava that reach three miles beyond the front of the Boulder Glacier.

Another sizable post-glacial eruption centered at Schriebers Meadow, low on Baker's south flank, where a cinder cone blanketed the vicinity with thick layers of reddish scoria and black ash, and produced an andesite lava flow that traveled about seven miles down Sulphur Creek valley, displacing a stream from the Easton Glacier. Sulphur Creek and Rock Creek now pursue new courses marginal to the lava. The Schriebers Meadow tephra was deposited some time between 10,350 and 6900 years ago.

Since the last ice age ended, numerous mudflows have created fills or terraces in most valleys heading on Baker. The largest occurred about 6000 years ago, when a mass of altered rock avalanched down the southwest side of the cone, possibly from Sherman Peak, to form a mudflow that poured more than 17 miles down the Middle Fork of the Nooksack River. A second arm of the mudflow, moving in a wave nearly 700 feet high, plastered a veneer of mud and rocks on the divide between Easton and Deming glaciers before streaming across Schriebers Meadow and down Sulphur Creek valley. Perhaps little more than 450 years ago, a similar avalanche, swept down the mountain's northeast flank into the Rainbow Creek valley. A still younger avalanche, originating in a rockfall from the south valley wall near the snout of Rainbow Glacier, deposited debris 600 feet high on the opposite valley wall, perhaps within the last century.

This east-side view of Mt. Baker shows both the lower Sherman Peak (left), and the main summit, Grant Peak (10,778 feet). The thermally active Sherman Crater lies between the two peaks. —Photo by Bob and Ira Spring

During the last 10,000 years, large clayey avalanches and mudflows have occurred, on average, once every 1300 years. Four of the eight major post-glacial mudflows happened during the last 600 years. The acceleration of events in recent centuries may be due to progressive alteration of the summit rock, which may have reached a point of crucial instability. It may also mean that events that trigger avalanches, such as earthquakes, steam explosions, or eruptions, have in recent centuries occurred with greater frequency.

The Summit Area and Recent Fumarolic Activity

Baker's summit area includes two peaks, separated by the Sherman Crater. The higher, northern peak terminates in a flat, ice-covered plateau about the length of a football field. The lower horn, Sherman Peak, forms the southern rim of crater. Aerial photographs taken in the summer of 1940, when snow accumulation was unusually low, reveal what seems to be part of a crater rim on the summit. The northwest walls of Sherman Crater show layers of lava slanting upward toward their source near the northern crest.

An ice stream descending from the north peak partly fills Sherman Crater. Before 1975, two major clusters of steaming fumaroles perforated the crater icepack, one on the southwest side, the other on the east side, perhaps through glacial ice. Thermal action was also intense along the western rim, where swarms of hissing jets sent up

Sherman Crater, looking west. Plumes of steam rise from numerous active fumaroles. —Photo by Bob and Ira Spring

impressive columns of vapor. Before the large new vent opened at the base of Lahar Lookout, the noisiest fumaroles were within the eastern perforation, just inside the deep cleft in the eastern crater rim. Acidic snow melt from Sherman Crater channeled into a single stream through the east gap in the crater wall, and disappeared under Boulder Glacier. A roaring fumarole on the stream bank blew vapor and water from sulphur-encrusted crevices. Typical temperatures were about 180 degrees Fahrenheit, approximately the boiling point of water at that altitude.

Long exposure to heat and acidic gases has converted much rock in the Sherman Crater to soft, clayey material extremely susceptible to sliding. At least six times between 1958 and 1973, avalanches of snow, ice, altered rock, and mud have fallen from Sherman Peak. All six avalanches followed nearly identical paths down the Boulder Glacier but none reached its end. Lahar Lookout, at the northern edge of the breach, is a similar mass of unstable, heat-altered rock and clay. The threat of a major avalanche from this area now poses the principal volcanic hazard.

The Steam Caves

Although the recent thermal changes have already altered their size and extent, the steam caves, approximately three-quarters of a mile in total length, remain one of the most interesting features of Sherman Crater. Like those in Rainier's summit craters, The Baker caves formed as heat melted a series of passageways and chambers between the crater floor and the overlying snowpack, which largely fills the bowl-shaped vent.

Although a few climbers may have observed their entrances before then, it was not until August, 1974, that the caves' existence and extent were established. Eugene Kiver led an expedition, of which the author was a member, to explore and map the caves. Even before the recent increase in thermal activity, air inside the caves was so murky that flashlight beams could penetrate only a few feet. The rotten egg stench of hydrogen sulphide was overpowering. That and other gases, including carbon dioxide and carbon monoxide, made gas masks a necessity. The 1974 party discovered an opening in the snowcover near the northwest crater that led into a long passage that apparently extended across the crater and connected with the large ice perforation near the gap in the eastern crater wall. Unfortunately, the passage floor was so muddy near its terminus that we could not exit from the eastern end. We saw several small pools of standing water, but no lake of any consequence. Kiver's second exploration, in August, 1975, revealed that the passage from the western rim then ended at the lake of warm meltwater in the central pit.

Baker's Historic Eruptions

The generally brief and incomplete written observations make it difficult to determine exactly when Baker's latest activity began, or the number and duration of individual eruptions. Although the volcano seems to have been quiet when Quimper and Vancouver first sighted it, the log of a Spanish expedition reconnoitering Bellingham Bay in June, 1792, indicates that the mountain was then active. According to an anonymous narrative published in 1802, the crews of the vessels *Sutil* and *Mexicana* recognized in "the ominous rumbling and flashes of fire to the east that continued day and night signs of a volcanic eruption."

The sudden increase in heat and gas emission that began in March, 1975, caused rapid melting of the Sherman Crater ice fill, exposing vigorous new steam vents. —Photo courtesy of Fred Munich

Chart of Postglacial Events at Mount Baker

Event	Approximate age
Increased fumarolic activiity at Sherman Crater	Present (1975)
Several small avalanches and mudflows poured down Boulder Glacier	Recent past to present
Eruption of altered rock debris from Sherman Crater	Within the last few centuries
Two or more mudflows poured short distances down Sulphur Creek valley	Do.
At least two mudflows poured 7 miles down Boulder Creek valley	Do.
Avalanche moved at least 5 miles down Rainbow Creek valley	Do.
A mudflow poured 9 miles down Park Creek valley	500
A mudflow poured almost 2 miles down Middle Fork Nooksack River valley	Between 6,000 & 300 yrs. ago
Eruption of tephra	Between 6,600 and 500
A mudflow poured more than 6 miles down Sulphur Creek valley	Between 6,600 and 300
A mudflow poured at least 18 miles down Middle Fork Nooksack River valley	6,000
A mudflow poured 9 miles down Park Creek valley	6,650
Eruption of tephra	Between 10,000 and 6,600
A mudflow poured at least 5 miles down Sulphur Creek valley	Do.
A lava flow moved 8 miles down Sulphur Creek valley	Do.
Eruption of scoria in Sulphur Creek valley	Do.
Pyroclastic flows, mudflows, and two lava flows moved down Boulder Creek valley, some reached Baker River valley	About 8,7000 (?)
A mudflow reached at least 4 miles down Sulphur Creek valley	10,340

After Hyde and Crandell, 1978

For decades after Baker's discovery, few literate observers were in the vicinity to record any activity, although Indian lore tells of prehistoric cataclysms. According to Edmund Coleman, who first climbed Baker in 1866, "an old Indian" recalled that when he was a "boy" "the mountain [burst] out with a terrible fire and great smoke," poisoning the fish in the Skagit River. In his compendium of Baker's historic activity, Majors placed the eruption "about 1810" and correlated it with the testimony of another Northwest Indian, John Hiaton, who stated that he had watched outbursts of both Rainier and Baker "about the year 1820."

The first authenticated 19th century eruption, and possibly the most severe, was that of 1843. In his review of the Pacific Northwest's physical geography, George Gibbs referred to Baker as an active volcano, noting that:

> . . . it would seem to have only recently resumed its activity, as I am informed, both on the authority of officers of the Hudson Bay Company, and also of Indians, that the eruptions of 1843 was the first known. It broke out simultaneously with St. Helens' [sic.] and covered the whole country with ashes.

Gibbs also cited Indian reports that the ". . . Skagit River was obstructed in its course, and all the fish died. This was, in substance, what they assured me on my visit to the river, adding that the country was on fire for miles round." Gibbs was certain that the fish ". . . were destroyed by the quantity of cinders and ashes brought down by the Hukullum [Baker River]. Since the above date, smoke is frequently seen issuing from the mountain." This volcanic decimation of the rivers' fish population so impressed local Indians and trappers that they recounted it to travelers for decades afterwards.

Mid-19th century visitors to Baker's east flank reported finding considerable devastation. Gibbs cited "some miners" who described "a level plain two or three miles wide, of black volcanic rock and sand, upon which were vast piles of half-burned timber, apparently swept down by a current of, as they supposed, lava, but more probably water." The miners also said they saw lava streams on the plain and mountainside, with "sulphur . . . scattered over its surface." They also noted "smoke ascending" from the east side "about two-thirds the distance above the snow line . . ." The supposed "lava" was probably landslide debris that had fallen from the walls of Sherman Crater and cascaded down the Boulder Glacier, much as smaller debris flows did between 1958 and 1973.

A number of pioneer accounts refer to extensive changes in Baker's summit configuration. According to F.G. Plummer, John Hiaton recounted ". . . a tradition of his race . . . that this mountain was formerly much higher and that a tremendous explosion threw down the entire south side." Plummer also quoted Father De Smet that " . . . in the year 1846 [1843?] Mounts Saint Helens and Baker became volcanoes, the latter immediately preceding the time of writing had undergone considerable changes on the south side where the crater was formed." These are the first known allusions to conspicuous alterations of the summit, which became a recurring theme in news paper accounts of eruptions in the 1860s. While sizable avalanches of altered rock may have taken place at or near Sherman Peak or other parts of the crater walls, there is no recognized independent evidence of any major summit collapse dating from the 19th century.

In 1896, when Plummer compiled a catalogue of reported eruptions in Alaska and the Pacific Northwest, he did not include the reputed 1820 activity of either Baker or Rainier. He did, however, note 1842 and 1846 as marking eruptions on Baker, omitting reference to the outburst of 1843. In faraway Belgium, Alexis Perrey recorded eruptions on Baker for both 1842 and 1843. As in the case of St. Helens, eyewitness and second-hand accounts are typically contradictory about dates. It seems likely that "eruption of great force" which supposedly occurred in 1846 was confused with the 1843 activity.

In his catalogue of active American volcanoes, Coombs accepted, 1843, 1854, 1858, and 1870 as dates of eruptions. But Baker seems to have been mildly active on other occasions. The *Oregon Spectator* of March 21, 1850, announced that both St. Helens and Baker were then ". . . in a state of eruption." In *The Canoe and The Saddle*, Theodore Winthrop noted that "Mt. Baker was active in 1852, sending up flame and smoke for several days." An earthquake occurred in October and, according to Plummer's 1893 article, the following January "persons living down [Puget] Sound could distinctly see a long black streak on the southwest slope of Mount Baker, which was variously estimated at from 1000 to 2000 feet in width. It was several months before the mass of lava (as it undoubtedly was) had cooled so as to receive the falling snow." U.S. Geological Survey investigators have recognized no lava flows that recent, so it seems that a landslide or mudflow was mistaken for lava.

The most convincing account of Baker's historic eruptions is that of George Davidson. In a long letter to *Science* he wrote.

> In 1854 I was one day observing at the trigonometrical station on Obstruction Island in the Rosario Strait, Washington Sound: I had finished the measure for horizontal direction of the summit of Mount Baker, and was commencing a series of measures of the vertical angles for elevation, when I found the whole summit of the mountain suddenly obscured by vast rolling masses of dense smoke, which in a few minutes

A steam plume rises from Sherman Crater, a thermally active vent just south of Baker's summit. The Black Buttes appear on the left.
—U.S. Geological Survey photo by Robert Krimmel

reached an estimated height of two thousand feet above the mountain, and soon enveloped the higher parts. Baker was distant thirty-nine geographical miles from my station . . . The weather was fine, and we hoped to see a brilliant display at night; but unfortunately the sky clouded, and we could not see the light at night, nor the mountain the next day; when the weather cleared, the eruption had ceased; and instead of the white mountain mass, we discovered the snow covering it was apparently melted away for two or three thousand feet below the two heads. Of course the snow may not have been melted, but only covered with ashes and scoriae; and we had not the means of deciding the question at that distance.

That Davidson was taking scientific measurements of Baker with his whole attention focused on the mountain when without warning he saw the summit enveloped in "rolling masses of dense smoke" heightens his credibility.

Davidson also noted that ". . . the crater was not on the summit, or on the secondary peak to the south-eastward, but on the flank of the higher peak, and opening towards the south or southwest. In subsequent years we occasionally saw small volumes of smoke issuing from this crater."

Four years later, observers were treated to a night display that bad weather denied to Davidson in 1854. Friends in Victoria, British Columbia, informed the scientist that ". . . night clouds over Mount Baker were brilliantly illuminated by the light from an eruption."

During November and December of 1859 a similarly "brilliant" eruption occurred. According to the Olympia *Pioneer and Democrat* of November 25, "two large bright jets of flame were seen having the appearance as if issuing from separate fissures or openings. It was seen but a few days and not accompanied by quantities of dark smoke." The newspaper commented that such eruptions were very unusual. The following December 3 brought another show with both sound and light effects. "Bright flashes of lightning with report like heavy thunder proceed from Mount Baker, which was in a state of eruption. On December 4 the volcano has shown a dense cloud of smoke all day." While such natural fireworks may have been the result of genuinely molten fragments being ejected from the crater, it seems more likely that the "flashes as of lightning" were simply electrical displays.

Baker seems to have remained spasmodically active throughout the first half of the following decade. Perrey recorded an eruption of April 26, 1860, probably a continuation of the 1859 activity. As usual, Perrey gave no details. He is likely to have received his information from reprints in eastern newspapers. On December 28, 1860, for example, the *Pioneer and Democrat* reported that "Mt. Baker is in a state of eruption, throwing off clouds of steam and smoke." A similar item appeared in the Washington *Sentinel* of August 18, 1863; "Mt. Baker is reported in a state of eruption," which may reflect equally brief reports in Victoria publications, such as the *Daily British Colonist's* statement on July 27 that "the volcano is beyond all doubt in a state of eruption, the flames were plainly seen from Beacon Hill." Except for noting the bare fact that Baker was allegedly erupting, these frontier papers manifest a dearth of scientific curiosity.

In any event, Baker's 19th century activity was sufficiently intense to inspire wildly exaggerated tales about the destruction of its summit. A correspondent for the *Scientific American* reprinted an *Oregonian* story of the previous April: "Mt. Baker, it is said, is rapidly sinking in. It is asserted that the mountain has fallen 1000 to 1500 feet, and that its summit, which was formerly a sharp point, is now much flattened. This peak has been for some time in a state of active eruption. Dense clouds of smoke have of late issued from it.' Correspondents of the California papers speak of the same phenomenon, one of whom asserts that the emission of steam is immense, and that 1200 feet of the summit has fallen in." Davidson's article in *Science* states that he made an exact drawing of the volcano's profile, ". . . the more particularly because rumors had found their way into the newspapers, asserting that the summit of Mount Baker had fallen in. On the contrary, I was perfectly satisfied, from my years of familiarity with its features, that no such catastrophe had taken place between 1852 and 1870; nor was I able to detect any changes in 1877, when I was daily in sight of Baker for some time."

Davidson's conscientious observations are confirmed by the climbing party of E.T. Coleman, which in 1868 first ascended Baker. On

September 1 of that year, the *Oregonian* carried an interview with members of the group, which established that the familiar two peaks of the mountain remained in place. In a mountaineering article for *Harper's* published the following year, Coleman described the "summit plateau" of the higher peak as about a "quarter of a mile in diameter . . . the white surface of the snow was unrelieved by a single rock." One of the climbers, David Ogilvey, explored the crater and concluded that ". . . the existence of the volcano is established beyond a doubt, the crater being about 300 [yards] wide, and at least 600 feet deep, from which puffs of sulphurous vapor are being emitted. The crater lies between two high peaks of the mountain, where the summits form a plateau quite bare and free from snow." Stratton, another member of the party described "three extinct craters," which may have been subsidiary vents responsible for some of the historic eruptions. On the Baker River side, Coleman said, "300 feet of the lip had been torn away where successive layers of lava had flowed out and cooled." He also commented that ". . . no traces of fire were visible by daylight but smoke was plainly observed. Fire must be slumbering beneath as there is no snow on the lava."

Davidson witnessed another apparent eruption in 1870. He was about 60 miles away when he ". . . beheld great volumes of smoke projected from the higher peaks." If this was a genuine outburst, it seems to have been the only historic one to have issued "from the highest peak," which is now completely buried in ice and manifests none of the thermal anomalies so abundant in the Sherman Crater. It is possible that Davidson mislocated the active vent. When J.S Diller, an eminently reliable authority, referred to the authenticated eruptions in 1854, 1858, and 1870, he undoubtedly had Davidson's published observations in mind.

Plummer's catalogue listed two more eruptions, in December, 1880, but gave no details. Several other sources alluded to unusual activity in 1880, beginning in April, when Baker was observed "smoking." The following September 7, a steamer captain ascending the Skagit River witnessed a "violent" eruption, with "flames streaming up from the summit and large volumes of smoke." By the time of his return trip downriver the activity had somewhat subsided. Although "smoke" continued to be seen rising from the peak, 1880 apparently marked the end of Baker's last eruptive period.

It appears that Baker produced numerous, but relatively minor steam eruptions for nearly a century. At least one threw altered rock from Sherman Crater, forming a deposit that now lies in areas east-northeast of the volcano. Steam explosions may have triggered avalanches from Sherman Peak or other parts of the Crater wall, some of which probably descended Boulder Glacier, devastating part of

Baker River and transporting enough sediment to kill salmon in the Skagit River. Frequent references to "fire" and "light" suggest that some of the ejecta may have been incandescent. It is also possible that the light came from burning volcanic gases, such as those noted at St. Helens in 1980, or lightning generated by static electricity in the eruption cloud.

Thermal activity persisted into the early 20th century. When C.E. Rusk climbed the east side in 1903, his party "saw large volumes of smoke rolling from between two peaks." Rusk saw from the crater rim "the most thrillingly weird spectacle [he] had ever seen."

> In the bowl-like depression immediately between the two peaks was a great orifice in the snow. It was perhaps fifty feet across, although the western side was partly blocked with snow so that the opening had somewhat the shape of a half-moon. At a distance of possibly two hundred feet a semicircular crevasse swept halfway around it. From the unknown depths of this abyss the black smoke rolled. It drifted away, shifting with the wind, until it was finally dissipated in the rarefied air. The wild, unearthly loneliness of the scene impressed us profoundly, for its counterpart perhaps does not exist on earth.

A photograph taken "about 1900" shows the crater "in practically the same condition" as it was when Rusk visited it. But by August, 1906, thermal activity had apparently declined for there was then "no sign either of orifice in the snow or smoke." Apparently heat and vapor emission in the Sherman Crater had varied considerably during the 20th century.

Present and Future Hazards

Although Baker is still discharging quantities of superheated steam and hydrogen sulphide, it is too early to determine whether this activity will culminate in an eruption. If Baker does erupt, beware of mudflows from Sherman Peak, Lahar Lookout and other nearby formations. Even a relatively small rockfall could dam the eastern outlet from Sherman Crater and permit a lake of considerable size to form. A sudden failure of the dam could then release large volumes of water, which might pour down Boulder Glacier, gather additional stored water within the crater ice and the glacier, and possibly create an outburst flood in upper Boulder Valley.

A really large mudflow could sweep down Boulder Glacier and theaten two large reservoirs, Baker Lake and Lake Shannon, which have combined lengths of 18 miles. The dams would probably survive but a spillover caused by large volumes of mud flowing into the lakes

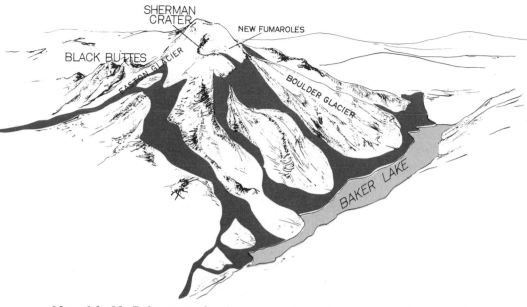

Map of the Mt. Baker area, showing zones potentially endangered by renewed volcanic activity. The darkened area east of the volcano has been covered by ashfalls several times during the last several thousand years, most recently in the mid-19th century. Valley floors east and south of the volcano have been repeatedly inundated by avalanches and mudflows. The Baker Lake area was decimated by pyroclastic flows when Sherman Crater was formed.
—After Hyde and Crandell, 1978

could mean serious difficulties for persons living in valleys southeast of the mountain.

Andesite lava flows move too slowly to directly threaten human life. But lava erupted onto glacial ice could cause flooding of valleys below. If a lava flow does occur it will probably emerge somewhere on the flanks of the mountain. If erupted below the snowline, hot lava might start a forest fire among the thick stands of fir, cedar, and hemlock that carpet the volcano's base.

Prevailing westerly winds would probably carry ash into unpopulated areas east of the volcano. No post-glacial ash deposit thicker than four inches a few miles east of the crater has been recognized. Even a temporary reversal of the usual wind patterns would produce only dense haze and light dusting near Bellingham, the largest city in the area. A pyroclastic flow erupted through the east or west gaps in the Sherman Crater rim would probably destroy large areas of forest. The series of pyroclastic flows that descended the east flank of the cone traveled at least as far as Baker Lake, now a popular camping site.

281

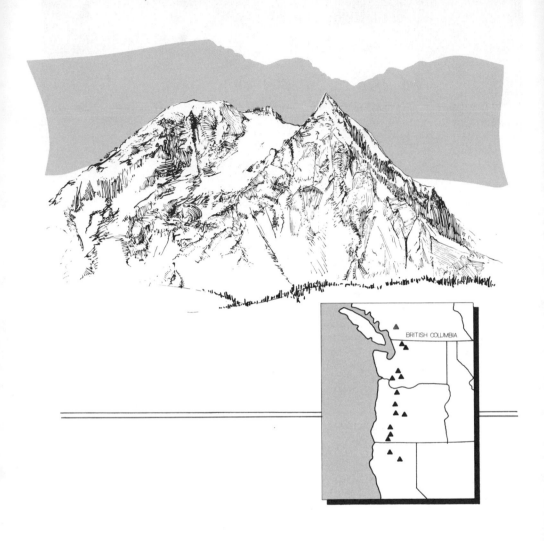

Location of Mount Garibaldi showing its location in the Cascade Range (Coast Mountains) of southern British Columbia.

Mt. Garibaldi is the only known major Pleistocene strato-volcano in North America to have been partly built atop an active glacier. Melting of the underlying glacier caused collapse of much of Garibaldi's original cone.
—Photo by Ed Cooper

XX
Mt. Garibaldi:
The Volcano that Overlapped a Glacier

Canada produced the only large Ice Age volcano known to have been built partly atop an active glacier. Metaphorically raised by fire on top of flowing ice, Garibaldi now dominates the rugged scenery of British Columbia's Garibaldi Provinicial Park.

About 40 miles north of Vancouver, B.C., Garibaldi, 8787 feet, is one in a chain of at least 18 volcanic centers that formed in the Coast Mountains of southern British Columbia. The Garibaldi Belt of volcanoes, which extends about 95 miles north-northwest from Watts Point on Howe Sound to Salal Glacier near the head of the Bridge River, continues the High Cascade line of volcanic peaks.

Compared with Cascade giants like Hood, Shasta, or Rainier, the Canadian volcanoes are unimpressive. Few reach or exceed 8800 feet in height and thus do not tower above the surrounding peaks of crystalline rock. The volcanic cones seem lost amid the steep ridges of the Coast Mountains, many of which are loftier and more spectacular than the volcanoes. Garibaldi is an exception; its twin-peaked, glacier-laden cone rises majestically above the head of Howe Sound.

The largest composite cone in the southern part of the Garibaldi Belt, Garibaldi is one of only two Cascade stratovolcanoes composed entirely of dacite. The other is Glacier Peak. Garibaldi is a deeply eroded complex of domes, pyroclastic flow deposits, and a few lava flows. It consists of such loosely compacted debris that large blocks are constantly tumbling into the adjacent valleys.

Garibaldi's Eruptive History

Garibaldi was built in three distinctive eruptive episodes, the first two of which were separated by long dormant intervals. In the first phase of activity, an ancestral volcano erected a broad composite cone of dacite flows and breccia. Remnants of this edifice form the upper 800 feet of Brohm Ridge and the lower northern and eastern flanks of the mountain. A series of coalescing dacite domes then erupted at Columnar Peak on the southeastern flank, possibly also on the northern edge of the structure, at Glacier Pikes. Potassium-argon dates indicate that the ancestral cone formed about 250,000 years ago.

The volcano was dormant for a long period, during which the Cheekye River cut a deep valley into the western flank of the ancestral pile. The valley later filled with glacial ice derived at least in part from a glacier in the Cheakamus Valley.

While glaciers still filled the Cheekye basin, but after the ice had already attained its maximum extent and had begun to recede, the modern volcano was born. Garibaldi began life as a viscous spire of dacite lava that emerged through a ridge completely surrounded by an ice sheet several thousand feet thick. Like that of Lassen Peak, Garibaldi's late Pleistocene growth was accompanied by the crumbling and avalanching of the rising lava mass. Large banks of talus formed around a solid core. Avalanches of incandescent lava fragments, accompanied by clouds of gas and hot ash, repeatedly swept down the flanks of the volcano. Apparently laid down in quick succession, the pyroclastic flows soon built a fragmental cone with a volume of 1.5 cubic miles. Garibaldi's original slopes were gentle, averaging only 12 to 15 degrees. Subsequent erosion has made the mountain much steeper.

Where the growing cone rested on the ice that filled old valleys, or on ridges only slightly above the general ice level, the pyroclastic flow deposits extend at least three miles away from their source.

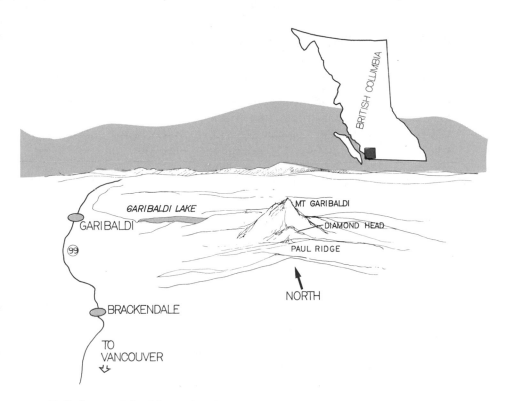

Relief map of the Mount Garibaldi.

Where the cone banked against higher ground, the volcano's slopes are correspondingly shorter and more abrupt. On the west, and to a lesser degree, the south, the volcano significantly overlapped the edge of a massive ice sheet. Melting of glacial ice during this eruptive cycle caused water to pond against the southern arm of Brohm Ridge. Ash flows deposited underwater formed volcanic sandstones atop Brohm Ridge, nearly 2000 feet above the present floor of the Cheekye drainage basin.

After the main Peléan eruptions had ceased, the surrounding ice surface rose to an elevation of 5500 feet. Whether this resurgence was merely a return to the pre-eruption levels after the great pyroclastic thaw, or a result of general climatic cooling is not known. In any event, the lowland ice sheet finally began to melt, with disastrous effect on Garibaldi. As the glacier shrank, Garibaldi's western flank began to collapse in a series of avalanches and mudflows.

Where the layer of ice between the pyroclastic flow deposits and bedrock was thin, the melting of the ice produced only minor breaks in the surface of the cone. The steep scarp at Cheekye Ridge is probably an example. However, where the volcano overlay ice hundreds or thousands of feet thick, as in the Cheekye Basin, radical disruption of the cone occurred. As the underlying ice melted, Garibaldi's broken and over-steepened western flank disintegrated into the rock slides that carried nearly half the original cone into the Squamish Valley.

A large volume of debris derived from Garibaldi fills much of the Squamish Valley, an extensive depression at the head of Howe Sound. Approximately 10 square miles of the valley floor are underlain by an average thickness of about 300 feet of dacite blocks and rubble that once composed the western slope of Garibaldi. Not all this material, approximately 0.6 cubic mile, poured into the Squamish Valley at once. Several fans of debris now form terraces upvalley toward the ruined cone, recording a series of slides or avalanches that occurred as the glacier wasted.

After most, or all, of the ice sheet had thawed, Garibaldi again became active. This time Garibaldi erupted quietly. Streams of liquid dacite issued from a crater north of the Atwell Peak plug dome, and flowed down the north and northeastern flanks of the mountain. One dacite stream extended several thousand feet west over the landslide scar on the volcano's western face. This late flow traveled down a slope of 30 to 35 degrees, a grade far steeper than the original 12 to 15 degree angle formed by the earlier pyroclastic flow deposits. Partial destruction of this flow by the additional slumping of the fragmental material on which it lies suggests that the lava was erupted shortly after the ice sheet had receded.

Garibaldi's southern peak is the pyramidal Atwell spire from which many of the pyroclastic flows were derived. The slightly higher northern summit is the Dalton Dome, associated with the last dacite flows. The third stage of activity produced about 0.15 cubic mile of dacite lava that covers the north and part of the western sides of the volcano with a thin layer of solid rock. Slightly more than half, about 0.8 of a cubic mile, of the original pyroclastic cone remains in place; an almost equal quantity of material now lies in the Squamish Valley.

The Garibaldi Lake Volcanic Field

Although the highest, Garibaldi is by no means the only intriguing volcano in the area. Three and a half miles north stands The Table, one of the most unusual volcanic structures anywhere. This conspicuous, flat-topped, steep-walled pile of andesitic dacite rises several hundred feet above the immediate terrain. From a distance it resembles a mesa similar to those in the southwestern United States;

from other angles it resembles the famous Devil's Tower in Wyoming. Closer inspection reveals that its mass is horizontally layered, a pile of lava flows that spread one atop the other like a stack of poker chips. A few curtains of congealed lava cling to the almost vertical sides of the pile.

The Table was built by the repeated flooding of lava into a more or less cylindrical vent thawed through the Cordilleran ice sheet which, in late Pleistocene time, mantled the entire region. The lava flows that plaster the nearly vertical sides of The Table like frosting on a cake are thought to have formed where molten rock flowed down into the gap thawed between the dacite mass and the encircling walls of glacial ice.

The Garibaldi area abounds in uncommon formations. The Black Tusk, a massive spire of dark lava shaped rather like a walrus tusk, seems to defy classification. It is so extensively eroded that its origin and nature are not clear. It may have been an intra-glacial vent like The Table, although the flows composing its summit are not horizontal. The Black Tusk's north-south elongation, parallel to the local direction of ice movement, may be the result of its having erupted along a north trending fissure, by ice movement at the time of eruption, or simply by later glacial erosion.

A series of sheer pinnacles on the west side of the Squamish Valley, opposite the town of Squamish, are also puzzling. The highest of these, The Castle, terminates in an impressive spire. It is also distinguished by a semicylindrical groove extending up its southern end "like a flue up some huge split smokestack." This spire, which does not seem to conform to any known geological category, may in fact be a lava spine, much as those extruded by Mont Pelée in the West Indies.

Another formation worth a visit is the slumped and asymmetrical Cinder Cone, a 500-foot-high mound of pyroclastics, which stands between two arms of Helmet Glacier on the flanks of Garibaldi. This cone, topped by a crater that contains an ice-melt lake much of the summer, as well as numerous other vents and associated lava flows, attests to the wealth of volcanic activity in this region during early post-glacial time.

Some Coast Range Volcanoes North of Garibaldi

In the Garibaldi Belt, most of the volcanoes occur in groups of three to seven. Besides the Garibaldi and Garibaldi Lake volcanic centers in the south and the Silverthorne complex in the north, they include the Mt. Cayley and Meager Creek volcanic fields in the center of the chain. Cayley is a composite dacite cone, smaller than Garibaldi, but similar in structure, and probably active during the same period.

Meager Mountain, an extensively eroded volcanic complex west of Meager Creek and Lillooet River, formed during at least four major eruptive cycles over a period of at least 4 million years. Activity began in late Pliocene time, and continued even after the last ice age. As the volcano evolved, the center of activity migrated northward, producing a series of dacite and andesite flows and pyroclastic deposits.

A vent on Meager Mountain's north side produced the most recent volcanic eruption in Canada. This explosive outburst ejected a large volume of tephra, the Bridge River ash, which forms layers up to 70 feet thick on some ridgetops near Plinth Peak. Blocks of rhyodacite pumice up to about 188 feet in diameter lie along the north side of Fall Creek. Overlying the airfall pumice are welded pyroclastic flow deposits up to four miles long with an aggregate thickness of about 475 feet. The Bridge River tephra and welded pyroclastic flows probably erupted from a vent near the headwall of Fall Creek about 2400 years ago.

To See Garibaldi

Drive Highway 99 northward from Vancouver, B.C., past the town of Squamish. The turn-off to Diamond Head in Garibaldi Park is sign-posted; watch along the east side of the highway between Squamish and Brackendale for the sign. Drive to Diamond Head base camp and hike the jeep trail 6 miles to Diamond Head Lodge. There is no automobile travel from the base camp to the lodge. From Diamond Head Lodge to the Gargoyles is about 2 more miles. The elevation gain from the parking lot to the Gargoyles is 2000 feet. From the base of the Gargoyles the view of Garibaldi's south side is unobstructed.

XXI
Volcanic Fire and
Glacial Ice:
Now and in the Future

Our peak-by-peak tour of the principal Western volcanoes has told us much of their past, but what of the future? Besides St. Helens, which of the fire-mountains are likely to erupt during our lifetime? What kinds of eruptions can we expect a particular volcano to produce? And what of the glaciers? Are these rivers of ice now destined to shrink until they disappear altogether? Or are some Western glaciers now advancing—the harbingers of a new Ice Age?

While final answers to these questions are not yet possible, geologists have made some educated guesses about the future behavior of both the volcanic fire and glacial ice that fashioned the Cascade mountainscape we know today. Because glacial fluctuations have been more consistently observed and precisely measured for longer periods of time, we will examine them first.

The Glaciers

Most people find the shimmering ice that encases the high peaks of Rainier, Shasta, Hood, Baker, or the Three Sisters their most appealing feature. Besides their beauty, and their value as a source of water and hydroelectric power, these frozen reservoirs also remind us of a time not long past when vastly larger glaciers covered not only most of the mountains, but part of the adjacent lowlands as well. When glaciers advance, as some are now doing, it is easy to believe that another Ice Age might be on the way.

In reviewing the current status of the Cascade glaciers we might first briefly summarize their behavior during the last few thousand years. After the last ice age ended, 10,000 to 12,000 years ago, there followed an unusually warm period that lasted until about 3000 years ago. Except for those on the highest peaks, few of the Cascade glaciers survived this warm interval.

New glaciers grew more than once since the end of the last great ice age, some as recently as the mid-18th or 19th centuries. On Rainier, various glaciers attained their maximum size between approximately 1350 and 1850 A.D. Following this period of growth, glaciers throughout the Cascade Range began a recession that made their ultimate disappearance seem possible. Rapid melting was particularly evident in the decades following 1900.

Twentieth century glacial activity at Rainier provides a fair example of glacial trends, even though there is no precise consistency of glacial events throughout the Cascades. During the mid-1800s, the ice on Rainier reportedly covered an estimated 45 to 55 square miles. By 1913, the area had shrunk to about 40 square miles, leaving a section of bare rock exposed for the first time in many hundreds of years. By 1950 glacial ice mantled no more than 37 square miles. Since 1950, Rainier's glaciers have responded to the cooler and wetter weather that began in the mid-1940s. The total volume of ice on Rainier is now about 160 billion cubic feet. Whereas some of the smaller glaciers, particularly those at lower elevations, are stable or receding, many of the large glaciers that originate at high altitudes have been expanding vigorously during recent decades.

The Emmons Glacier, which covers approximately four square miles of Rainier's eastern flank has been advancing since 1953. After the massive rockfall from Little Tahoma Peak buried its lower part in December, 1963, insulating the ice, the rate of growth accelerated. Emmons Glacier advanced 601 feet between 1967 and 1973. By 1980 the terminus had advanced an additional 885 feet.

Nisqually Glacier, which sweeps down Rainier's south slope and skirts the Paradise Visitor Center, fluctuates rapidly, perhaps because its southern exposure makes it particularly sensitive to climatic changes. The location of Nisqually's snout has been recorded annually since 1918. After 34 years of unbroken retreats, the glacier had melted back half a mile, leaving a residue of dirty, stagnant ice that masked the actual terminus. Beginning in 1953, however, a revitalized glacier overrode the stagnant ice and pushed forward until 1969. Between 1969 and 1974 the Nisqually Glacier again lost ground, retreating 197 feet, but then reversed direction and advanced downvalley about 560 feet between 1974 and 1980.

A major flood that swallowed the lower portion of the Kautz Glacier in 1947, sending a bouldery mudflow down Kautz Creek canyon, left the glacier significantly smaller. After 1966, this small south-side glacier began to regain some of its lost volume, advancing 438 feet by 1980.

South Tahoma Glacier, on Rainier's southwest flank, was advancing until the summer of 1967, when a large flood burst from it at about the 7400 foot level, possibly because active steam vents opened beneath the glacier. Despite being almost severed from the upper part of the glacier, the South Tahoma terminus continued to advance, moving ahead about 600 feet from 1967 to 1969. Between 1969 and 1974, it advanced 250 feet, only 50 feet between 1974 and 1976. Meanwhile, the lower 3000 feet of the glacier, virtually detached by the outburst floods, began to stagnate and melt.

Tahoma Glacier flows from the summit icecap in a magnificent icefall down the volcano's western face, then divides into two ice lobes that descend different valleys. The Tahoma Creek lobe advanced about 350 feet between 1967 and 1977, building a moraine from which it had retreated about 147 feet by 1980. The Puyallup River lobe has consistently moved ahead, gaining about 580 feet between 1967 and 1980.

Heading at the Willis Wall, the largest glacial cirque on Mt. Rainier, the Carbon glacier is the longest, thickest, and largest ice stream in the lower 48 states. It advanced about 425 feet between 1965 and 1973. Winthrop Glacier, which descends Rainier's north flank, has hardly changed since about 1910, probably because its protective covering of rockfall debris reduces melting.

Since 1945 the Cascades have experienced several winters of abnormally high snowfall. The average snow-pack in the North Cascades was about 50 percent greater in the 1960s and 1970s than during the 1920s and 1930s. The winter of 1969 to 1970 and the years immediately following set new snow records. The late 1970s, however, brought near-drought conditions. Many high-altitude Pacific Northwest glaciers were still advancing in 1980, probably because of heavy snowfall in the 1960s through mid-1970s.

Some Rainier glaciers have shown little or no change in the last several decades, while others expanded or shrunk. Why should glaciers on the same mountain behave so differently? First, each has its own position on the volcano. Furthermore, different combinations of steepness, ice thickness, and shape of the bedrock channel all affect the glacier's behavior. In addition, glaciers respond at different rates to changes in climate. The region of ice accumulation may lie several thousand feet above the glacier's terminus, so it can take many years for heavy snowfall on the upper part of the glacier to cause the glacier's front to advance.

Glacier Activity at Other Cascade Peaks

Although statistics are not so complete for other major Cascade volcanoes as for Rainier, it appears that a similar non-synchronous pattern of growth, stabilization, and shrinkage is affecting other parts of the range. Several glaciers in the north Cascades are advancing steadily while others are stagnant or retreating. Although no single factor is entirely responsible, some general conclusions can be drawn: most of the expanding ice streams face north or northeast, where they are shadowed much of the day. Most of the advancing glaciers have higher mean altitudes than nearby glaciers that are not growing. In general, the higher the glacier heads on its host peak, the more likely it is to grow.

U.S. Geological Survey reports show that at least six glaciers on Baker were growing at the close of the 1960s. Between 1953 and 1969, the Boulder Glacier, on Baker's east flank, advanced 2135 feet. Between 1949 and 1968, the Coleman and Roosevelt glaciers, on the volcano's northwest slope, moved ahead more than 1300 feet. During roughly the same period the Park, Deming, and Easton glaciers also advanced more than 1000 feet. Meanwhile, the Sholes Glacier, low on Baker's north side, shrank.

Glacier Peak supports at least eight glaciers that have advanced since 1949. Between 1949 and 1956, the Chocolate Glacier advanced 1100 feet, while the Cool, Kennedy, Scimitar, North Guardian, Vista, Dusty and Ermine glaciers all pushed ahead several hundred feet. Meanwhile, the White Chuck Glacier, southwest of the volcano, and the Honeycomb Glacier to the southeast continued to retreat. Observations made in the early 1980s suggest that even glaciers at high elevations are now slowing their growth.

St. Helens' Glaciers

The catastrophic eruptions of St. Helens in 1980, which destroyed 70 percent of the mountain's ice cover, provided a unique opportunity to study the effects of the holocaust on the small surviving glaciers. The 3.5 billion cubic feet of ice destroyed in the eruption included most of the Forsyth, Leschi, Loowit, and Wishbone glaciers. Only a narrow tongue of the Forsyth glacier survived, while the Nelson, Ape, and Shoestring glaciers were left beheaded, their cross-sections exposed along the eastern crater rim. The surviving glaciers were further reduced by the hot pyroclastic flows that raced over their surfaces. Hot ash melted about 20 feet of snow and firn ice from the Shoestring Glacier, initiating the destructive mudflows that poured down the valleys of Muddy River and Pine Creek, eventually into Swift Reservoir. Three to ten feet of volcanic debris buried St. Helens' remaining glaciers.

During the late summer and early fall of 1980, floods and mudflows repeatedly burst from the Shoestring Glacier. From the end of August through early October at least one mudflow, averaging seven to ten feet thick and with a volume of 170,000 cubic feet, originated on Shoestring Glacier each day. Numerous mudflows also poured from other glaciers on the west and south sides of the volcano, well below the lost summit.

Studies of the effects of an ash cover on the rate of snowmelt show that a trace of ash increases the melting rate by as much as 20 to 30 percent. A cover of about one tenth of an inch of ash creates a maximum enhancement of melting—90 percent. Conversely, ash depths greater than an inch or more insulate the snow from external heat and reduce melting. Because the ash cover on the Shoestring Glacier averages 51 inches, it was not surprising that August, 1980, found nearly 200 inches of the previous winter's snow unmelted on the lower glacier.

The future of most of St. Helens' glaciers is doubtful. The removal of the volcano's former summit deprives Forsyth, Nelson, and Shoestring glaciers of their areas of snow accumulation. With no new waves of ice descending to feed them, the surviving portions may eventually melt. But their thick blankets of ash and mudflows may preserve the remaining ice for years.

Some Oregon and California Glaciers.

Several of the Oregon Cascade glaciers have been monitored during the past few decades. Eliot Glacier, on Hood's north flank, is the second largest in the state and the most fully studied. Like many others, Eliot Glacier shrank so greatly after 1900 that it seemed in danger of vanishing. In 1958, however, it was rejuvenated when an "ice wave" from the upper glacier—the result of increased snowfall in the late 1940s and after—reached the terminus. The annual increase in glacial thickness has since been fairly constant. Between September 1970 and September 1971, Eliot Glacier gained a volume of ice equivalent to about 147 acre feet of water. This is good news to the farmers and fruit growers in the Hood River valley, who depend upon Hood's ice fields for much of their water supply.

Once mighty Collier Glacier, which descends from the Middle Sister to cut across the North Sister's western flank, is apparently still retreating. With a maximum thickness of about 300 feet, it remains the deepest ice mass on the Three Sisters. In 1981, a survey revealed that the South Sister's crater ice-fill is about 200 feet deep. Two ice fields on Thielsen's shaded north flank may be tiny glaciers. McLoughlin's northeastern cirque is now empty of glacial ice, although a sizable glacier still occupied the basin as late as 1900.

The southernmost Cascade glaciers are on Shasta, which received several record snowfalls before the droughts of the late 1970s and 1980s. The winter of 1973-74 deposited more than 40 feet of new snow on the volcano's slopes. Rejuvenation of the larger glaciers on Shasta's northern and northeastern flanks was then underway. The Whitney Glacier advanced approximately 1500 feet between 1944 and 1972, to cover an area larger than most Sierra glaciers.

Survey of Present Hazards

Potential loss of life and property from future eruptions has increased markedly during this century as recreational, industrial, and economic use of our volcanic areas continues to grow. Pressure from a rapidly expanding population ensures that more people will settle in the Cascades. Thousands of summer homes, scores of lumber camps, hydroelectric projects, ski resorts, and dams already exist. Millions of fishermen, hikers, campers, and tourists visit the sleeping volcanic giants. Most are unaware of the menace. Despite the devastating mudflows that swept down the North Fork Toutle River valley, overwhelming approximately 140 houses and other buildings on May 18, 1980, most people do not yet seem to realize that similar events can strike lowlands bordering almost any western volcano.

Most media-inspired fears can be dismissed. No major city situated west of the Cascade Range is likely to suffer the fate of Pompeii. None is as close to a potentially active volcano as were Pompeii and Herculaneum, which are only four or five miles from Vesuvius. Prevailing westerly winds from the Pacific would ordinarily direct ashclouds eastward, away from the most populous areas. This was the case during the May 18 eruption of St. Helens, as it had been during the majority of its previous eruptions.

As demonstrated on May 25, June 12, and October 16, 1980, however, reversals of the normal wind patterns can coincide with explosive eruptions. On these occasions, winds carried St. Helens' ashfalls to towns in western Washington and Oregon. While west-side ash-dustings are mercifully rare, towns east of the Cascades could receive damaging fallout from ash plumes. Large sections of eastern Washington, northern Idaho, and western Montana experienced almost total blackouts during the passage of St. Helens' 1980 eruption cloud. Communities at or near the base of volcanoes may be in grave peril from pyroclastic flows and mudflows. Many more communities lie downstream from reservoirs which, if overtopped by a large mudflow or flood, could cause disaster in the lowlands below. Dozens of settlements below Rainier are in range of large mudflows.

A few towns or alpine resorts actually on the flanks of large volcanoes are vulnerable to the most frightful of all volcanic phenomena, pyroclastic flows. These swift-moving "firestorms" can incinerate large areas in a few minutes.

Fortunately, volcanoes usually give ample warning of an impending eruption. Most eruptions begin on a small scale. Unmistakable warning signs would almost certainly persuade authorities to order mass evacuation. Increased steam emission, swarms of small earthquakes centered beneath the volcano, the appearance of hot spots, even swelling of the volcanic cone would give clear indication of an imminent eruption.

Nevertheless, it is possible that a volcano might explode before there was time to evacuate the surrounding areas. Or, one might erupt on a scale vastly greater than any it had known in the past. Rainier, Adams, Hood or some other could produce a cataclysm like the one that destroyed Mazama 6900 years ago. The probability of such an event happening in any one year is very low, but such an eruption would be a regional disaster.

There are a few places in the Cascades, such as Rainier, where massive avalanches of rock occur without warning. A sudden thaw, an earthquake, or an unheralded steam explosion could at any time send millions of tons of rock crashing down, as happened at Little Tahoma Peak in 1963. Chaos Crags, near the north base of Lassen Peak, pose an avalanche threat. The danger is so real that in 1974 the National Park Service moved the entire Lassen visitor center, museum, and hotel and dining accommodations from near Mazanita Lake to a new site.

A Summary of Volcanic Hazards

To estimate the damage a Cascade volcano might do if it erupted, it is necessary to know as much as possible about its past conduct. Unfortunately, the life histories of only a few of the major peaks have been analyzed with the required thoroughness. Except for Lassen and St. Helens, no Cascade volcano has erupted since about the middle of the last century.

In assessing the hazards posed by a particular volcano, geologists generally assume that its future eruptions will resemble those of the past; a volcano is expected to conform to its former habits. The obvious weakness in this assumption is that a volcano may erupt in new ways, as did St. Helens.

The following survey of potential volcanic hazards is based on our limited knowledge of each volcano's past activity.

Mono Lake-Long Valley Region: Next to St. Helens, the U.S. Geological Survey considers this area the most likely source of the next volcanic eruption in the lower 48 states. A tabular sheet of magma (dike) reaching the surface could produce a long chain of vents erupting almost simultaneously at the Mono and/or Inyo Craters. The eruptive cycle would probably occur in three stages: (1) steam blast explosions ejecting towering columns of ash that will blanket large areas many miles downwind; (2) ejection of moderate volumes of rhyolitic pumice, some in the form of pyroclastic flows; (3) after the magma has been degassed, the quiet effusion of pasty lava erecting steep domes to seal the vents. If eruptions take place close to communities like Mammoth Lakes, property losses may be high.

Lassen Peak - Chaos Crags: Almost any part of Lassen National Park is susceptible to renewed volcanic activity. Lassen is likely to produce silicic lavas that are potentially explosive and dangerous to people and property in the immediate vicinity. Future pyroclastic flows at Lassen or Chaos Crags will follow existing topographical depressions, such as the Manzanita Creek and Lost Creek valleys. Approximately 1100 years ago a series of rhyodacite pumice flows poured about 13 miles down these stream valleys, while a horizontal blast of steam and hot rock fragments traveled 4.5 miles down Lassen Peak into Hat Creek valley in 1915. The same year a bouldery mudflow extended 20 miles down both Lost and Hat creeks, burying farm and pasture land. Events of this nature can be expected to occur sporadically in the future.

The Chaos Jumbles area is especially vulnerable to sizable rockfalls and avalanches triggered by earthquakes or steam explosions that could occur without warning. Lava flows may occur at any of the nearby volcanoes, such as Prospect Peak, or Cinder Cone. Ash from these or other vents in the Park, though likely to be of relatively small volume, may blanket an area several miles downwind.

Shasta: Large tephra eruptions are not common at Shasta, but ash derived from collapsing domes could mantle the volcano's flanks to a depth of a foot or more. Although activity has centered at the Hotlum Cone for the last 9000 years, new vents could open anywhere. Extrusion and collapse of domes high on Shasta's west flank could send hot pyroclastic flows sweeping over the towns of Weed and Mt. Shasta, which are partly built on earlier pyroclastic deposits. Mudflows have accompanied virtually every eruptive episode, and buried valley floors many tens of miles from their source. McCloud stands on mudflow deposits, some less than 200 years old. Rapid melting of Shasta's ice-cover could initiate large-scale flooding of the upper Sacramento River, threatening Shasta Lake and Dam.

McLoughlin: This small volcano has apparently not been explosive since the early stages of its construction. Future eruptions are likely to produce relatively quiet emissions of andesite lava along the mountain's flanks, although renewed pyroclastic activity is possible. Except during seasons when McLoughlin is snow covered, there would be little danger from floods or mudflows.

Crater Lake: Since Mazama's collapse about 6900 years ago, all activity has been confined to the caldera, where eruptive episodes produced dacite and rhyodacite flows and domes, as well as Wizard Island. Mazama erupted both rhyodacite and basaltic andesite during the caldera eruption, so it is possible that basaltic andesite may appear again, probably in the form of cinder cones and associated flows. Larger eruptions of ash and pyroclastic flows are possible.

Thielsen and Union Peak: Probably extinct.

Newberry Volcano: A broad spectrum of magma types and kinds of eruptions have marked the more recent eruptions of this enormous volcano. Lava flows have reached as far as Bend, while ash blanketed Oregon's high lava plains many tens of miles to the east. New cinder cones, flows, domes, and pyroclastic flows could erupt anywhere inside the central depression or on the outer slopes. Renewed activity along the Northwest rift zone could send basalt flowing across Highway 97, even into Bend. Cinder cones similar to Lava Butte might form anywhere along the fracture zone.

Bachelor: This volcano formed since the last ice age maximum. Future lava flows would probably move sluggishly enough to allow plenty of time to evacuate the affected area, although sudden melting of snow and ice during winter could generate mudflows potentially endangering the north-side ski resort, Cascade Lakes Highway, and adjacent summer homes. Lava or mudflow damming of local streams could cause flooding and even create new lakes.

Brokentop: Probably extinct.

South Sister: Most recently active of the Three Sisters, about 1900 years ago, it produced flank eruptions of pumice and blocky lava flows that dammed Fall Creek to form the Green Lakes while a chain of obsidian domes erupted along the south flank. Neither pyroclastic flows nor extensive mudflows have characterized South Sister's post-glacial history. The highly silicic composition of the more recent lavas suggests that future eruptions could be explosive.

Middle Sister: The volcano's post-glacial behavior is not well enough known to provide a basis for assessing its potential.

North Sister: Probably extinct, although the numerous cinder cones and lava flows on its eroded flanks suggest that new cinder cones could form.

Belknap Crater: Large flows of blocky basalt can be expected in the future, as well as moderate tephra eruptions from summit and parasitic cinder cones. The McKenzie Pass Highway and McKenzie River valley might be affected by slow-moving lava flows.

Washington, Black Butte, and Three Fingered Jack: All three have probably been extinct for 100,000 to 200,000 years, but small pyroclastic eruptions near their bases could erect cinder cones and produce basaltic flows similar to those found throughout the Oregon Cascades.

Jefferson: Jefferson has not erupted since before the last two Pleistocene glaciations, perhaps 140,000 years or more ago, and may be extinct.

Hood: Perhaps the next Pacific Northwest volcano to erupt, Hood will probably adhere to the behavior pattern it set during the last 12,000 years, producing silicic domes, pyroclastic flows with accompanying ashclouds, and extensive mudflows. If future eruptions center south of the summit ridge behind Crater Rock, material avalanching from growing domes will be largely deflected into the White River drainage to the southeast and into the Sandy River basin on the west. Hood's last major activity (about 1760-1810) sent floods and/or mudflows down the Sandy River all the way to the Columbia and down the White River into the Deschutes, inundating valley floors where thousands of people now live. Ash fallout from crumbling domes and hot avalanches could blanket wide areas many miles downwind.

Adams: Most of Adams' bulky cone above the 7000-foot-level was erected between about 25,000 and 10,000 years ago. Since then, Adams has erupted at least seven times, producing voluminous lava flows from vents below the summit but little pyroclastic material. Future eruptions are also likely to be quietly effusive. Only one sizable mudflow is known to have affected a presently inhabited area during the last 7000 years.

St. Helens: This most consistently explosive of Cascade peaks has repeatedly produced large ash eruptions, pyroclastic flows, and mudflows throughout the last 40,000 years. Several thousand square miles have been seriously affected by ashfalls: lava flows, hot debris flows, and mudflows traveled as far as eight to ten miles. Failure of the debris plug now damming the North Fork Toutle River could create additional floods and mudflows. As long as the crater remains open to the north, future pumice flows and debris from shattering domes will probably be directed into the Spirit Lake and North Fork Toutle River areas. More eruptions as large as that of May 18 are possible in the next few decades, but most geologists think it more likely that continued dome activity and small eruptions may partially rebuild St. Helens' decapitated cone. Flows of andesite lava and extrusion of a new summit dome may complete the current eruptive cycle. But St. Helens is dangerously unpredictable and may yet stage a paroxysmal outburst as violent as any in the past.

Rainier: Immense mudflows from this glaciated giant have repeatedly extended as far as the Puget Lowland during the last 7000 years, burying townsites where 60,000 people now live. Canyons heading on the east, south, and west sides of the mountain have been most frequently affected. Ash eruptions, although relatively common, have generally been small or moderate. Post-glacial eruptions of lava have occurred at the summit, but flows could erupt on any flank of the volcano. Hot rock ejected on snow and ice would quickly generate meltwater floods and mudflows.

An earthquake or steam blast could trigger avalanches of hydrothermally decayed rock from the steep cliffs of the Sunset Amphitheater high on Rainier's west face, sending a wave of rock debris hundreds of feet high speeding many miles down adjacent valley floors. A repetition of the Osceola mudflow that inundated at least 125 square miles near Puget Sound about 5700 years ago is theoretically possible, but future activity will more likely create debris flows the size of the Electron Mudflow, which streamed 35 miles down the Puyallup River valley 600 years ago.

Glacier Peak: Some of the largest late-glacial and post-glacial pyroclastic eruptions in the Northwest originated at Glacier Peak. In spite of its remote location and relatively small stature, Glacier Peak's eruptions devastated areas tens of miles downwind and downvalley, as far as the Puget Lowland. Future eruptions on this scale would bury valley heads on the mountain and severely damage or destroy low-lying towns, farms, and other settlements.

Baker: Unlike Glacier Peak, Baker has typically erupted only small-to-moderate volumes of pyroclastics during the last 10,000 years. Explosive eruptions, block-and-ash flows, and lava flows have affected the east flank of the mountain and the Baker River drainage. Intense heat and steam emission continues in Sherman Crater, the walls of which contain large sections of chemically altered, unstable rock particularly susceptible to avalanching. Because Sherman Crater is breached on the east, meltwater from the crater icepack could pour eastward down Boulder Glacier, possibly generating mudflows extending as far as Baker Lake. Hot pyroclastic flows, melting even more snow and ice, would probably follow a similar route. Lava flows might erupt on any flank of the cone, perhaps disrupting glaciers and sending floods and mudflows downvalley. The increasing frequency of avalanches and mudflows during the last few centuries—plus the numerous historic eruptions—suggests that Baker will erupt again, perhaps within our lifetimes.

Garibaldi: Although Garibaldi apparently has not erupted since the close of the Fraser Glaciation, it may not be extinct. If it follows its earlier pattern, this dacite volcano will produce pyroclastic flows and thin, relatively short lava flows, all limited to within three or four miles of their source. Because of its glacier cover, some flooding and mudflows can be expected.

Forecasting the Next Eruption

Local weather forecasting is a chancy affair, and trying to outguess Vulcan's next move is almost infinitely more so. Using the recent past as a clue to the future suggests that most Cascade eruptions will occur at the large composite cones. The Mono-Inyo Craters' record during the past 2000 years indicates that within the next few decades new vents will erupt explosively.

The big composite cones—high, steep, and glacier-covered—represent the more complex puzzle and greater threat. Most have been dormant for long periods of time. Gas-charged silicic magma chambers may form under several whose last eruptions produced dacite or rhyodacite lava. Most are so erratic that predicting their next eruption is virtually impossible. A long-dormant volcano, its internal plumbing blocked by solidified lava from the last eruption, may blast open new vents on the flanks of the cone, triggering massive avalanches and producing laterally directed pyroclastic surges. The sudden collapse of St. Helens' north side and the ground-hugging horizontal blasts at St. Helens and Lassen Peak demonstrate the variety of behavior of which our volcanoes are capable.

Once a volcano has cleared its vent, its activity may be more predictable. Nearly all of St. Helens eruptions after May, 1980, were predicted. Until the vent clears, forecasting the time and nature of an imminent eruption is not possible. Gas concentrates in the upper part of the magma column and typically erupts explosively when it reaches the surface. Because its timing and size can not be pinpointed in advance, the initial blast is commonly the most deadly.

XXII
When Mt. Shasta Erupts

(The following fictional account describes what might happen were Shasta to become the third Cascade volcano to erupt during the 20th century. The events are based on the volcano's known behavior during the past 10,000 years. The narrative presents the kinds of dangerous activity we can expect when Shasta awakes from its present slumber. In them we see that the social and scientific responses to a volcanic crisis may prove as humanly decisive as the event itself.)

Prologue

Mt. Shasta shows signs of restlessness. On September 2, 199__, climbers reported having seen "wisps of steam" issuing from the crater of Shastina, the large secondary cone on Shasta's western flank. Forest Service Rangers who investigated the story could find no new steam vents, but noted a strong sulfurous odor. The United States Geological Survey will fly over the volcano to take infrared images that might reveal new hot spots.

In a report dated September 6, the U.S. Geological Survey states that its flight detected no increase in heat emission on Shastina or the summit of Hotlum cone. Because all known eruptive activity has centered at the Hotlum vent for about the last 9000 years, geologists expect most future eruptions to occur there, although spokesman cautioned that a new vent could open anywhere on the mountain.

On September 7, a sharp earthquake jolts the towns of Dunsmuir, McCloud, Weed, and Mt. Shasta City. Seismologists place the focus of the shock about a mile west of Shasta's crest at a depth of 12 miles. Although felt in Yreka, the tremor, which registered a magnitude of about 4.0 on the Richter Scale, is much milder there.

A second earthquake, scoring 4.7 on the Richter Scale, rolls through parts of Siskiyou County the night of September 7-8. That morning two lighter shocks, between nine and ten miles underground, are recorded in Weed. A more severe jolt follows on September 8, which causes mild alarm as far south as Redding and as far north as Yreka. This magnitude of 5.1 is centered at a depth of only seven miles. The increasing shallowness of the earthquakes may indicate that magma is rising, possibly heralding an eruption. Shasta's last activity occurred in 1786.

During the next 48 hours, the Weed seismograph records at least 320 earthquakes, only a dozen of which are strong enough to be felt. A second U.S. Geological Survey flight discovers two large hot spots on Shastina, both inside the summit crater. The larger is close to the narrow gash that heads Diller Canyon.

By the evening of September 10, residents of Weed, Mt. Shasta City and McCloud report an almost constant rattling of windows and dishes. Timbers in old farm houses creek eerily as the earth shifts beneath them. After consulting with representatives from the Geological Survey, the California State Board of Emergency Preparedness alerts the local police and Forest Service to a possible mass evacuation. The Governor has already placed the State Police on alert and requested help from the National Guard should the speedy evacuation of several thousand people become necessary.

These announcements are immediately opposed by delegations of local businessmen, as well as public relations agents for the timber and railroad companies who own land around the mountain. These spokesmen remind legislators and other state officials of the "unconscionable" loss of business occasioned by the "unnecessary" volcanic hazard warnings issued for the Mammoth Lakes region in the early 1980s. To date, the spokesmen point out, despite the swarms of volcanic earthquakes at the Mammoth resort, no eruption has occurred. Employees of northern California's depressed logging industry cannot afford loss of work and income such "irresponsible" restrictions would impose, nor, add local entrepreneurs, is it fair to stem the profitable influx of tourists eager to see an active volcano. The Governor promises "not to act hastily."

After a brief lull, another swarm of earthquakes, now focused a scant two miles beneath the volcano's western flank strikes the

Mount Shasta region shortly before dawn September 12. By noon, the seismographs are receiving almost continual shocks. At 12:35 p.m. an awesome column of dark ash suddenly bursts from the top of Shastina and rises an estimated 12,000 feet skyward. A low rumbling, reminiscent of distant thunder, accompanies the cloud, which swells and unfurls like a great black umbrella. Fortunately, prevailing winds direct the plume toward the sparsely inhabited country east of the volcano, where fine ash begins to dust the Medicine Lake Highland, 35 miles distant.

A Forest Service airplane radios that Mount Shasta's snowfields are blackened by the ashfall. Large fragments of solid rock litter glaciers and summit icefields. The material thrown out thus far seem to have been cold, for it lies on snowdrifts without melting the underlying ice. Most of the ejecta, a Geological Survey spokesman explained, is probably derived from the solid lava plug that blocked the volcano's throat.

Small to moderate steam eruptions occur spasmodically for more than a week, while the small earthquakes beneath the peak decrease in number, but increase in average magnitude. Because of hazards from laterally directed explosions, rockfalls, and avalanches, the general public is barred from the mountain's vicinity. Citizens of Weed, Mount Shasta City, McCloud, and Dunsmuir, as well as ranchers and summer residents near the base of the volcano are warned to be ready to leave immediately if the eruptions grow more violent. The Yakohovians and Seekers of Light, two of the many local religious cults, formally protest this restriction, which they claim infringes upon their constitutional freedom. The prohibition against climbing the mountain will interfere with the international festival of mystics and psychics which they had planned to celebrate on Shasta's heights.

Columnists in the San Francisco *Chronicle* complain that the volcano is the biggest disappointment since the Comet Kahoutek failed to show in the 1970s. A federal agency issues an official press release pointing out that since the last ice age Shasta has produced only one large explosive eruption, which formed the Red Banks pumice deposit between about 9700 to 9600 years ago. "Considering the laws of probability," one geologist observes, "smaller events rather than large ones are to be expected."

Shortly before the dinner hour, September 19, local conjecture about Shasta's intentions ends with a roar heard for a hundred miles in every direction. As a column of charcoal gray ash soars 70,000 feet above Shastina, boulders large enough to be seen from miles away are tossed high into the air and crash downslope. The whole mountain trembles with these detonations, triggering avalanches of freshly erupted material and dislodging old rock formations.

By sunset the eruption has reached its height. A ruddy glare shows fitfully amid the inky billows towering above Shastina's crest. High altitude northerly winds now bear the ash plume over the head of the Sacramento Valley. A steady rain of powdery ash falls at Shasta Dam, 45 miles south of the volcano. Radio and TV stations in Redding, nine miles farther south, report that fine dust has begun to drift through the sultry air, compounding the usual "smaze" of this agricultural region.

Roused from his bed in Sacramento, the Governor of California issues a state-of-emergency proclamation. The curtain of warm ash spreading over the Shasta reservoir, which, following an exceptionally wet winter, still holds an estimated 3.8 million acre feet of water, has alarmed Forest Service and U.S Geological Survey officials, who have been up all night evaluating measures to ensure the safety of persons living within the endangered area. In spite of protests from owners of large ranches in the Sacramento Valley, the Governor orders the sluices of Shasta Dam opened to permit a large flow down the Sacramento River. Because of the increasing temperatures of the magmatic eruptions, authorities fear that quantities of the snow and glacial ice mantling Shasta may melt and generate catastrophic mudflows. If a mudflow were to reach Shasta Lake, it could displace millions of acre feet of water over the dam crest, and inundate the upper Sacramento Valley.

At midnight the explosive eruption of frothy pumice continues unabated, but the winds above the mountain have virtually ceased. Instead of being directed to the south-southeast as before, the giant mushroom cloud hangs suspended in stagnant air above Shasta. Ash and walnut to sand-sized pumice fragments are now showering down over the entire mountain and adjacent terrain.

State authorities have barricaded Route 97 at Dorris, a tiny hamlet just south of the California-Oregon border. Interstate 5 closed before nightfall between Yreka in the north and Redding in the south. With visibility reduced to near zero, traffic crawls through the turgid air, transporting the last evacuees. The fate of persons remaining in isolated farms is unknown, but radio bulletins urge those trapped by the choking ashfall to remain indoors.

Hours after sunrise, September 20, northern California is a land in twilight. It is almost noon before the titanic cauliflower cloud begins to fall. Although it will take days for all the ash to settle, the slowly clearing air reveals a thick blanket of beige to orange pumice. Preliminary estimates indicate that pumice lies nearly a foot thick over Mount Shasta City and Weed, with progressively thinner layers extending over Dunsmuir and McCloud.

With the Shastina vent now spouting an ash plume only a few thousand feet high, aerial surveys reveal that numerous pyroclastic flows swept down the volcano's flanks. Several, three to five miles in length, descended Diller Canyon to the outskirts of Weed. One of the largest poured through a newly-formed breach in the northeast rim of Shastina's crater and across Whitney Glacier, the large ice stream in the saddle between Shastina and the main cone. Meltwater from glacial ice, mixed with incandescent pyroclastic material, formed a sizable mudflow that streamed down Whitney Creek, inundating a stretch of Highway 97 at Shasta's northern foot.

Almost miraculously, no lives were lost during the explosive activity, but property damage is estimated in the tens of millions of dollars. Roofs throughout McCloud, Weed, and Mt. Shasta City collapsed under the weight of the pumice fall while thousands of evergreens were totally defoliated.

As an early autumn night closes over northern California, a lurid glare fitfully glows through the dust-charged darkness. Intermittent explosions spray molten fragments over the cone.

Thousands of tourists and sensation-seekers flocking to witness the spectacle have impeded the evacuation. Closure of all roads between Chico, Redding, and the Oregon boundary continues, but rash people with four-wheel drive vehicles continue to circumvent the barricades. Fine ash sucked up into their motors soon stalls most roadblock runners forcing them to abandon their vehicles and walk.

By September 24 the eruption appears to be over and displaced residents at Dunsmuir, Weed, McCloud, and Mt. Shasta clamor to return to their homes. However, scientists are reluctant to advise civil authorities that the volcano is again safe, because seismographs have begun to record earthquakes different in nature from the sharp jolts and harmonic tremors preceded and accompanied the explosive eruption. When pressed by TV reporters to interpret the significance of the seismic phenomena, a University of California seismologist said it was "too soon" to decipher Shasta's message and that scientists could "only wait and see" what would happen next.

Because Shastina is now only mildly steaming, most state and federal officials agree that those who so wish should return home and begin digging out the ash and mud. A bearded young seismologist from a private university, however, emphasized to a local TV newswoman that the current earthquake swarm, though producing events of much lower magnitude than those that heralded the pumice eruption, are steadily increasing in frequency.

"An alarmist," declared Mr. James Fitz-Brown, president of the Mt. Shasta Summer Development Association. He went on to say, "We understand the concern which the recent little blow-off can inspire, but we have been assured by several world-famous geologists that whatever pressure had built up inside Mt. Shasta is now relieved. We who have lived and worked near this wonderful mountain know that it is fundamentally a recreational attraction, not a menace. We don't need nervous nellies who needlessly worry people or keep them away from their jobs and homes." A timber company executive added that the financial loss to his company was already great and that it was "vital" to the area's economic welfare to salvage the denuded trees surrounding Mt. Shasta before boring insects made the timber worthless.

As the debate to reopen the Shasta area continues, aerial surveys of the peak show that a large dark mass has risen into Shastina's crater, almost filling the western half of the vent. One prominent local timberman argues that this proves that the volcano's throat is now plugged. Some scientists are skeptical. Pre-dawn flights on September 25, indicated that heat emission has increased tremendously. The new spires of viscous lava have split and fractured and are crumbling rapidly, undermined by small steam explosions around their bases.

For the next five days Shastina rumbles ominously, throwing out puffs of steam black with ash and sending small avalanches of hot lava blocks down its steep slopes. Refugees from the earlier eruption are eager to get back to home and work. But U.S Geological Survey scientists, remembering the lethal suddeness with which St. Helens exploded, remain reluctant to sound an all clear.

Some petroleum geologists, brought in by real estate brokers headquartered in Los Angeles, speak acidly of scientific "hypercaution" and announce that it is as safe to reoccupy Weed, McCloud, Mt. Shasta and other settlements as it is to live in Miami or Galveston during the average winter. The mathematical probability of a killer hurricane or second major eruption is equally low.

Two presidentially appointed scientists confer with federal geologists, then announce that, "The government cannot in conscience continue to disrupt people's lives or interfere with their free movement unless there is a clear and immediate threat to life, which we believe to be highly improbable at this particular volcano." When

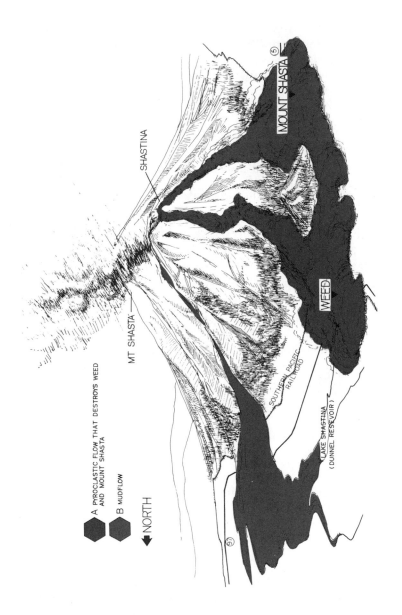

Map showing areas affected by Mt. Shasta's hypothetical eruption. A. Pyroclastic flow that destroyed the towns of Weed and Mt. Shasta. B. Mudflow.

asked about the young geologist who had stated that he felt as if he "were living in the eye of a hurricane, a moment of calm between two catastrophes," the official explained that the young man had been reassigned to studying 700,000 year-old ash flow deposits in the Nevada desert.

Welcoming the opportunity to lift a politically unpopular closure, the California Governor announces that people may return to McCloud, Weed, and Mount Shasta City. Residents had already gone back to Dunsmuir on the assurance that too little snow remains on the southwest side of the cone to form a mudflow large enough to reach town.

Another conference of Geological Survey administrators with a contingent of field geologists leads to a significantly different view. The concerned scientists urge that federal and state authorities allow people who work in the area to enter the danger zone only during daylight hours. Roadblocks are maintained at all major access routes, but scores of people continue to use the many logging roads.

One grizzled logger and his family expressed their confidence that state and federal authorities had made the right decision. "We'd starve to death if we weren't allowed back in the woods." "Besides," his wife added, "all the experts wouldn't let us back in even for work, if there was any real danger." "We need to get in there and cut before they turn the whole place into another damn national monument," was the forthright analysis of a timber company executive.

Near midnight, September 29, a Geological Survey group in the fire lookout west of the volcano reports that a towering column of incandescent matter is spiraling upward from Shastina. The roaring of the volcano and the accompanying earthquakes herald a new stage in the eruption.

The Climax

At exactly 6:21 on the morning of September 30, a violent explosion shatters the great dome that had almost filled Shastina's crater. An immense avalanche of seething gas and molten blocks of the dome's interior, sweeps down the west slope of Shastina, spreading out over the volcano's western flank. In moments the towns of Weed and Mt. Shasta are engulfed in the pyroclastic flow and totally consumed. The entire area bursts into flames with temperatures in the center of the holocaust approaching 2000 degrees Fahrenheit. Iron roofs, steel frame buildings, and glass windows melt into unrecognizable heaps. Most wooden structures simply vaporize. Smoke from burning forest and buildings darkens the dense haze of ash settling from the glowing cloud.

From their vantage point west of Shasta, survey geologists have a panoramic view of the catastrophe. The grotesquely convoluted hot ash cloud rising up and spreading beyond the basal pyroclastic flow is rent by lightning bolts as it rolls and boils upward like the mushroom from an atomic blast. Perhaps the most awesome aspect of the eruption is the near silence. The cloud sounds like wind rushing through the treetops.

By October 3, five more pyroclastic flows derived from dome collapse have shot from the vast gap in the Shastina crater's western lip. At least two of these are as large as the first. To the small towns in their path the subsequent eruptions matter little.

Conservative estimates place the numbers of people consumed in the cataclysm at several hundred. Most bodies were cremated to ash, which mingled indistinguishably with the general debris. Many others were apparently vaporized. Least fortunate were those caught on the fringes of the cloud of superheated ash and steam. Some managed to walk or crawl from the affected area only to perish a few hours or days later.

Two California towns have met the same volcanic fate that decimated St. Pierre, leading city on the Caribbean Island of Martinique. In May 1902, the collapse of a dome growing in the summit crater of Mont Pelée propelled a seething ashcloud through St. Pierre, incinerating the entire city and killing all but four of its 30,000 inhabitants.

Bibliography

I. Our Western Volcanoes: An Introduction

Atkeson, Ray, 1969, *Northwest Heritage: The Cascade Range*, Charles H. Belding, Portland, Oregon, 181 p.

Baldwin, Ewart M., 1981, *Geology of Oregon*, 3rd ed., Kendall-Hunt Publishing Co., Dubuque, Iowa, 179 p.

Begét, J.E., 1982, Glacier Peak: An Active Volcano: *Science*, vol. 215, pp. 1389-1390.

Bullard, Fred M., 1984, *Volcanoes of the Earth*, 2nd rev. ed., University of Texas Press, Austin, 629 p.

Crandell, D.R., Mullineaux, D.R., and Miller, C.D., 1979, Volcanic Hazard Studies in the Cascade Range of the Western United States, in, Sheets, P.D., and Grayson, D.K., eds., *Volcanic Activity and Human Ecology*, Academic Press, New York, pp. 195-219.

Decker, Robert, and Decker, Barbara, 1980, *Volcano Watching*, Hawaii Natural History Association, Hawaii Volcanoes National Park, 80 p.

Decker, Robert, and Decker, Barbara, 1981, *Volcanoes*, W.H. Freeman and Company, San Francisco, 244 p.

Easterbrook, Don J., 1982, *Potential Geologic Natural Landmarks, Cascade Range Province*, Department of Geology, Western Washington University, Bellingham, Washington, 277 p.

Folsom, M.M., 1970, Volcanic Eruptions: The Pioneers' Attitude on the Pacific Coast from 1800 to 1875: *The Ore Bin*, vol. 32, pp. 61-71.

Gibbs, George, 1955, George Gibbs' Account of Indian Mythology in Oregon and Washington Territories, ed. Ella Clark: *Oregon Historical Quarterly*, vol. 56, pp. 293-326.

Harris, Stephen L., 1983, In the Shadow of the Mountains: *Pacific Northwest*, vol. 17, no. 1, pp. 24-33.

Harris, Stephen L., 1986, The OTHER Cascade Volcanoes: Historic Eruptions at Mount St. Helens' Sister Peaks, in Keller, S.A.C., ed., *Mount St. Helens: Five Years Later*, Eastern Washington University Press, Cheney.

Harris, Stephen L., 1983, Volcanic Hazards in the West: *American West*, vol. 20, no. 6, pp. 30-39.

Johnston, D.A. and Donnelly-Nolan, Julie, 1981, *Guides to Some Volcanic Terranes in Washington, Idaho, Oregon, and Northern California: U.S. Geological Survey Circular 838*, 189 p.

Kiver, E.P., 1982, The Cascade Volcanoes: Comparison of Geological and Historic Record, in Keller, S.A.C., ed., *Mount St. Helens: One Year Later*, Eastern Washington University Press, Cheney, pp. 3-12.

Lipman, P.W., and Mullineaux, D.R., 1981, *The 1980 Eruptions of Mount St. Helens, Washington: U.S. Geological Survey Professional Paper 1250*, 844 p.

Martin, R.C., and Davis, J.F., 1982, *Status of Volcanic Prediction and Emergency Response Capabilities in Volcanic Zones of California, Special Publication 63*, California Department of Conservation, Division of Mines and Geology, 275 p.

Simkin, T., Siebert, L., McClelland, L., Bridge, D., Newhall, C., and Latter, J.H., 1981, *Volcanoes of the World. A Regional Directory, Gazeteer, and Chronology of Volcanism during the Last 10,000 Years*, Smithsonian Institution, Hutchinson Ross Publishing Co., Stroudsburg, Pennsylvania, 233 p.

Swanson, D.A., 1982, Volcanic Studies in the Pacific Northwest, in Leviton, A.E., Rodda, P.U., Yochelson, E., and Aldrich, M.L., *Frontiers of Geological Exploration of Western North America, Pacific Division*, American Association for the Advancement of Science, San Francisco, California, pp. 193-207.

Williams, John, 1912, *Guardians of the Columbia*, John H. Williams, Tacoma, 144 p.

II. Our Place in the Ring of Fire

Alt, David D., Hyndman, Donald W., 1984, *Roadside Geology of Washington*, Mountain Press Publishing Co., Missoula, Montana, 282 p.

Barrash, Warren, and Venkatakrishnan, Ramesh, 1982, Timing of Late Cenozoic Volcanic and Tectonic Events along the Western Margin of the North American Plate: *Geological Society of America Bulletin*, vol. 93, no. 10, pp. 977-987.

Beck, Myrl, Cox, Allan, and Jones, David L., 1980, Mesozoic and Cenozoic Microplate Tectonics of Western North America: *Geology*, vol. 8, no. 9, pp. 454-456.

Ben-Avraham, Z., Nur, A., Jones D.L., and Cox, A., 1981, Continental Accretion: from Oceanic Plateaus to Allochthonous Terranes: *Science*, vol. 213, pp. 47-74.

Brantley, Steve, and Power, John, 1985, Reports from the U.S. Geological Survey's Cascades Volcano Observatory at Vancouver, Washington: *Earthquake Information Bulletin*, vol. 17, no. 1, pp. 20-32.

Brantley, Steve, and Topinka, Lyn, 1984, Volcanic Studies at the U.S. Geological Survey's David A. Johnston Cascades Volcano Observatory, Vancouver, Washington: *Earthquake Information Bulletin*, vol. 16, no. 2, pp. 1-121.

Chadwick, W.W., Jr., Iwatsubo, E.Y., Swanson, D.A., and Ewert, J.W., 1985, *Measurements of Slope Distances and Vertical Angles of Mount Baker and Mount Rainier, Washington, Mount Hood and Crater Lake, Oregon, and Mount Shasta and Lassen Peak, California, 1980-1984*, U.S. Geological Survey Open File Report 85-205, 96 p.

Coney, P.J., Jones, David L., and Monger, J.W., 1980, Cordilleran Suspect Terranes: *Nature*, vol. 288, pp. 329-333.

Denlinger, R.P., Riley, F.S., Boling, J.K., and Carpenter, M.C., 1985, Deformation of Long Valley Caldera between August 1982 and August 1983: *Journal of Geophysical Research*, vol. 90, no. B13, pp. 11,199-11,209.

Dickinson, W.R., 1979, Mesozoic Forearc Basin in Central Oregon: *Geology*, vol. 7. pp. 166-170.

Dickinson, W.R., Ingersoll, R.V., Cowan, D.S., Helmold, K.P., and Suczek, C.A., 1982, Provenance of Franciscan Graywackes in Coastal California: *Bulletin of the Geological Society of America*, vol. 93, pp. 95-107.

Dzurisin, Daniel, Cashman, K., and Sylvester, A.G., 1982, *Tilt Measurments at Long Valley Caldera, California, May-August, 1982: U.S. Geological Survey Open File Report 82-893*, 34 p.

Dzurisin, Daniel, Johnson, D.J., Murray, T.L., and Myers, B., 1982, *Tilt Networks at Mount Shasta and Lassen Peak, California: U.S. Geological Survey Open File Report 82-670*, 42 p.

Dzurisin, Daniel, Westphal, J.A., and Johnson, D.J., 1983, Eruption Prediction Aided by Electronic Tiltmeter Data at Mt. St. Helens: *Science*, vol. 221, no. 4618, pp. 1381-1382.

Gore, Rick, 1985, Our Restless Planet Earth: *National Geographic*, vol. 168, no. 2, pp. 142-181.

Jones, David L., Howell, D.G., Coney, P.J., and Monger, H.W.H., 1983, Recognition, Character and Analysis of Tectonostratigraphic Terranes in Western North America: *Journal of Geological Education*, vol. 31, pp. 295-303.

Jones, David L., Silberling, N.J., and Hillhouse, J., 1977, Wrangellia—A Displaced Terrane in Northwestern North America: *Canadian Journal of Earth Sciences*, vol. 14, pp. 2565-2577.

Malone, Stephen D., Boyko, C., and Weaver, C.S., 1983, Seismic Precursors to the Mount St. Helens Eruptions of 1981 and 1982: *Science*, vol. 221, no. 4618, pp. 1376-1378.

Robyn, Thomas L., and Hoover, J.D., 1982, Late Cenozoic Deformation and Volcanism in the Blue Mountains of Central Oregon: Microplate Interactions? *Geology*, vol. 10, no. 11, pp. 572-576.

Sullivan, Walter, 1985, Pieces of a Global Jigsaw Puzzle: *Smithsonian*, vol. 15, no. 10, pp. 66-74.

Swanson, D.A., Casadevall, T.J., Dzurisin, D., Malone, S.D., Newhall, C.G., and Weaver, C.S., 1983, Predicting Eruptions at Mount St. Helens, June 1980 through December 1982: *Science,* vol. 221, no. 4618, pp. 1370-1376.

Swanson, D.A., Casadevall, T.J., Dzurisin, D., Holcomb, R.T., Newhall, C.G., Malone, S.D., and Weaver, C.S., 1985, Forecasts and Predictions of Eruptive Activity at Mount St. Helens, USA: 1975-1984: *Journal of Geodynamics*, vol. 3, no. 314, pp. 75-89.

Tabor, R.W., Frizzell, V.A., Jr., Vance, J.A., and Naeser, C.W., 1984, Ages and Stratigraphy of Lower and Middle Tertiary Sedimentary and Volcanic Rocks of the Central Cascades, Washington: Application to the Tectonic History of the Straight Creek Fault: *Geological Society of America* Bulletin, vol. 95, no. 1, pp. 26-44.

Tilling, Robert I., 1982, *Eruptions of Mount St. Helens: Past Present, and Future*: U.S. Geological Survey series of general interest publications, 46 p.

Vogt, B.F., ed., 1985, Scientists Report on Five Years of Mount St. Helens Studies: *Oregon Geology*, vol. 47, no. 9, p. 109.

III. The Fire: How a Volcano Works

Armstrong, R.L., 1975, Episodic Volcanism in the Central Oregon Cascade Range; Conformation and Correlation with the Snake River Plain: *Geology*, vol. 8, no. 7, pp. 356-357.

Bullard, Fred M., 1984, *Volcanoes of the Earth*, 3rd ed., University of Texas Press, Austin and London, 579 p.

Citron, G.P., Kay, R.W., Mahlburg, Kay S., Snee, L.W., Sutter, L.W., 1980, Tectonic significance of early Oligocene plutonism on Adale Island, central Aleutian Islands, Alaska: *Geology*, vol. 8, no. 8, p. 375.

Coombs, H.A., and Howard, A.D., 1960, United States of America: In *Catalogue of the Active Volcanoes of the World, Part 9*, International Volcanological Association, Naples, Italy.

Cotton, C.A., 1952, *Volcanoes As Landscape Forms*, 2nd ed., John Wiley and Sons, Inc., New York, 416 p.

Cox, Allen, ed., 1973, *Plate Tectonics and Geomagnetic Reversals, W.H., Freeman & Co., San Francisco*.

Decker, Robert and Decker, Barbara, 1981, *Volcanoes*, W.H. Freeman & Co., San Francisco, 244 p.

Dole, H.M., ed., 1968, *Andesite Conference Guidebook: Oregon Dept. of Geology and Mineral Industries Bulletin 62*, 107 p.

Eaton, J.P., and Murata, K.J., 1960, How Volcanoes Grow: *Science*, vol. 132, no. 3432, pp. 925-938.

Eichelberg, J.C., 1975, Origin of Andesite and Dacite: Evidence of Mixing at Glass Mt. in California and at other Circum-Pacific Volcanoes: *Geol. Soc. of Amer. Bull.*, vol. 86, p. 1381-1391.

Erikson, E.H., Jr., 1969, Petrology of the Composite Snoqualmie Batholith, Central Cascade Mountains, Washington: *Geol. Soc. Am. Bull.*, vol. 80, pp. 2213-2239.

Fisher, R.V., 1977, Erosion by Volcanic Base-surge Density Currents: U-shaped channels: *Geol. Soc. Amer. Bull.. vol. 88, pp. 1287-1297.*

Francis, Peter, 1976, *Volcanoes*, Penguin Books, Ltd., 368 pp.

Green, Jack, and Short, Nicholas M., 1971, *Volcanic Landforms and Surface Features*, Springer-Verlag, New York.

Hildreth, Wes, 1980, The Valley-of-Ten-Thousand-Smokes Tuff, II. Compositional and Isotopic Variability of the Ejecta [Abst]: *Geol. Soc. of America Abstracts*, vol. 12, no. 3, p. 111.

Hughes, John M., Stoiber, Richard E., and Carr, Michael J., 1980, Segmentation of the Cascade Volcanic Chain: *Geology*, vol. 8, no. 1, pp. 15-17.

Johnson, D.A., and Hildreth, Wes, 1980, The Valley-of-Ten-Thousand Smokes Tuff, I. Emplacement, Compaction and Welding [Abst]: Geol. Soc. of American Abstracts, *vol. 12, no. 3, p. 113.*

Kennett, J.P., Thunell, R.C., and McBirney, A.R., 1975, Correlation of Neogene Volcanic Episodicity Recorded in Marine and Terrestrial Deposits, Western North America [Abs]: *Geol. Soc. America Abstracts*, vol. 7, no. 7, p. 1143.

Kruger, Christoph, ed., 1971, *Volcanoes*, G.P. Putnam's Sons, New York, 168 p.

Laursen, J.M., and Hammond, P.E., 1974, Summary of radiometric ages of Oregon and Washington rocks, through 1972: *Isochron/West*, no. 9, 32 p.

Lyman, William D., 1963, *The Columbia River: Its History, Its Myths, Its Scenery, Its Commerce*, rev. ed., Portland, Binfords & Mort, Pubs., 367 p.

Macdonald, Gordon A., 1972, *Volcanoes*, Prentice-Hall, Englewood Cliffs, N.J., 510 p.

Macdonald, Gordon A., and Abbott, A.T., 1970, *Volcanoes in the Sea–the Geology of Hawaii*, University of Hawaii Press, Honolulu, 441 p.

McBirney, Alexander R., 1968, Petrochemistry of the Cascade Andesite Volcanoes, in Doles, H.M., ed., *Andesite Conference Guidebook: Oregon Dept. of Geology and Mineral Industries,* Bulletin 62, pp. 101-107.

McBirney, A.R., 1978, Volcanic Evolution of the Cascade Range: Ann. Rev. Earth Planet Science, vol. 6, pp. 437-456.

McBirney, A.R., Sutter, J.F., Naslund, H.R., Sutton, K.G., and White, C.M., 1974, Episodic Volcanism in the Central Oregon Cascade Range: *Geology*, vol. 2, pp. 585-589.

Oakeshott, Gordon B., 1976, *Volcanoes and Earthquakes: Geologic Violence*, McGraw Hill Earlth Science Paperback Series, 143 p.

Ollier, Cliff, 1969, *Volcanoes*, the MIT Press, Cambridge and London, 177 p.

Peck, D.L., Griggs, A.B., Schlicker, H.G., Wells, F.G., and Dole, H.M., 1964, *Geology of the Central and Northern parts of the Western Cascade Range in Oregon: U.S. Geological Survey Professional Paper 449*, 56 p.

316

Rittmann, A., 1962, *Volcanoes and Their Activity*, Wiley-Interscience Publishers, New York, 305 p.

Robinson, P.T., and Brem, G.F., 1979, The John Day Formation of Oregon: A Record of Early Cascade Volcanoes [Abs]: *Geol. Soc. Am. Abstr. Programs*, vol. 11, no. 3, p. 125.

Simkin, T., Siebert, L., McClelland, L., Bridge, D., Newhall, C., and Latter, J.H., 1981, *Volcanoes of the World: A Regional Directory, Gazeteer, and Chronology of Volcanism during the Last 10,000 Years*, New York: Academic Press, 240 p.

Smith, A.L. and Carmichael, I.S.E., 1968, Quaternary Lavas from the Soutern Cascades, Western U.S.A.: *Contributions to Minerology and Petrology*, vol. 19, pp. 212-238.

Smith, J.G., Sawlan, M.G., and Katcher, A.C., 1980, An Important Lower Oligocene Welded-Tuff Marker Bed in the Western Cascade Range of Southern Oregon [Abst]; *Geol. Soc. of America Abstracts*, vol. 12, no. 3, p. 153.

Swenson, David H., 1973, *Geochemistry of Three Cascade Volcanoes*: Master's 1973 New Mexico Meeting.

Vance, J.A., 1977, Early and Middle Cenozoic Magmatism and Tectonics, Cascade Mountains [Abst]; *EOS*, vol. 58, no. 12, p. 1247.

Walker, G.W., Dalrymple, G.B., Lanphere, M.A., 1974, Index to Potassium-Argon Ages of Cenozoic Volcanic Rocks of Oregon: *U.S. Geological Survey Misc. Field Studies Map MF-569*, scale 1:1,000,000, text, table, 2 p.

Weissenborn, A., ed., and Cater, F.W., 1966, The Cascade Mountains, In U.S. Geological Survey, *Mineral and Water Resources of Washington*, Washington Division of Mines and Geology Reprint 9, pp. 27-37.

White, C.M., and McBirney, A.R., 1978, Some Quantitative Aspects of Orogenic Volcanism in the Oregon Cascades, *Geol. Soc. Amer. Mem.*, 1979.

Williams, Howel, 1962, *Ancient Volcanoes of Oregon*, 3rd ed., Condon Lectures, University of Oregon Press, Eugene, Oregon, 68 p.

Williams, Howel and McBirney, Alexander R., 1979, *Volcanology*, Freeman, Cooper & Co., San Francisco, 397 p.

Wise, W.S., 1964, A Guide to the Volcanic Rocks of the Cascade Range: *Mazama*, vol. 46, no. 13, pp. 23-25.

Wise, William S., 1970, *Cenozoic Volcanism in the Cascade Mountains of Southern Washington*, Washington Division of Mines and Geology, Bull. No. 60.

IV. The Ice: How a Glacier Works

Brugman, M.M., and Post, Austin, 1981, *Effects of Volcanism on the Glaciers of Mount St. Helens: U.S. Geological Survey Circular 850 D*, 11 p.

Chorlton, Windsor, 1983, *Ice Ages*, Time-Life Books, Alexandria, Virginia, 176 p.

Crandell, D.R., 1965, Glacial History of Western Washington and Oregon: In Wright, H.E., and Frey, D.G., eds., *Quaternary of the United States,* Princeton University Press, Princeton, N.J., pp. 341-354.

Crandell, D.R., 1971, *Postglacial Lahars from Mount Rainier Volcano, Washington: U.S. Geological Survey Professional Paper 677*, 73 p.

Crandell, D.R., 1983, *The Geologic Story of Mount Rainier*, rev. ed.: *U.S. Geological Survey Bulletin 1292*, 40 p.

Crandell, D.R., and Miller, R.D., 1974, *Quaternary Stratigraphy and Extent of Glaciation in the Mount Rainier Region, Washington: U.S. Geological Survey Professional Paper 847*, 59 p.

Cummins, John, 1981, *Mudflows Resulting from the May 18, 1980, Eruption of Mount St. Helens, Washington: U.S. Geological Survey Circular 850-B*, 16 p.

Driedger, Carolyn L., 1981, *The Effect of Ash Thickness on Snow Ablation:* U.S. Geological Survey Preliminary Report.

Driedger, Carolyn L., 1986, *A Visitor's Guide to Mount Rainier Glaciers,* Longmire, Washington: Pacific Northwest National Parks and Forests Association.

Driedger, Carolyn L., and Kennard, P.M., 1986, *Ice Volumes on Cascade Volcanoes: Mount Rainier, Mount Hood, Three Sisters, and Mount Shasta: U.S. Geological Survey Professional Paper 1365.*

Easterbrook, D.J., 1969, Pleistocene Chronology of the Puget Lowland and San Juan Islands, Washington: *Geological Society of America Bulletin*, vol. 80, pp. 2273-2286.

Heliker, C.C., Johnson, Arthur, and Hodge, S.M., 1983, *The Nisqually Glacier, Mount Rainier, Washington, 1857-1879: U.S Geological Survey, Open File Report 83-541*, 20 p.

Kennard, P.M., 1983, *Volumes of Glaciers on Cascade Volcanoes: unpublished Ph.D. Thesis, University of Washington, Seattle, 151 p.*

McKee, Bates, 1972, *Cascadia: The Geologic Evolution of the Pacific Northwest,* McGraw-Hill Book Co., New York, 394 p.

Paterson, W.S.B., 1981, *The Physics of Glaciers*, Pergamon Press, New York.

Porter, S.C., 1971, Fluctuations of Late Pleistocene Alpine Glaciers in Western North America: In Turkeian, K.K., ed., *The Late Cenozoic Glacial Ages*, Yale University Press, pp. 307-329.

Post, A., Richardson, D., Tangborn, W., Rosselot, F., *Inventory of Glaciers in the North Cascades, Washington: U.S. Geological Survey Professional Paper 705-A, 26 p.*

Russell, I.C., 1897 *Glaciers of Mount Rainier: U.S. Geological Survey Annual Report 18,* Part 2, pp. 349-415.

Stagner, Howard R., 1966, *Behind the Scenery of Mount Rainier National Park*, The Mount Rainier Natural History Association, Longmire, Wash. 64 p.

Veatch, Fred M., 1969, *Analysis of a 24-Year Photographic Record of Nisqually Glacier, Mount Rainier National Park, Washington: U.S. Geological Survey Professional Paper 631, 52 p.*

V: The Mono Lake-Long Valley Region

Bailey, Roy A., 1982, Other Potential Eruption Centers in California: Long Valley-Mono Lake, Cosco, and Clear Lake Volcanic Fields, in Martin, R.C., and Davis, J.F., *Status of Volcanic Prediction and Emergency Response Capabilities in Volcanic Hazard Zones of California, Special Publication 63*, California Department of Conservation: Division of Mines and Geology, pp. 17-28.

Bailey, Roy A., Dalrymple, G.B., and Lanphere, M.A., 1976, Volcanism, Structure and Geochronology of Long Valley Caldera, Mono County, California: *Journal of Geophysical Research*, vol. 81, pp. 725-744.

Cramer, C.H., and Toppozada, T.R., 1980, *A Seismological Study of the May 1980 and Earlier Earthquake Activity near Mammoth Lakes, California*: California Division of Mines and Geology Special Report 150, pp. 91-136.

Coash, John R., 1981, *Guidebook to the Glacial Geology, Volcanoes, and Earthquakes of Mammoth, Field Guide, NAGT-FWS Fall Field Conference, Mammoth, Sept. 27-28, 1980.*

Denlinger, R.P., Riley, F.S., Boling, J.K., and Carpenter, M.C., 1985, Deformation of Long Valley Caldera between August 1982 and August 1983: *Journal of Geophysical Research*, vol. 90, no. B13, pp. 11,199-11,209.

Fink, J.H., 1985, Geometry of Silicic Dikes beneath the Inyo Domes, California: *Journal of Geophysical Research*, vol. 90, no. B13, pp. 11,127-11,134.

Hill, David P., 1976, Structure of Long Valley Caldera, California, from a Seismic Refraction Experiment: *Journal of Geophysical Research*, vol. 81, pp. 745-753.

Hill, David P., Bailey Roy A., and Ryall, Alan S., 1985, Active Tectonic and Magmatic Processes Beneath Long Valley Caldera, Eastern California: An Overview: *Journal of Geophysical Research*, vol. 90, no. B13, pp. 11,111-11,120.

Hill David P., Kissling, E., Luetgert, J.H., and Kradolfer, U., 1985, Constraints on the Upper Crustal Structure of Long Valley-Mono Craters Volcanic Complex, Eastern California, from Seismic Redaction Measurements: *Journal of Geophysical Research*, vol. 90, no. B13, pp. 11,135-11,150.

Hill, David P., Wallace, R.E., and Cockerham, R.S., 1985, Review of Evidence on the Potential for Major Earthquakes and Volcanism in the Long Valley-Mono Craters-White Mountains Regions of Eastern California: *Earthquake Predict. Res.*, vol. 3, Terra Scientific Publishing Company, Tokyo, Japan, pp. 571-594.

Izett, G.A., Obradovich, J.D., and Mehnert, H.H., 1982, *The Bishop Ash Bed and Some Older Closely Related Ash Beds in California, Nevada, and Utah: U.S. Geological Survey Open-File Report 82-584*, 60 p.

Julian, Bruce R., and Sipkin, S.A., 1985, Earthquake Processes in the Long Valley Caldera Area, California: *Journal of Geophysical Research*, vol. 90, no. B13, pp. 11,155-11,170.

Kilbourne, Richard T., 1982, Chronology of Eruptions in California during the Last 2,000 Years, in Martin, R.C., and Davis J.F., *Status of Volcanic Prediction and Emergency Response Capabilities in Volcanic Hazard Zones of California*, California Department of Conservation: Division of Mines and Geology, Special Publication 63, pp. 29-40.

Kilbourne, Richard T., and Anderson, Catherine L. 1981, Volcanic History and "Active" Volcanism in California, *California Geology*, vol. 34, no. 8, pp. 159-168.

Lide, Chester S., and Ryall, A.S., 1985, Aftershock Distribution Related to the Controversary Regarding Mechanisms of the May 1980 Mammoth Lakes, California, Earthquakes: *Journal of Geophysical Research*, vol. 90, no. B13, pp. 11,151-11,154.

Metz, J.M., and Mahood, G.A., 1985, Precursors to the Bishop Tuff Eruption: Glass Mountain, Long Valley, California: *Journal of Geophysical Research*, vol. 90, no. B13, pp. 11,121-11,126.

Miller, C. Dan, 1985, Holocene Eruptions of the Inyo Volcanic Chain, California—Implications for Possible Eruptions Long Valley Caldera: *Geology*, vol. 13, pp. 14-17.

Miller, C.D., Mullineaux, D.R., Crandell, D.R., and Bailey, R.A., 1982, Potential Hazards from Future Volcanic Eruptions in the Long Valley-Mono Lake Area. East-Central California and Southwest Nevada—A Preliminary Assessment: Geological Survey Circular 877.

Rinehart, C. Dean, and Huber, N. King, 1965, The Inyo Crater Lakes—A Blast in the Past: *Mineral Information Service, California Division of Mines and Geology*, vol. 18, no. 9, pp. 169-180.

Rinehart, C. Dean, and Smith, Ward C., 1982, *Earthquakes and Young Volcanoes along the Eastern Sierra Nevada*, William Kaufmann, Inc., Los Altos, California.

Ryall, F., and Ryall, A., 1982, Propagation Effects and Seismicity Associated with Magma in Long Valley Caldera, Eastern California: *Earthquake Notes*, vol. 53, no. 1, pp. 46-47.

Sieh, Kerry and Bursik, Marcus, 1986, *Most Recent Eruption of the Mono Craters, Eastern Central California:* Journal of Geophysical Research (in press).

Wood, Spencer, 1977, Distribution, Correlation and Radiocarbon Dating of Late Holocene Tephra, Mono and Inyo Craters, Eastern California: *Geological Society of American Bulletin,* vol. 88, pp. 89-95.

VI. Lassen Peak

Anonymous, 1850, Volcanic Eruptions, *Daily Pacific News,* San Francisco, August 21, 1850, p. 1, col. 2.

Anonymous, 1859, An Active Volcano in Calif., *Sacramento Daily Union*, March 18, 1859, p. 3.

Anderson, C.A., 1933, *Tuscan Formation of Northern California, University of California Publications: Bulletin of the Dept. of Geological Science*, vol. 23, no. 7, pp. 215-276.

Anderson, C.A., 1935, Alteration of the Lavas Surrounding the Hot Springs in Lassen Volcanic National Park: *American Mineralogist, vol. 20, no. 4, pp. 240-252.*

Anderson, C.A., 1940, The Hat Creek Lava Flow: *American Journal of Science*, vol. 238, pp. 477-492.

Bowen, P.A., 1978, Pleistocene Volcanism near Lassen Peak, Calif.: *Geol. Soc. Am., Abstr. Programs*, vol. 10, no. 3, p. 97.

Carlson, Kenneth and Wheeler, G., 1980, Renaming "Quartz Basalts" of Cinder Cone, Lassen Volcanic National Park: *California Geology*, vol. 33, no. 5, p. 101.

Chesterman, Charles W., 1971, Volcanism in California: *California Geology*, California Division of Mines and Geology, vol. 24, no. 8, pp. 139-147.

Clynne, Michael A., 1984, *Stratigraphy and Major Element Geochemistry of the Lassen Volcanic Center, California: U.S. Geological Survey Open-File Report 84-224*, 168 p.

Coombs, H.A., and Howard, A.D., 1960, United States of America: *In Catalog of Active Volcanoes of the World, Part 9*. International Volcanological Association, Naples, Italy.

Crandell, D.R., 1972, Glaciation Near Lassen Peak, Northern California: *U.S. Geological Survey Professional Paper* 800-C, pp. C179-C188.

Crandell, D.R., Mullineaux, D.R., Sigafoos, R.S., and Rubin, Meyer, 1974, Chaos Crags Eruptions and Rockfall-Avalanches, Lassen Volcanic National Park, California: *Journal of Research U.S. Geological Survey*, vol. 2, no. 1, pp. 49-59.

Crandell, D.R., Mullineaux, D.R., and Bath, G.D., 1970, Late Glacial and Postglacial Dacitic Volcanism Near Lassen Peak, California: *Geological Society of America Abstracts with Programs 1970, vol. 2, no. 2, Cordilleran Section, pp. 83-84.*

Day, A.L. and Allen, E.T., 1925, *The Volcanic Activity and Hot Springs of Lassen Peak: Carnegie Inst. Washington Pub. no. 360*, 190 p.

Decker, R.W., and Harlow, David, 1970, *Microearthquakes at Cascade Volcanoes: American Geophysical Union Transaction, vol. 51, p. 351.*

Diller, J.S., 1887, The Latest Volcanic Eruption in Northern California: *American Journal of Science*, 3rd Ser., vol. 33, pp. 45-50.

Diller, J.S., 1889, *Geology of the Lassen Peak District*: U.S. Geological Survey, 1886-87, 8th Annual Report, pp. 395-432.

Diller, J.S., 1914, The Eruptions of Lassen Peak, California: *Mazama*, vol. 4, no. 3, pp. 54-59.

Diller, J.S., and Hillebrand, W.F., 1896, The Lassen Peak Region: *U.S. Geological Survey Bulletin 148*, pp. 191-193.

Eppler, D.B., 1984, *Characteristics of Volcanic Blasts, Mudflows and Rock-fall Avalanches in Lassen Volcanic National Park, California* (Ph.D. Thesis), Arizona State University, Tempe, 261 p.

Eppler, D.B., 1987a, the May 1915 Eruptions of Lassen Peak, California, I: Characteristics of Events Occurring on May 19: *Bulletin of Volcanology* (in press).

Eppler, D.B., 1987b, The May 1915 Eruptions of Lassen Peak, II: May 22 Volcanic Blast Effects, Sedimentology and Stratigraphy of Blast and Lahar Deposits, and Characteristics of the Blast Cloud: *Journal of Volcanology and Geothermal Research,* vol. 31, pp. 65-85.

Eppler, D.B., Fink, J., and Fletcher, R., 1987, Rheologic Properties and Kinematics of Emplacement of the Chaos Jumbles Rockfall Avalanche, Lassen Volcanic National Park, California: *Journal of Geophysical Research,* vol. 92, no. B5, pp. 3623-3633.

Finch, R.H., and Anderson, C.A., 1930, The Quartz-Basalt Eruptions of Cinder Cone, Lassen Volcanic National Park, California: *University of California Publication Bulletin of the Dept. of Geological Science*, vol. 19, pp. 245-273.

Finch, R.H., 1937, A Tree Ring Calendar for Dating Volcanic Events, Cinder Cone, Lassen National Park, California: *American Journal of Science*, vol. 33, pp. 140-146.

Harkness, H.W., 1875, A Recent Volcano in Plumas County: *California Academy of Sciences, Proceedings*, vol. 5, 1873-74, pp. 408-412.

Heath, J.P., 1960, Repeated Avalanches at Chaos Jumbles, Lassen Volcanic National Park: *American Journal of Science*, vol. 258, pp. 744-751.

Heath, J.P., 1967, Primary Conifer Succession, Lassen Volcanic National Park: *Ecology*, vol. 48, pp. 270-275.

Heiken, G., and Eichelberger, J.C., 1980, Eruptions at Chaos Crags, Lassen Volcanic National Park, Calif. in Gordon A. Macdonald memorial volume (McBirney, A.R. Editor), *J. Volcanol. Geotherm. Res.*, vol. 7, no. 3-4, p. 443-481.

Hill, Mary, 1970, "Mount Lassen Is In Eruption and There Is No Mistake About That:" *Mineral Information Service*, California Division of Mines and Geology, vol. 23, no. 11, pp. 211-224.

Hinds, N.E.A., 1952, Cascade Range: *In Evolution of the California Landscape. California Division of Mines and Geology, Bulletin 158, pp. 119-142.*

James, David E., 1966, Geology and Rock Magnetism of Cinder Cone Lava Flows, Lassen Volcanic National Park, California: *Geological Society of America Bulletin*, vol. 77, pp. 303-312.

Klein, Fred W., 1979, Earthquakes in Lassen Volcanic National Park California, *Seismol. Soc. Am. Bull.*, Vol. 69, no. 3, pp. 867-875.

Loomis, Benjamin Franklin, 1926, *Pictorial History of the Lassen Volcano* (rev. ed., by Schultz, 1948), Loomis Museum Association, Mineral, Calif., 100 p.

Macdonald, G.A., 1963, Geology of the Manzanita Lake Quadrangle, California: *U.S. Geological Survey Quadrangle Map GQ-248.*

Macdonald, G.A., and Katsura, Takashi, 1965, Eruption of Lassen Peak, California, in 1915: Example of Mixed Magmas: *Geological Society of America Bulletin*, vol. 76, pp. 475-482.

Macdonald, G.A., 1966, Geology of the Cascade Range and Modoc Plateau: *In Geology of Northern California, California Division of Mines and Geology Bulletin 190*, pp. 65-96.

Moxham, R.M., 1970, Thermal Features at Volcanoes in the Cascade Range, as Observed by Aerial Infrared Surveys: *Bulletin Volcanologique*, vol. 34, pp. 77-106.

Oakshott, Gordon B., 1971, California's Changing Landscapes: *A Guide to Geology of the State*, McGraw-Hill Book Company.

Russell, Israel C., 1910, *Volcanoes of North America*, pp. 28-33.

Schulz, P.E., 1952, *Geology of Lassen's Landscape*, Ann Arbor, Mich., Edwards Brothers, Inc., p. 98

Strong, Douglas, 1973, *"These Happy Grounds": A History of the Lassen Region*, National Park Service and Loomis Museum Association, 101 p.

Unger, J.D., and Coakley, J.M., 1971, Microearthquakes Near Lassen Peak, California: *U.S. Geological Survey Professional Paper 750-C*, pp. C156-C157.

Whitney, J.D., 1865, Geological Survey of California; Geology, vol. 1, Report of progress and synopsis of the field work from 1860 to 1864. *California Geol. Survey (Pubs.)*, 498 p.

Williams, Howel, 1928, A Recent Volcanic Eruption Near Lassen Peak, California: *California University Dept. of Geol. Sci. Bulletin*, vol. 17, no. 7, pp. 241-263.

Williams, Howel, 1929, The Volcanic Domes of Lassen Peak and Vicinity, California: *American Journal of Science*, vol. 18, pp. 313-330.

Williams, Howel, 1931, The Dacites of Lassen Peak and Vicinity, California: *American Journal of Science,* vol. 22, pp. 385-403.

Williams, Howel, 1932a, *Geology of the Lassen Volcanic National Park, California, University of California Publications in Geological Science*, vol. 21, no. 8, pp. 195-385.

Williams, Howel, 1932b, The History and Character of Volcanic Domes: *University of California Dept. of Geological Science Bulletin,* vol. 21, pp. 51-146.

Willendrup, A.W., 1984, *The Lassen Peak Eruptions and Their Lingering Legacy*: Occasional Publication No. 8 Association for Northern California Records and Research.

VII: Mt. Shasta

Brantley, S., and Glicken, H., 1986, Volcanic Debris Avalanches: *Earthquakes and Volcanoes,* vol. 18, no. 5, pp. 195-206.

Chesterman, Charles W., 1971, Volcanism in California: *California Geology*, California Division of Mines and Geology, vol. 24, no. 8, pp. 139-424.

Chesterman, Charles W., 1982, Potentially Active Volcanic Zones in California, in, Martin, R.J. and Davis, J.F., eds., *Status of Volcanic Prediction and Emergency Response Capabilities in Volcanic Hazard Zones of California*, California Division of Mines and Geology Special Publication 63, pp. 9-16.

Christiansen, Robert L., 1982, Volcanic Hazard Potential in the California Cascades, in Martin, R.J. and Davis, J.F., eds. *Status of Volcanic Prediction and Emergency Response Capabilities in Volcanic Hazard Zones of California*, California Division of Mines and Geology Special Publication 63, pp. 43-59.

Christiansen, Robert L., and Miller, C. Dan, 1976, Volcanic Evolution of Mt. Shasta, California, *Abstracts with Programs, Geological Society of America, Cordilleran Section Meetings*, vol. 8, no. 3, pp. 360-361.

Condie, K.C., and Swenson, D.H., 1973, Compositional Variation in Three Cascade Stratovolcanoes: Jefferson, Rainier, and Shasta: *Bulletin Volcanologique*, vol. 37, no. 2, pp. 205-230.

Coombs, H.A., and Howard A.D., 1960, United States of America: *In Catalogue of the Active Volcanoes of the World, Part 9*, International Volcanological Association, Naples, Italy.

Crandell, D.R., 1973, Hot Pyroclastic-Flow Deposits of Probable Holocene Age West of Mount Shasta Volcano, California: *Geological Society of America, Abstracts with Programs*, 69th Annual Meeting, Cordilleran Section, vol. 5, no. 1, p. 28.

Crandell, D.R., Miller, C.D., Glicken, H.X., Christiansen, R.L., and Newhall, C.G., 1984, Catastrophic Debris Avalanche from Ancestral Mount Shasta Volcano, California: *Geology*, vol. 12, pp. 143-146.

Crandell, D.R., Mullineaux, D.R., and Miller, C.D., Volcanic Hazard Studies in the Cascade Range of the Western United States, in Sheets, P.D., and Graysen, D.K., eds., *Volcanic Activity and Human Ecology:* Academic Press, New York, pp. 195-219.

Crandell, D.R., Waldron, H.H., 1969, Volcanic Hazards in the Cascade Range: In Olson, R.A., and Wallace, M.M., eds., Geologic Hazards and Public Problems, Office of Emergency Preparedness, pp. 5-18.

Diller, J.S, 1895, Mount Shasta: A Typical Volcano: *National Geographic Society Monograph*, vol. 1, no. 8, pp. 237-268.

Diller, J.S., 1915, Mount Shasta—Some of Its Geologic Aspects: *Mazama*, vol. 4, pp. 11-16.

Eichelberger, J.C., and Gooler, R., 1975, Banded Andesitic Bombs of Mt. Shasta [Abst.]: *Geol. Soc. Am. Abstr. Programs*, vol. 7, no. 7, p. 1065-1066.

Eichorn, Arthur F., 1957, *The Mount Shasta Story*, rev. ed., the Mount Shasta Herald, Mount Shasta, Calif.

Finch, R.H., 1930, Activity of a California Volcano in 1786: *The Volcano Letter*, no. 308, p. 3

Hardesty, William P., 1915, Physical Geography of Mount Shasta Region: *Mazama*, vol. 4, no. 4, pp. 17-18.

Hill, M., and Eganhoff, E.L., 1976, A California Jokulhaup, *California Geology*, vol. 29, no. 7, pp. 154-158.

Hinds, Norman E.A., 1952, *Evolution of the California Landscape: California Division of Mines Bulletin 158*, pp. 119-130.

Kilbourne, R.T., 1982, Chronology of Eruptions in California during the Last 2000 Years, in Martin, R.J. and Davis J.F., eds., *Status of Volcanic Prediction and Emergency Response Capabilities in Volcanic Hazard Zones of California*: California Division of Mines and Geology Special Publication 63, pp. 29-40.

Kilbourne, Richard T., and Anderson, Catherine L., 1981, Volcanic history and "Active" Volcanism in California, *California Geology*, vol. 34, no. 8, pp. 159-168.

LaFehr, T.R., Gravity, Isostasy, and Crustal Structure in the Southern Cascade Range: *Journal of Geophysical Research*, vol. 70, no. 22, pp. 5581-5597.

Macdonald, Gordon A., 1966, Geology of Cascade Range and Modoc Plateau: In Bailey, Edgar H., ed., *Geology of Northern California, California Division of Mines and Geology Bulletin 190*, pp. 65-95.

Miller, C. Dan, 1978, Holocene Pyroclastic-Flow Deposits from Shastina and Black Butte, West of Mount Shasta, California: *Journal of Research*, U.S Geological Survey, vol. 6, no. 5, pp. 611-624.

Miller, C. Dan, 1980, Potential Hazards from Future Eruptions in the Vicinity of Mounta Shasta Volcano, Northern California: *Geological Survey Bull. 1503*, 43 p.

Miller, C. Dan, and Crandell, Dwight R., 1975, Postglacial Pyroclastic-Flow Deposits and Lahars from Black Butte and Shastina, West of Mt. Shasta, California, *Abstracts with Programs, Geological Society of America, Cordilleran Section*, pp. 347-348.

Moxham, R.M., 1970, Thermal Features at Volcanoes in the Cascade Range, as Observed by Aerial Infrared Surveys: *Bulletin Volcanologique*, vol. 34, no. 1, pp. 79-106.

Muir, John, 1918, *Steep Trails*, Houghton Mifflin, Boston and New York.

Oakshott, Gordon B., 1971, *California's Changing Landscape: A Guide to the Geology of the State*, McGraw-Hill Book Co.

Russell, Israel C., 1897, *Volcanoes of North America*, Macmillan, New York, pp. 225-228.

Shelton, John, 1966, *Geology Illustrated*, W.H. Freeman and Co., San Francisco, pp. 64-65.

Williams, Howel, 1932, Mount Shasta, A Cascade Volcano: *Journal of Geology*, vol. 40, no. 5, pp. 417-429.

Williams, Howel, 1934, Mount Shasta, California, *Zeitschrift Vulkanologie*, vol. 15, no. 4, pp. 225-253.

Williams, Howel, Geology of the Macdoel Quandrangle: *California Division of Mines Bulletin 151*, pp. 7-60.

Medicine Lake Volcano

Anderson, C.A., 1933, Volcanic History of Glass Mountain, Northern California, *American Journal of Science*, vol. 26, pp. 485-506.

Anderson, C.A., 1941, Volcanoes of the Medicine Lake Highland, California: *University of California Publications, Bulletin of the Dept. of Geological Sciences*, vol. 25, no. 7, pp. 347-422.

Donnelly-Nolan, J.M., 1987, Medicine Lake Volcano and Lava Beds National Monument, California: *Geological Society of America Centennial Field Guide, Cordilleran Section, 1987*, pp. 289-294.

Donnelly-Nolan, J.M., and Champion, D.E., 1987, Geologic Map of Lava Beds National Monument, Northern California: *U.S. Geological Survey Map I-1804*.

Donnelly-Nolan, J., Ciancanelli, E.V., Eichelberger, J.C., Fink, J.H., and Heiken, G., 1981, Road Log for Field Trip to Medicine Lake Highland, in Johnston D., and Donnelly-Nolan, J., *Guides to Some Volcanic Terranes in Washington, Idaho, Oregon, and Northern California: Geological Survey Circular 838*, pp. 141-149.

Donnelly-Nolan, J.M., and Nolan, K.M., 1986, Catastrophic Flooding and Eruption of Ash Flow Tuff at Medicine Lake Volcano, California: *Geology*, vol. 14, pp. 875-878.

Eichelberger, J.C., 1975, Origin of Andesite and Dacite: Evidence of Mixing at Glass Mountain in California and at Other Circum-Pacific Volcanoes: *Geologic Society of America Bulletin,* vol. 86, pp. 1381-1391.

Eichelberger, J.C., 1981, Mechanism of Magma Mixing at Glass Mountain, Medicine Lake Highland Volcano, California, in Johnston, D., and Donnelly-Nolan J., *Guides to Some Volcanic Terranes in Washington, Idaho, Oregon, and Northern California: Geological Survey Circular 838,* pp. 183-189.

Heiken, G., 1978, Plinian-Type Eruptions in the Medicine Lake Highland, California, and the Nature of the Underlying Magma: *Journal of Volcanology and Geothermal Research,* vol. 4, pp. 375-402.

Heiken, G., 1981, Holocene Plinian Tephra Deposits of the Medicine Lake Highland, California, in Johnston, D., and Donnelly-Nolan, J., *Guides to Some Volcanic Terranes in Washington, Idaho, Oregon, and Northern California: Geological Survey Circular 838,* pp. 177-181.

Mertzman, S.A., Jr., 1977, The Petrology and Geochemistry of the Medicine Lake Volcano, California: *Contributions to Mineralogy and Petrology,* vol. 62, pp. 221-247.

Mertzman, S.A., 1981, Pre-Holocene Silicic Volcanism on the Northern and Western Margins of the Medicine Lake Highland, California, in Johnston, D., and Donnelly-Nolan, J., *Guides to Some Volcanic Terranes in Washington, Idaho, Oregon, and Northern California: Geological Survey Circular 838,* pp. 163-169.

Waters, A.C., 1981, Captain Jack's Stronghold (The Geologic Events That Created a Natural Fortress), in Johnston, D., Donnelly-Nolan, J., eds., *Guides to Some Volcanic Terranes in Washington, Idaho, Oregon, and Northern California: Geological Survey Circular 838,* pp. 151-161.

VIII. Mt. McLoughlin

Crandell, D.R., 1965, Glacial History of Western Washington and Oregon: In Write, H.E., and Frey, D.G. eds., *Quaternary of the United States*, Princeton University Press, Princeton, N.J., pp. 341-354.

Emmons, Arthur B., 1886, Notes on Mount Pitt [McLoughlin]: *California Academy of Science Bulletin*, vol. 1, pp. 229-234.

Lynch, Bill, 1973, Monitoring Our Snoring Mountains: *Eugene* (Oregon) *Register-Guard Emerald Empire*, March 18, 1973, pp. 3-4.

Mcdonald, Gordon A., 1972, *Volcanoes*, Prentice-Hall, Englewood Cliffs, N.J., 510 p.

Maynard, Leroy G., 1974, *Geology of Mt. McLoughlin*, unpublished Master of Science thesis, University of Oregon, Eugene, Oregon.

Montague, Malcolm J., 1973, The Little Glacier That Couldn't: *Mazama*, vol. 40, no. 13, pp. 73-75.

Phillips, Kenneth, 1939, Farewell to Sholes Glacier: *Mazama*, vol. 21, no. 12, pp. 37-40.

Russell, Israel C., 1897, *Volcanoes of North America*, Macmillan, New York, p. 236.

Williams, Howel, 1942, *The Geology of Crater Lake National Park, Oregon, with a Reconnaissance of the Cascade Range Southward to Mount Shasta: Carnegie Institution Publication 540*, pp. 18-20.

IX. Crater Lake

Allen, J.E., 1936, Structures in the Dacite Flows at Crater Lake, Oregon: *Journal of Geology*, vol. 44, pp. 737-744.

Allen, John E., 1979, Speculations on Oregon Calderas, Known and Unknown: *Oreg. Geol.*, vol. 41, no. 2, p. 31-32.

Allison, I.S., 1966, *Fossil Lake Oregon: Its Geology and Fossil Fauna*, Oregon State University, Corvallis, 48 p.

Atwood, W.W., Jr., 1935, The Glacial History of an Extinct Volcano, Crater Lake National Park: *Journal of Geology*, vol. 43, pp. 142-168.

Bacon, Charles, 1983, Eruptive History of Mount Mazama and Crater Lake Caldera, Cascade Range, U.S.A.: *Journal of Volcanology and Geothermal Research*, vol. 18, pp. 57-115.

Bacon, Charles, 1985, Implications of Silicic Vent Patterns for the Presence of Large Crustal Magma Chambers: *Journal of Geophysical Research,* vol. 90, pp. 11,243-11,252.

Bacon, Charles, 1986, Magmatic Inclusions in Silicic and Intermediate Volcanic Rocks: *Journal of Geophysical Research,* vol. 91, pp. 6091-6112.

Bacon, Charles, 1987, Mount Mazama and Crater Lake Caldera, Oregon: *Geological Society of America Centennial Field Guide – Cordilleran Section, 1987,* pp. 301-306.

Bacon, Charles, and Druitt, T.H., 1988, Compositional Evolution of the Zoned Calcalkaline Magma Chamber of Mount Mazama, Crater Lake, Oregon: *Contributions to Mineralogy and Petrology,* vol. 98, pp. 224-256.

Blackwell, D.D., Bowen, R.G., Hull, D.A., Riccio, J. and Steele, J.L., 1982, Heat Flow, Arc Volcanisms and Subductors in Northern Oregon: *Journal of Geophysical Research*, vol. 87, pp. 8735-8754.

Blakely, R.J., Jackens, R.C., Simpson, R.W., and Couch, R.W., 1985, Tectonic Setting of the Southern Cascade Range As Interpreted from Its Magnetic and Gravity Fields: *Geological Society of American Bulletin*, vol. 96, no. 1, pp. 43-48.

Blank, H. Richard, 1968, Aeromagnetic and Gravity Surveys of Crater Lake Region, Oregon: In Dole, H.M., ed., *Andesite Conference Guidebook: Oregon Dept. of Geology and Mineral Industries Bulletin 62*, pp. 42-52.

Blinman, Eric, Mehringer, Peter J., Jr., and Sheppard, John C., 1979, Pollen Influx and the Deposition of Mazama and Glacier Peak Tephra, in Sheets, Payson D., and Grayson, Donald K., *volcanic Activity and Human Ecology*, New York: Academic Press, pp. 393-425.

Clark, Ella E., 1966, *Indian Legends of the Pacific Northwest*, University of California Press, Berkeley.

Crandell, D.R., and Mullineaux, D.R., 1967, *Volcanic Hazards at Mount Rainier, Washington: U.S. Geological Survey Bulletin 1238.*

Crandell, D.R., Mullineaux, D.R., Miller, C.D., and Rubin, M., 1962, Pyroclastic Deposits of Recent Age at Mount Rainier, Washington: *U.S. Geological Survey Professional Paper 450-D*, pp. 64-68.

Crandell, D.R., and Waldron, H.H., 1969, Volcanic Hazards in the Cascade Range: In Olson, R.A., and Wallace, M.M., eds., *Geologic Hazards and Public Problems*, Office of Emergency Preparedness, pp. 5-18.

Cranson, K.R., 1983, *Crater Lake–Gem of the Cascades: The Geological Story of Crater Lake National Park*, rev. ed., K.R. Cranson Press, Lansing, Michigan, 119 p.

David, P.P., 1970, Discovery of Mazama Ash in Saskatchewan, Canada: *Canadian Journal of Earth Science*, vol. 7, pp. 1579-1583.

Davis, J.O., 1978, Quaternary Tephra-chronology of the Lake Lahontan Area, Nevada and California: *Nevada Archaeological Survey*, Res. Paper, no. 7, 137 pp.

Diller, J.S., and Patton, H.B., 1902, *The Geology and Petrography of Crater Lake National Park: U.S. Geological Survey Professional Paper 3, Part I: Geology.*

Druit, Timothy H., and Bacon, Charles R., 1986, Lithic Breccia and Ignimbrite Erupted during the Collapse of Crater Lake Caldera, Oregon: *Journal of Volcanology and Geothermal Research*, in press.

Dutton, Major C.D., 1886, Crater Lake, A proposed National Park: *Science*, vol. 7, pp. 179-182.

Eichelberger, J.C., and Crowe, B., 1978, Climactic Eruptions of Crater Lake: Evidence for Early Stage of Mixing Within a Pluton [abst.]: *Geol. Soc. Am. of Abstr. Programs*, vol. 10, no. 3, p. 104.

Fryxell, R., 1965, Mazama and Glacier Peak Volcanic Ash Layers: Relative Ages: *Science*. vol. 147, pp. 1288-1290.

Gorman, N.W., 1897, The Discovery and Early History of Crater Lake: *Mazama*, vol. 1, no. 2, pp. 150-161.

Grayson, Donald K., 1979, Mount Mazama, Climatic Change, and Fort Rock Basin, in Sheets, Payson, D., and Grayson, Donald K., *Volcanic Activity and Human Ecology*, Academic Press, pp. 427-457.

Hansen, H.P., 1942, Post-Mazama Forest Succession on the East Slope of the Central Cascades of Oregon: *American Midland Naturalist*, vol. 27, pp. 523-534.

Harris, Stephen L., 1985, Western Geology Illustrated: American Indian Legends II: The Battle of Llao and Skell: *American West*, vol. 22, no. 6, pp. 10-11.

Haward, M.E., and Youngberg, C.T., 1969, Soils from Mazama Ash in Oregon: Identification, Distribution and Properties: In Pawluk, S., ed., *Pedology and Quaternary Research*, University of Alberta, Edmonton, Alberta, pp. 163-178.

Hildreth, Wes, 1981, Gradients in Silicic Magma Chambers: Implications for Lithospheric Magmatism: *Journal of Geophysical Research*, vol. 86, pp. 10,153-10,192.

Horberg, Leland, and Robie, R.A., 1955, Postglacial volcanic Ash in the Rocky Mountain Piedmont, Montana and Alberta: *Geological Society of America Bulletin*, vol. 66, pp. 949-955.

Kirk, Ruth, 1975, *Exploring Crater Lake Country*, University of Washington Press, Seattle, 74 p.

Kettleman, L.R., 1973, Mineralogy, Correlation and Grain-Size Distribution of Mazama Tephra and Other Postglacial Pyroclastic Layers, Pacific Northwest: *Geological Society of America Bulletin, vol. 84, p. 2957-2980.*

Leeman, W.P., 1979, Petrology and Evolution of Mount Mazama: *A Geochemical Study* [abst.]: *Geol. Soc. Am., Abstr. Programs*, vol. 11, no. 3, p. 88

Lidstrom, J.W., 1971, *A New Model for the Formation of Crater Lake Caldera, Oregon* Ph.D. Thesis, Oregon State University, Corvallis.

Macdonald, Gordon A., 1972, *Volcanoes*, Prentice-Hall, Englewood Cliffs, N.J., 510 p.

Mack, Richard N., Okazakj, Rose, and Valastro, Sam, 1979, Bracketing Dates for Two Ash Falls from Mount Mazama: *Nature*, vol. 279, no. 5711, p. 228-229.

Mason, Ralph S., 1961, Did Mt. Mazama Collapse in June or January? *Mazama*, vol. 43, no. 13, p. 31.

McBirney, Alexander R., 1968a, Compositional Variations of the Climactic Eruption of Mount Mazama: In Dole, H.M., ed., *Andesite Conference Guidebook: Oregon Dept. of Geology and Mineral Industries Bulletin 62*, pp. 53-56.

McBirney, Alexander, R., 1968b, Petrochemistry of Cascade Andesite Volcanoes: In Doles, H.M., ed., *Andesite Conference Guidebook: Oregon Dept. of Geology and Mineral Industries Bulletin 62*, pp. 101-107.

Mehringer, P.J. Jr., Blinman, Ed, and Peterson, K.L., 1977, Pollen Influx and Volcanic Ash: *Science*, vol. 198, 257-261.

Merriam, J.C., 1933, Crater Lake: A Study in Appreciation of Nature: *American Magazine of Art,* vol. 26, pp. 357-361.

Moore, B.N., 1934, Deposits of Possible *Nuee Ardente* Origin in the Crater Lake Region, Oregon: *Journal of Geology*, vol. 42, pp. 358-385.

Nasmith, H., Mathews, W.H., and Rouse, G.E., 1967, Bridge River Ash and Some Recent Ash Beds in British Columbia: *Canadian Journal of Earth Science*, vol. 4, pp. 163-170.

Nelson, C.H., Larsen, M.C., Thor, D.R., and Wright, A.A., 1980, Sedimentary History of the Floor of Crater Lake Oregon: A Restricted Basin with Non-Channelized Turbidities, *Geological Society of America, Abstracts with Programs*, vol. 12, no. 7, p. 491.

Place, Howard and Marian, 1974, *The Story of Crater Lake National Park*, The Caxton Printers, Ltd., Caldwell, Idaho, 84 p.

Powers, H.A., and Wilcox, R.E., 1964, Volcanic Ash from Mount Mazama (Crater Lake) and from Glacier Peak: *Science*, vol. 144, no. 3624, pp. 1334-1336.

Purdom, W.B., 1963, *The Geologic History of the Diamond Lake Area, Umpqua National Forest, Douglas County*, Oregon: U.S. Dept. of Agriculture, Forest Service, and Douglas County Park Dept.

Rai, D., 1971, *Stratigraphy and Genesis of Soils from Volcanic Ash in the Blue Mountains of Eastern Oregon* Ph.D. Thesis, Oregon State University, Corvallis, 136 p.

Ritchey, J.L., 1979, *Origin of Divergent Magmas at Crater Lake, Oregon*, Ph.D. Thesis, University of Oregon, Eugene, Oregon.

Royce, C.F., Jr., 1967, Mazama Ash from the Continental Slope off Washington: *Northwest Science*, vol. 41, no. 3, pp. 103-109.

Rubin M., and Alexander, C., 1960, U.S. Geological Survey Radiocarbon Dates: *A.J.S. Radiocarbon Supplement*, 2, pp. 129-185.

Shelton, John S., 1966 *Geology Illustrated*, W.H. Freeman, San Francisco.

Smith, J.G., Sawlan, M.G., and Katcher, A.C., 1980, An Important Lower Oligocene Welded-Tuff Marker Bed in the Western Cascade Range of Southern Oregon (Abst.): *Geol. Soc. of America Abstracts*, vol. 12, no. 3, p. 153.

Smith, W.D., and Swartzlow, C.R., 1936, Mount Mazama: Explosion Versus Collapse: *Bulletin of the Geological Society of America*, vol. 47, pp. 1809-1830.

Taylor, E.M., 1967, Accidental Plutonic Ejecta at Crater Lake Oregon: *Geological Society of America 1967 Annual Meeting, Program, New Orleans*, p. 221.

Westgate, J.A., and Dreimanis, A., 1967, Volcanic Ash Layer of Recent Age at Banff National Park, Alberta, Canada: *Canadian Journal of Earth Science*, vol. 4, pp. 155-161.

Wilcox, R.E., 1965, Volcanic-ash Chronology: In *Quarternary of the United States*, Princeton University Press, Princeton, N.J., pp. 807-816.

Williams, Howel, 1941a, Calderas and Their Origin: *University of California Publications, Bulletin of the Department of Geological Science*, vol. 25, pp. 239-346.

Williams, Howel, 1941b, *Crater Lake: The Story of Its Origin*, University of California Press, 97 p.

Williams, Howel, 1942, *The Geology of Crater Lake National Park, Oregon: Carnegie Institution Publication 540*, 162 p.

Williams, Howel, 1957, *A Geologic Map of the Bend Quadrangle, Oregon, and a Reconnaissance Geologic Map of the Central Portion of the High Cascade Mountains*, Oregon Dept. of Geology and Mineral Industries in coop. with U.S. Geological Survey.

Williams, Howel, 1961, The Floor of Crater Lake, Oregon: *American Journal of Science*, vol. 259, pp. 81-83.

Williams, Howel, 1962, *Ancient Volcanoes of Oregon*, 3rd ed., Condon Lectures, University of Oregon Press, 68 p.

Williams, Howel, and Goles, Gordon, 1968, Volume of the Mazama Ash Fall and The Origin of Crater Lake Caldera: In Dole, H.M., ed., *Andesite Conference Guidebook: Oregon Dept. of Geology and Mineral Industries Bulletin 62*, pp. 37-41.

X. Mt. Thielsen

Barnes, C.G., 1978, *The Geology of the Mount Bailey Area, Oregon* (unpublished Master of Science Thesis), University of Oregon, Eugene, 123 p.

Cummins, William S., 1964, New Routes on Three Fingered Jack: *Mazama*, vol. 46, no. 13, pp. 52-53.

Davie, Ellen I., 1980, *The Geology and Petrology of Three Fingered Jack, A High Cascade Volcano in Central Oregon*, (unpublished Master of Science Thesis), University of Oregon, Eugene, 137 p.

Diller, J.S., 1884, Fulgurite from Mt. Thielsen, Oregon: *American Journal of Science*, vol. 128, pp. 252-258.

Diller, J.S., and Patton, H.B., 1902, *The Geology and Petrography of Crater Lake National Park: U.S. Geological Survey Professional Paper No. 3*.

Hall, Don Alan, 1975, *On Top of Oregon*, Golden West Press, Corvallis, Oregon, 180 p.

Lathrop, T.G., 1968, Return of the Ice Age? *Mazama*, vol. 50, no. 13, pp. 34-36.

Nafziger, R.H., 1971, Oregon's Southernmost Glacier: A Three Year Report: *Mazama*, vol. 53, no. 13, pp. 30-33.

Purdom, W.B., 1963, *The Geologic History of the Diamond Lake Area, Umpqua National Forest, Douglas Co., Oregon*, published by U.S. Dept. of Agriculture, Forest Service, and Douglas Co. Park Dept.

Scott, W.E., 1977, Quaternary Glaciation and Volcanism, Metolius River Area, Oregon: *Geological Society of America Bulletin*, vol. 88, pp. 113-124

Taylor, Edward M., 1968, Roadside Geology, Santiam and McKenzie Pass Highways, Oregon: In Dole, H.M., ed. *Andesite Conference Guidebook, Oregon Dept. of Geology and Mineral Industries Bulletin 62*, pp. 3-33.

Taylor, Edward M., 1981, Central High Cascade Roadside Geology, Bend, Sisters, McKenzie Pass, and Santiam Pass, Oregon: in Johnston, D.A., and Donnelly-Nolan, Julie, eds., *Guides to Some Volcanic Terranes in Washington, Idaho, Oregon, and Northern California: U.S. Geological Survey Circular 838*, pp. 55-83.

Williams, Howel, 1933, *Mount Thielsen, A Dissected Cascade Volcano: University of California Publication, Dept. of Geological Science Bulletin*, vol. 23, pp. 195-213.

Williams, Howel, 1957, *A Geologic Map of the Bend Quadrangle, Oregon and a Reconnaissance Geologic Map of the Central Portion of the High Cascade Mountains*, Oregon Dept. Geology and Mineral Industries, in coop. with U.S. Geological Survey.

Williams, Howel, 1962, *Ancient Volcanoes of Oregon*, 3rd edition, Condon Lectures, University of Oregon Press, p. 68.

Williams, Ira, 1921, Mount Thielsen: *Mazama*, vol. 6, pp. 19-25.

XI. Newberry Volcano

Anonymous, 1981, Newberry Well Is Hottest Geothermal Prospect Yet Reported in Oregon, *Oregon Geology*, vol. 43, no. 9, p. 126.

Chitwood, L.A., Jensen, R.A., and Groh, E.A., 1977, The Age of Lava Butte: *The Ore Bin*, vol. 39, no. 10, pp. 155-165

Brogan, Philip F., n.d., *The Lava Butte Geological Area,* U.S. Dept. of Agriculture Forest Service, Pacific Northwest Region, 12 p.

Friedman, Irving, 1971, Obsidian Hydration Dates in the Newberry Volcano Area, Oregon: *Geological Survey Research for 1971: In U.S. Geological Survey Professional Paper 750-A*, p. A117.

Friedman, Irving, 1977. Hydration Dating of Volcanism at Newberry Crater, Oregon: *Journal Research U.S. Geol. Survey*, vol. 5, no. 3, pp. 337-342.

Higgins, M.W., 1969, Air-fall Ash and Pumice Lapilli Deposits from Central Pumice Cone, Newberry Caldera, Oregon: *U.S. Geological Survey Professional Paper 650-D*, pp. D26-D32.

Higgins, M.W., 1973, Petrology of Newberry Volcano, Central Oregon: *Geological Society of America Bulletin*, vol. 84, no. 2, pp. 455-488.

Higgins, M.W., and Walters, A.C., 1967, Newberry Caldera, Oregon—A Peliminary Report: *The Ore Bin*, vol. 29, no. 3, pp. 37-60.

Higgins, M.W., and Walters, A.C., 1968, Newberry Caldera Field Trip: In Dole, H.M., ed., *Andesite Conference Guidebook. Oregon State Dept. of Geology and Mineral Industries, Bulletin 62*, pp. 59-77.

Higgins, M.W., and Walters, A.C., 1970, A Re-evaluation of Basalt-Obsidian Relations at East Lake Fissure, Newberry Caldera, Oregon: *Geological Society of America Bulletin*, vol. 81, no. 9, pp. 2835-2842.

MacLeod, N.S., 1978, Newberry Volcano, Oregon, Preliminary Results of New Field Investigations [abst.]: *Geol. Soc. Am., Abstr. Programs*, vol. 10, no. 3, p. 115.

MacLeod, Norman S., Chitwood, Lawrence A., and McKee, Edwin H., 1979, Fieldtrip Guide for Newberry Volcano, Oregon, PNAGU 1979 Meeting, Bend, Oregon.

MacLeod, Norman S., Sherrod, David R., Chitwood, Lawrence A., and McKee, Edwin H., 1981, Newberry Volcano, Oregon, in *Guides to Some Volcanic Terranes in Washington, Idaho, Oregon, and Northern California, Geological Survey Circular 838*, pp. 85-103.

Oregon Dept. of Geology and Mineral Industries, 1965, Articles on Recent Volcanism in Oregon, reprinted from *The Ore Bin, Oregon Dept. of Geology and Mineral Industries–Misc. Paper 10.*

Peterson, N.V., and Groh, E.A., eds., 1965, *State of Oregon Lunar Geological Field Conference Guide Book: Oregon Dept. of Geology and Mineral Industries Bulletin 57*, 51 p.

Peterson, N.V., and Groh, E.A., 1969, The Ages of Some Holocene Volcanic Eruptions in the Newberry Volcano Area, Oregon, *The Ore Bin*, vol. 31, no. 4, pp. 73-87.

Sherrod, P.R., and MacLeod, N.S., 1979, The Last Eruptions at Newberry Volcano, Central Oregon [abst.] *Geol. Soc. Am., Abstr. Programs*, vol. II, no. 3, p. 127.

Williams, Howel, 1935, Newberry Volcano, Central Oregon: *Geological Society of America Bulletin*, vol. 46., no. 2, pp. 253-304.

Williams, Howel, 1941, Calderas and Their Origin: *University of California Publicatins, Bulletin of the Department of Geological Sciences*, vol. 25, pp. 239-346.

Williams, Howel, 1962, *Ancient Volcanoes of Oregon*, 3rd ed. Condon Lectures, University of Oregon Press, 68 p.

XII. Three Sisters

Armstrong, R.L., Taylor, E.M., Hales, P.O., and Parker, D.J., 1975, K-Ar Dates for Volcanic Rocks, Central Cascade Range of Oregon, *Isochron/West*: no. 13, pp. 5-10

Bacon, Charles R., 1985, Implications of Silicic Vent Patterns for the Presence of Large Crustal Magma Chambers: *Journal of Geophysical Research*, vol. 90, no. B13, pp. 11,243-11,252.

Baldwin, Edward M., 1981, *Geology of Oregon*, 3rd edition, Kendall/Hunt Publishing Co., Dubuque, Iowa, 170 p.

Brogan, P.F., 1964, *East of the Cascades*, Binfords and Mort, Portland, Oregon, 304 p.

Clark, James G., 1983, *Geology and Petrology of South Sister Volcano, High Cascade Region, Oregon* Ph.D. Thesis, University of Oregon, Eugene.

Crowe, B., and Nolf, B., 1977, Composite Cone Growth Modeled after Broken Top, a Dissected High Cascade Volcano: *Geological Society of America Abstracts with Programs*, vol. 9, pp. 940-941.

Driedger, C.L., and Kennard, P.M., 1986, *Ice Volumes on Cascade Volcanoes: Mount Rainier, Mount Hood, Three Sisters, and Mount Shasta*: U.S. Geological Survey Professional Paper 1365.

Fairbanks, Harold Wellman, 1901, Notes on the Geology of the Three Sisters, Oregon: *Journal of Geology*, vol. 9, p. 73.

Hodge, Edwin T., 1925, *Mount Multnomah, Ancient Ancestor of the Three Sisters*: University of Oregon Publications, vol. 3, no. 2, 160 p.

Hopson, Ruth E., 1961, The Arctic Alpine Zone in the Three Sisters Region: *Mazama*, vol. 43, no. 13, pp. 14-27.

Hopson, Ruth E., 1960, Collier Glacier—A Photographic Record: *Mazama*, vol. 42, no. 13.

Hyslop, Robert S., 1971, South Sister's "Nordwand": *Mazama*, vol. 53, no. 13, pp. 21-23.

Miller, James P., 1853, Letter to the Rev. Joseph T. Cooper, D.D., July 4-12, 1853, in Cooper, Rev. Joseph T., *The Evangelical Repository of the Associate Presbyterian Church*, Oct., 1853, Philadelphia, Pa.

Mimura, Koji, and MacLeod, Norman S., 1982, Source Directions of Pumice and Ash Deposits near Bend, Oregon [abst.]

Newberry, John Strong, 1858, On the Parts of California and Oregon Explored: *American Journal of Science*, 2nd series, vol. 26, no. 76, pp. 123-127.

Peterson, N.V. and Groh, E.A., 1965, *Lunar Geological Field Conference Guidebook: Oregon Dept. of Geology and Mineral Industries Bulletin 57, 51 p.*

Robinson, Paul T., and Brem, G.F., 1981, Guide to Geologic Field Trip between Kimberly and Bend, Oregon with Emphasis on the John Day Formation, in *Guide to Some Volcanic Terranes in Washington, Idaho, Oregon, and Northern California: U.S Geological Survey Circular 838*, pp. 29-54.

Scott, W.E., 1983, Character and Age of Holocene Rhyodacite Eruptions at South Sister Volcano (abs.): *EOS, Trans. American Geophysical Union*, vol. 64 (45), pp. 899-900.

Scott, W.E., 1986, Holocene Rhyodacite Eruptions on the Flanks of South Sister Volcano, Oregon: *Geological Society of America Special Paper* (in press).

Scott, W.E., and Gardner, C.A., 1985, Late Pleistocene-Early Holocene Development of the Bachelor Butte Volcanic Zone, Oregon: *EOS, Trans. American Geophysical Union*, vol. 66, no. 46, p. 1141

Smith, Warren D., 1916, A Geologist's Thoughts on Returning from the Mazama Outing of 1916: *Mazama*, vol. 5, no. 1, pp. 24-28.

Stearns, James, 1912, Physiography of the Three Sisters: *Mazama*, vol. 4, no. 1, pp. 15-20.

Taylor, Edward M., 1965, Recent Volcanism between Three Fingered Jack and North Sister, Oregon Cascade Range: Oregon Dept. of Geology and Mineral Industries, *The Ore Bin*, vol. 27, no. 7, pp. 121-147.

Taylor, Edward M., 1968, Roadside Geology, Santiam and McKenzie Pass Highways, Oregon, in Dole, H.M., ed., *Andesite Conference Guidebook: Oregon Dept. of Geology and Mineral Industries Bulletin 62*, pp. 3-33.

Taylor, Edward M., 1978, *Field Geology of S.W. Broken Top Quadrangle, Oregon, Special Paper 2*, State of Oregon Dept. of Geology and Mineral Industries, 50 p.

Taylor, Edward M., 1981, Central High Cascade Roadside Geology, Bend, Sisters, McKenzie Pass, and Santiam Pass, Oregon, in *Guide to Some Volcanic Terranes in Washington, Idaho, Oregon, and Northern California: U.S. Geological Survey Circular 836*, pp. 55-83.

Williams, Howel, 1944, Volcanoes of the Three Sisters Region, Oregon Cascades: *University of California Dept. of Geological Sciences Bulletin*, vol. 27, no. 3, pp. 37-84.

Williams, Howel, 1957, *A Geological Map of the Bend Quandrangle, Oregon, and a Reconnaissance Geologic Map of the Central Portion of the High Cascade Mountains*, Oregon Dept. of Geology and Mineral Industries, in coop. with U.S. Geological Survey.

Williams, Howel, 1962, *Ancient Volcanoes of Oregon*, 3rd edition, Condon Lectures, University of Oregon Press.

Wozniak, Karl C., 1982, *Geology of the Northern Part of the South-east Three Sisters Quandrangle, Oregon*, M.S. Thesis, Oregon State University, Corvallis, Oregon.

Wozniak, K.C., and Taylor, E.M., 1981, Late Pleistocene Summit Construction and Holocene Flank Eruptions of South Sister Volcano, Oregon: *EOS, Trans. American Geophysical Union*, vol. 62, no. 6, p. 61.

XIII. Mt. Jefferson

Beg̃et, James E., 1981, Evidence of Pleistocene Explosive Eruptions of Mount Jefferson, Oregon: *EOS*, vol. 62, p. 1089.

Beg̃et, James E., 1982 Pleistocene Pyroclastic Deposits from Eruptions of Mount Jefferson, Oregon: *AMQUA*, Projects and Abstracts, p. 67.

Condie, K.C., and Swenson, D.H., 1973, Compositional Variation in Three Cascade Stratovolcanoes: Jefferson, Rainier, and Shasta: *Bulletin Volcanologique*, vol. 37, no. 2, pp. 205-230.

Green, Robert C., 1968, *Petrography and Petrology of Volcanic Rocks in the Mount Jefferson Area High Cascade Range, Oregon: U.S. Geological Survey Bulletin 1251-G.*

Hodge, Edwin T., 1925, Geology of Mount Jefferson: *Mazama*, vol. 7, no. 2 pp. 25-58.

Holden, Edward S., 1898, *A Catalogue of Earthquakes on the Pacific Coast 1769 to 1897: Smithsonian Miscellaneous Collection 1087*, page 226.

McBirney, A.R., 1978, Volcanic Evolution of the Cascade Range: *Ann, Rev. Earth Planet. Science*, vol. 6, pp. 437-456.

McBirney, A.R., Sutter, J.F., Naslund, H.R., Sutton, K.G., and White, C.M., 1974, Episodic Volcanism in the Central Oregon Cascade Range: *Geology*, vol. 2, no. 12, pp. 585-589.

Scott, W.E., 1977, Quaternary Glaciation and Volcanism, Metolius River Area, Oregon: *Geological Society of America Bulletin*, vol. 88, pp. 113-124.

Sutton, Kenneth, 1974, *Geology of Mt. Jefferson*, unpublished Masters of Science Thesis, University of Oregon, Eugene, Oregon.

Sutton, Kenneth, 1975, written communication, unpublished field reports, maps, drawings, etc., on Mt. Jefferson.

Thayer, T.P., 1937, Petrology of Later Tertiary and Quaternary Rocks of the North Central Cascade Mountains in Oregon: *Geological Society of America Bulletin*, vol. 48, no. 11, pp. 1611-1651.

Thayer, T.P., 1939, *Geology of the Salem Hills and the North Santiam River Basin, Oregon: Oregon Dept. of Geology ad Mineral Industries Bulletin 15*, 40 p.

Walker, G.W., Greene, R.C., and Pattee, E.C., 1966, *Mineral Resources of the Mount Jefferson Primitive Area, Oregon: U.S. Geological Survey Bulletin 1230-D*, 32 p.

Williams, Howel and McBirney, Alexander R., 1979, *Volcanology*, Freeman, Cooper & Company, San Francisco, 397 p.

Williams, Ira A., 1916, Some Little-Known Scenic Pleasure Places in the Cascade Range in Oregon: *Oregon Bureau of Mines and Mineral Resources of Oregon*, vol. 2, no. 1, 114 p.

XIV. Mt. Hood

Ayeres, F.D., and Creswell, A.E., 1951, The Mount Hood Fumaroles: *Mazama*, vol. 33, no. 13, pp. 33-39.

Brantley, S., Yamaguchi, D., Cameron, K., and Pringle, P., 1986, Tree-Ring Dating of Volcanic Deposits: *Earthquakes and Volcanoes*, vol. 18, no. 5, pp. 184-194.

Cameron, K.A., and Pringle, P.T., 1986, Post-glacial Lahars of the Sandy River Basin, Mount Hood, Oregon: *Northwest Science,* vol. 60, pp. 225-237.

Cameron, K.A., and Pringle, P.T., 1987, A Detailed Chronology of the Most Recent Major Eruptive Period at Mount Hood, Oregon: *Geological Society of America Bulletin,* vol. 99, pp. 845-851.

Clark, Ella E., 1966, *Indian Legends of the Pacific Northwest,* Univ. of California Press, Berkley and Los Angeles.

Courtney, W.F., 1902, Eruption of Mount Hood, *Everett Record,* May 17, 1902.

Crandell, Dwight R., 1980, *Recent Eruptive History of Mount Hood, Oregon, and Potential Hazards from Future Eruptions:* U.S. Geological Survey Bulletin 1492, 81 p.

Crandell, D.R., and Rubin M., 1977, Late-glacial and Postglacial Eruptions at Mt. Hood, Oregon [Abst.], *Geol. Soci. Am., Abstr. Programs,* vol. 9, no. 4, p. 406.

Crandell and Wise, written comm., 1973.

Driedger, C.L., and Kennard, P.M., 1986, Ice Volumes on Cascade Volcanoes: Mount Rainier, Mount Hood, Three Sisters, and Mount Shasta: *U.S. Geological Survey Professional Paper 1365.*

Folsom, M.M., 1970, Volcanic Eruptions: The Pioneers' Attitude on the Pacific Coast from 1800 to 1875: Oregon Dept. of Geology and Mineral Industries: *The Ore Bin,* vol. 32, pp. 61-71.

Grauer, Jack, 1975, *Mount Hood: A Complete History,* John Foerste "Jack" Grauer, 300 p.

Green, Susan Molly, Weaver, Craig S., Iyer, H.M., n.d., *Seismic Studies at the Mt. Hood Volcano, Northern Cascade Range, Oregon:* U.S. Geological Survey Open-File Report 79-1691, 27 p.

Hague, Arnold, 1871, Mt. Hood, *American Journal of Science,* series 3, vol. 1, pp. 165-167.

Hammond, Paul E., 1973, If Mount Hood Erupts: Oregon Dept. of Geology and Mineral Industries: *The Ore Bin,* vol. 35, no. 6, pp. 93-102.

Hazard, J.T., 1932, *Snow Sentinels of the Pacific Northwest,* Lowman and Hanford, Seattle.

Hinds, Franklin A., Letter of January 28, 1866, handwritten manuscript 1500: *Oregon Historical Society,* Portland, Oregon.

Hodge, E.T., 1931, Stadter Buried Forest: *Mazama,* vol. 13, pp. 82-86.

Hodge, E.T., 1934, Volcanic and Seismic History of Oregon: *Proceedings of the Fifth Pacific Congress:* vol. 3, *Division of physical sciences,* A7, *Seismology and Volcanology,* pp. 2451-2460.

Hodge, E.T., 1935, Mt. Hood, Geological Society of the Oregon Country: *Geologic News Letter,* vol. 1, no. 13, pp. 3-4.

Holden, Edward S., 1898, *A Catalogue of Earthquakes on the Pacific Coast 1769 to 1897: Smithsonian Miscellaneous Publication 1087.*

Jillson, Willard Rouse, 1917, The Volcanic Activity of Mount St. Helens and Mount Hood in Historical Time: *Geographical Review,* vol. 3, pp. 482-483.

Keith, Terry, E.C., 1981-1982, Written Commun. on Sandy Glacier Volcano.

Lawrence, D.B., Mount Hood's Latest Eruption and Glacier Advances: *Mazama,* vol. 30, no. 13, p. 22-29.

Lawrence, D.C., and Lawrence E.G., 1959, Radiocarbon Dating of events on Mount Hood and Mount St. Helens: *Mazama,* vol. 41, pp. 10-18.

McKee, Bates, 1972, *Cascadia: The Geologic Evolution of the Pacific Northwest,* McGraw-Hill Book Company, pp. 212-213.

Parker, Samuel, 1846, *Journal of an Exploring Party Beyond the Rocky Mountains*, Ithaca, New York.

Steel, William Gladstone, ed., 1899, Mountain Lore: *Oregon Native Son*, vol. 1, p. 276.

Steel, William Gladstone, ed., 1906, Mount Hood in Eruption: *Steel Points*, vol. 1, no. 1, p. 23.

Steel, William Gladstone, 1907a, Mount Hood: *Steel Points*, vol. 1, no. 3, pp. 89-99.

Steel, William Gladstone, ed., 1907b, Eruption of Mout Hood: *Steel Points*, vol. 1, no. 3, pp. 135-136.

Sylvester, A.H., 1908, Evidences of Recent Volcanic Activity and the Glaciers of Mt. Hood, Oregon: *Science*, vol. 27, p. 585.

Sylvester, A.H. 1908, Is our Noblest Volcano Awakening to New Life? *National Geographic Magazine*, vol. 19, pp. 515-525.

Symons, Thomas W., 1882, The Upper Columbia of the Great Plain of the Columbia, In *U.S. Senate, 47th Congress, Executive Document no. 186*, Ch. 9, pp. 98-104.

Trimble, D.E., 1963, Geology of Portland, Oregon, and Adjacent Areas: *U.S. Geological Survey Bulletin 1119*, 119 p.

White, Craig, 1980, *Geology and Geochemistry of Mt. Hood Volcano*, Special Paper 8, State of Oregon Department of Geology and Mineral Industries, 26 pp.

Williams, John H., 1912, *The Guardians of the Columbia*, J.H. Williams, Tacoma.

Wise, W.S., 1964a, The Geologic History of Mount Hood, Oregon: *Mazama*, vol. 46, pp. 13-22.

Wise, W.S., 1964b., A Guide to the Volcanic Rocks of the Cascade Range: *Mazama*, vol. 46, pp. 23-25.

Wise, William S., 1966, The Last Eruptive Phase of the Mt. Hood Volcano: *Mazama*, vol. 58, no. 13, pp. 14-19.

Wise, William S., 1968, Geology of the Mt. Hood Volcano: In Dole, H.M., ed., *Andesite Conference Guidebook, Oregon Dept. of Geology and Mineral Industries, Bulletin 62*, pp. 81-98.

Wise, William S., 1969, Geology and Petrology of the Mt. Hood Area: A Study of High Cascade Volcanism: *Geological Society of America Bulletin*, vol. 80, pp. 969-1006.

Wollenberg, Harold A. and others, February, 1979, Geochemical Studies of Rocks, Water, and Gases at Mt. Hood, Oregon: LBL (Lawrence Berkeley Lab., Energy and Environment Division), no. 7092, 57 p.

Writers Program of the Work Projects Administration in the State of Oregon, 1940, *Mount Hood: A Guide,* Oregon State Board of Control, 132 p.

XV. Mt. Adams

Beckey, Fred, 1973, *Cascade Alpine Guide: Climbing and High Routes, Columbia River to Stevens Pass.* The Mountaineers, Seattle.

Bylam, R.M., 1921, The Mount Adams Slide of 1921, *Mazama*, vol. 6, no. 2, pp. 44-46.

Church, S.E., Swanson, D.A., Williams, D.L., Clayton, G.A., Close, T.J., and Peters, T.J., 1983, *Mineral Resource Potential Map of the Goat Rocks Wilderness and Adjacent Roadless Areas, Lewis and Yakima Counties, Washinton*: U.S. Geological Survey Miscellaneous Field Studies Map MF-1653-A.

Cline, D.R., 1976, *Reconnaissance of the Water Resources of the Upper Klickitat River Basin, Yakima Indian Reservation, Washington:* U.S. Geological Survey Open-File Report 75-518, 54 p.

Crandell, D.R., and Miller, R.D., 1974, *Quaternary Stratigraphy and Extent of Glaciation in the Mount Rainier Region, Washington*: Geological Survey Professional Paper 847, 59 p.

Easterbrook, D.J., 1975, Mount Baker Eruptions, *Geology*, vol. 3, no. 12, pp. 679-682.

Elkman, Leonard C., 1962, *Scenic Geology of the Pacific Northwest*, Binford and Mort, Portland, Oregon, pp. 138-142.

Ellingson, Jack A., 1969, Geology of the Goat Rocks Volcano, Southern Cascade Mountains, Washington: *Geological Society of America Abstracts with Programs, 1969*, Part 3, Cordilleran Section, p. 15.

Ellingson, Jack A., 1972, The Rocks and Structure of the White Pass Area, Washington: *Northwest Science,* vol. 46, no. 1, pp. 9-24

Fiske, R.S., Hopson, C.A., and Waters, A.C., 1963, *Geology of the Mount Rainier National Park, Washington*: U.S. Geological Survey Professional Paper 444, 93 p.

Fowler, C.S., 1935, *The Origin of the Sulfur Deposits of Mount Adams* M.S. Thesis, State College of Washington, Pullman, 22 p.

Fowler, C.S., 1936, The Geology of the Mount Adams Country, *Geological Newsletter*, vol. 2, no. 1, pp. 2-5.

Frank, David, 1983, *Origin, Distribution, and Rapid Removal of Hydrothermally Formed Clay at Mount Baker, Washington*: Geological Survey Professional Paper 1022-E, 31 p.

Gibbs, George, 1855, *Report of the Geology of the Central Portion of Washington Territory: U.S. Pacific Railroad Exploration:* U.S. 33rd Congress, 1st Session, House Executive Document 129, vol. 18, part 1, 1:494-512.

Hammond, Paul E., 1979, A Tectonic Model for Evolution of the Cascade Range, in Armentrout, J.M., Cole, M.R., and TerBest, H., eds., *Cenozoic Paleogeography of the Western United States*, Society of Economic Paleontologists and Mineralogists, Los Angeles, pp. 219-237.

Hildreth, Wes, 1981, Gradients in Silicic Magma Chambers: Implications for Lithospheric Magmatism, *Journal of Geophysical Research,* vol. 86, pp. 10,153-10,192.

Hildreth, Wes, and Fierstein, Judy, 1983, Mount Adams Volcano and Its 30 "Parasites": *Geological Society of America Abstracts with Programs*, vol. 15, p. 331.

Hildreth, Wes, Fierstein, Judy, and Miller, M.S., 1983, *Mineral and Geothermal Resource Potential of the Mount Adams Wilderness and Contiguous Roadless Areas, Skamania and Yakima Counties, Washington:* U.S. Geological Survey Open-File Report 83-474, 49 p.

Hodge, E.T., 1934, Mount Adams Sulphur Deposits, Unpublished Report for the Glacier Mining Company, White Salmon, Washington, 35 p.

Hopkins, Kenneth D., 1969 Later Quaternary Glaciation and Volcanism on the South Slope of Mount Adams, Washington, *Geological Society of America Abstracts with Programs, 1969,* Part 3, Cordilleran Section, p. 27.

Hopkins, Kenneth D., 1976, *Geology of the South and East Slopes of Mount Adams Volcano, Cascade Range, Washington,* Ph.D. Dissertation, University of Washington, Seattle, 144 p.

Moxham, R.M., 1970, Thermal Features at Volcanoes in the Cascade Range, as Observed by Aerial Infrared Surveys, *Bulletin Volcanologique,* vol. 34, no. 1, pp. 77-106.

Phillips, K.N., 1941, Fumaroles of Mount Saint Helens and Mount Adams, *Mazama,* vol. 23, no. 12, pp. 37-42.

Rusk, C.E., 1919, *Mount Adams,* the Yakima Commercial Club, Yakima, Washington, 26 p.

Rusk, C.E., 1978, *Tales of a Western Mountaineer* (reprint), The Mountaineers, Seattle, 309 p.

Smutek, Ray, 1972, Mount Adams—A History, *Off Belay,* no. 3, pp. 11-20.

Swanson, Donald A., and Clayton, G.A., 1983, *Generalized Geologic Map of the Goat Rocks Wilderness and Roadless Areas (6036, parts A,C, and D), Lewis and Yakima Counties, Washington*: U.S. Geological Survey Open-File Report 83-357.

Throssell, W.I., 1940, The Massif: Mt. Adams in Southwestern Washington, *Rocks and Minerals,* vol. 15, no. 1, pp. 14-19.

Williams, John H., 1912, *The Guardian of the Columbia*, John H. Williams, Tacoma, Washington, 144 p.

XVI. Mt. St. Helens

Abelson, Philip H., 1980, Monitoring Volcanism: *Science,* vol. 209, no. 454, p. 343.

Ackerman, M., Lippens, C., and Lectievallier, M., 1980, Volcanic Material from Mt. St. Helens in the Stratosphere over Europe: *Nature,* vol. 287, October 16, 1980, pp. 614-615.

Adamson, Thelma, 1934, Folk-tales of the Coast Salish, *Memoirs of the American Folk-Lore Society,* vol. XXVII, p. 268.

Allen, A.J., ed., 1848, Ten Years in Oregon: *Travels and Adventures of Doctor E. White and Lady,* Ithaca, N.Y., p. 200.

Alpha, T.R., Moore, J.G., and Jones, D.R., 1980, Sequential Physiographic Diagrams of Mount St. Helens, Spring 1980: *U.S. Geological Survey,* Open-File Rep., No. 80-792.

Anonymous, 1854, *Sketches of Mission Life among the Indians of Oregon:* New York, Carlton & Phillips, Sunday-School Union, 200 Mulberry Street.

Anonymous, 1859, *Weekly Oregonian,* Portland, Oregon, August 20, 1859.

Anonymous, 1894, Ascent of Mt. St. Helens: *American Naturalist*, vol. 28., pp. 46-48.

Anonymous, 1854, G.S.O.C. Field Trip to Mt. St. Helens: *Geological News Letter,* vol. 20., no. 9, pp. 78-81.

Anonymous, The Once and Future Mountain, 1980, *Audobon,* vol. 82, no. 4, pp. 24-41.

Anonymous, 1980, Eruption of Mt. St. Helens: Seismology: *Nature,* vol. 285, June 19, 1980, pp. 529-531.

Anonymous, 1980, In the Wake of Mt. St. Helens: *Science News,* vol. 117, no. 23, pp. 355-356.

Anonymous, 1980, Volcano Log: Mount St. Helens 1980: *Earthquake Information Bulletin*, vol. 12, no. 4, pp. 142-149.

Anonymous, 1981, Forest Service Seeks Volcanic Study Area: *San Francisco Chronicle,* San Francisco, Calif., Feb. 9, 1981.

Anonymous, 1981, Mount St. Helens: *California Geology,* January 1981, pp. 17-19.

Bacon, Charles R., 1980, Goals are set for Research in Cascades: *Geotimes,* August 1980, pp. 16-18.

Battele Memorial Institute, Pacific Northwest Laboratory, 1980, *Preliminary report on physical, chemical, and mineralogical composition and health implications of ash from the Mount St. Helens eruption of May 18, 1980,* Rep. No. PNL-SA-8674, U.S Dept. Energy, Pacific Northwest Lab., Richland, WA, U.S.

Borchardt, Glenn A., Norgren, Joel A., and Harward, Moyle E., 1973, Correlation of Ash Layers in Peat Bogs of Eastern Oregon: *Geological Society of America Bulletin,* vol. 84, pp. 3101-3108.

Brantley, Steve, and Power, John, 1985, Reports from the U.S. Geological Survey's Cascades Volcano Observatory at Vancouver, Washington: *Earthquake Information Bulletin,* vol. 17, no. 1, pp. 20-32.

Brantley, Steve, and Topinka, Lyn, 1984, Volcanic Studies at the U.S. Geological Survey's David A. Johnston Cascades Volcano Observatory, Vancouver, Washington: *Earthquake Information Bulletin,* vol. 16, no. 2, pp. 44-122.

Brewer, Henry Bridgman, 1842, Daybook from Sept. 3, 1839 to Feb. 13, 1843: In Holmes, Kenneth L., 1955, Mount Saint Helens' Recent Eruptions: *Oregon Historical Quarterly*, vol. 56, pp. 197-210.

Burnett, Peter H., 1902, The Letters of Peter H. Burnett: *Oregon Historical Quarterly,* vol. 3, pp. 423-424.

Butzer, Karl W., 1980, Volcanism in Human History: *Science,* vol. 208, May 16, 1980, pp. 736-738.

Carithers, Ward, 1946, *Pumice and Pumicite Occurrences of Washington:* Washington Division of Mines and Geology, Report Investigation 15, 78 p.

Christiansen, Robert L., 1980, Eruption of Mt. St. Helens: Volcanology: *Nature,* vol. 285, June 19, 1980, pp. 531-533.

Christiansen, R.L., and Peterson, D.W., 1981, Chronology of the 1980 Eruptive Activity: In Lipman, P.W., and Mullineaux, D.R., eds., *The 1980 Eruptions of Mount St. Helens, Washington:* U.S. Geological Survey Professional Paper 1250, pp. 17-30.

Chadwick, W.W., Jr., Swanson, D.A., Iwatsubo, E.Y., Heilker, C.C., and Leighley, T.A., 1983, Deformation Monitoring at Mount St. Helens in 1981 and 1982: *Science,* vol. 221, no. 4618, pp. 1378-1380.

Casadevall, T., Rose, W., Gerlach, T., Greenland, L.P., Ewert, J., Wunderman, R., and Symonds, R., 1983, Gas Emissions and the Eruptions of Mount St. Helens Through 1982: *Science,* vol. 221, no. 4618, pp. 1383-1385.

Clark, Ella E., 1966, *Indian Legends of the Pacific Northwest*, Univ. of California Press, Berkeley and Los Angeles.

Coombs, H.A., and Howard, A.D., 1960, United States of America: In *Catalog of Active Volcanoes of the World, Part 9,* International Volcanological Association, Naples, Italy.

Crandell, D.R., Booth, B., Kusumadinata, K., Shimozuru, D., Walker, G.P.L., and Westercamp, D., 1983, *Sourcebook for Volcanic Hazards Zonation,* UNESCO, Paris, in press.

Crandell, D.R., and Mullineaux, D.R., 1978, Potential Hazards from Future Eruptions of Mount St. Helens Volcano, Washington: *U.S. Geological Survey Bulletin,* no. 1383-C, 26 p.

Crandell, Dwight R., 1969, Surficial Geology of Mount Rainier National Park, Washington, *U.S. Geological Survey Bulletin 1288*, 41 p.

Crandell, Dwight R., 1971, Postglacial Lahars from Mount Rainier Volcano, Washington: *U.S. Geological Survey Prof. paper 677,* 75 p.

Crandell, Dwight R., and Mullineaux, D.R., 1967, Volcanic Hazards at Mount Rainier, Washington: *U.S. Geological Survey Bulletin 1238,* 26 p.

Crandell, Dwight R., and Mullineaux, D.R., 1973, Pine Creek Volcanic Assemblage at Mount St. Helens, Washington: *U.S. Geological Survey Bulletin 1383-A,* 23 p.

Crandell, Dwight R., Mullineaux, D.R., Miller, R.D., and Rubin, Meyer, 1962, Pyroclastic Deposits of Recent Age at Mount Rainier, Washington: In Short Papers in Geology, Hidrology, and Topography, *U.S. Geological Survey Prof. Paper 450-D,* pp. D64-D68.

Crandell, D.R., Mullineaux, D.R., and Rubin, Meyer, 1975a, Mount St. Helens Volcano: Recent and Future Behavior, *The Ore Bin,* vol. 37, no. 3, p. 1.

Crandell, Dwight R., Mullineaux, Donal R., Rubin, Meyer, 1975b, Mount St. Helens Volcano: Recent and Future Behavior: *Science,* vol. 187, no. 4175, pp. 438-441.

Crandell, Dwight R., Mullineaux, Donal R., and Miller, C. Dan, 1979, Volcanic-Hazards Studies in the Cascade Range of the Western United States, In Sheets, P.D., and Grayson, D.K., eds., *Volcanic Activity and Human Ecology,* New York: Academic Press, pp. 195-219.

Crandell, D.R., and Waldron, H.H., 1969, Volcanic Hazards in the Cascade Range, In Olson, R.A., and Wallace, M.M., Geologic Hazards and Public Problems, Office of Emergency Preparedness, pp. 5-18.

Cummans, John, 1981, *Mudflows Resulting from the May 18, 1980, Eruption of Mount St. Helens, Washington: Geological Survey Circular 850-B,* 16 p.

Danna, James D., 1849, United States Exploring Expedition During the years 1838, 1839, 1840, 1841, 1842, under the Commands of Charles Wilkes, *U.S.N., vol. 10, Geology,* Philadelphia, p. 640.

Decker, Robert and Decker, Barbara, 1981, The Eruptions of Mount St. Helens, *Scientific America,* vol. 244, no. 3, pp. 68-80.

Decker, R.W., and Harlow, David, 1970, Microearthquakes at Cascade Volcaoes [abstract], *American Geophysical Union Transactions,* vol. 51, no. 4, p. 351.

del Moral, Roger, 1981, Life returns to Mount St. Helens: *Nature History,* May 1981, pp. 36-46.

De Mofras, Eugene Duflot, 1925, Extract from Exploration of the Oregon Territory, the Californias, and the Gulf of California, Undertaken during the Years 1840, 1841 and 1842, trans. and ed. by N.B. Pipes: *Oregon Historical Quarterly,* vol. 26, pp. 151-190.

Diller, J.S, 1899, Latest Volcanic Eruptions of the Pacific Coast: *Science,* new series, vol. 9 pp. 639-640.

Diller, J.S., 1915, The Relief of Our Pacific Coast: *Science,* vol. 41, pp. 48-57.

Drury, Clifford M., 1940, *Elkanah and Mary Walker,* Caldwell, Idaho.

Dryer, Thomas J., 1853, Ascent of Mount Saint Helens: *Weekly Oregonian,* Portland, Oregon, Sept. 3, 1853, p. 2.

Dryer, Thomas J., 1854, Eruption of Mount Saint Helens: *Weekly Oregonian,* Portland, Oregon, February 25, 1854, p. 2.

Dutton, C.E., 1885, Letter: *Science,* vol. 6, pp. 46-47.

Dzursin, Daniel, Westphal, J.A., and Johnson, D.J., 1983, Eruption Prediction Aided by Electronic Tiltmeter Data at Mount St. Helens: *Science,* vol. 221, no. 4618, pp. 1381-1383.

Elliott, C.P., 1897, Mount Saint Helens: *National Geographic Magazine,* vol. 8, pp. 226-230.

Emmons, S.F., 1877, The Volcanoes of the Pacific Coast of the United States: *Journal of the American Geographical Society,* vol. 9, pp. 45-46.

Erdmann, C.E., and Warren, W., 1938, Report on the Geology of Three Dam Sites on the Toutle River, Cowlitz and Skamania Counties, Washington: *U.S. Geological Survey Open-File Report,* 119 p.

Findley, Rowe, 1981, Eruption of Mount St. Helens, *National Geographic,* vol. 159, no. 1, pp. 3-65.

Folsom, M.M., 1970, Volcanic Eruptions: The Pioneers' Attitude on the Pacific Coast from 1800 to 1875: Oregon Dept. of Geology and Mineral Industries, *The Ore Bin,* vol. 32, pp. 61-71.

Forsyth, C.E., 1910, Mount St. Helens, *The Mountaineer,* vol. 3, pp. 56-62.

Foxworthy, B.L., and Hill, Mary, 1982, *Volcanic Eruptions of 1980 at Mount St. Helens: The First 100 Days:* U.S. Geological Survey Professional Paper 1249.

Frémont, John C., 1845, *Report of the Exploring expedition to the Rocky Mountains in the year 1842, and to Oregon and North California in the years 1843-44,* Washington, D.C., p. 193-194.

Fritz, William J., 1980, Stumps transported and deposited upright by Mount St. Helens Mudflows: *Geology,* vol. 8, no. 12, pp. 586-588.

Frost, Joseph H., 1934, The Journal of John H. Frost, ed. N.B. Pipes: *Oregon Historical Quarterly,* vol. 35, pp. 373-374.

Fruchter, Jonathan S. et al, 1980, Mt. St. Helens Ash from May 18, 1980 Eruption: Chemical, Physical, Mineralogical and Biological Properties: *Science,* vol. 209, no. 4461, p. 116.

Gairdner, Meredith, 1836, Letter. *The Edinburgh New Philosophical Journal,* vol. 20, no. 39, p. 206.

Gary, George, 1923, Diary of Rev. George Gary, ed. F.G. Young: *Oregon Historical Quarterly,* vol. 24, pp. 76-77.

Gibbs, George, 1855, Report on the Geology of the Central Part of Washington Territory, U.S. 33rd Congress, *1st session, House Document 129,* vol. 18, part 1, pp. 494-512.

Gibbs, George, 1872, Physical Geography of the Northwestern Boundary of the United States: *Journal of the American Geographical Society.*

Greeley, Ronald, and Hyde, J.H., 1972, Lava Tubes of the Cave Basalt, Mount St. Helens, Washington: *Geological Society of America Bulletin,* vol. 83, pp. 2397-2418.

Halliday, W.R., 1963, Features and Significance of the Mount St. Helens Cave Area: *National Parks Magazine,* vol. 37, no. 195, pp. 11-14.

Hammond, Paul E., 1980, Mt. St. Helens Adds Fireworks in the Cascades: *Geotimes,* vol. 25, no. 7, pp. 16-18.

Hammond, Paul E., 1980, Mt. St. Helens Blasts off 400 M: *Geotimes,* vol. 25, no. 8, p. 14.

Harris, Stephen L., 1980, A Rumbling in Our Midst, *Pacific Northwest,* vol. 14, no. 4, pp. 22-25.

Hawkins, L.L., 1903, Lava Caves of St. Helens: *Mazama,* vol. 2, no. 3, pp. 134-135.

Hoblitt, Richard P., Crandell, Dwight R., and Mullineaux, Donal R., 1980, Mount St. Helens Eruptive Behavior During the Past 1,500 years: *Geology,* vol. 8, no. 11, pp. 555-559.

Hoblitt, Richard P., and Miller, C. Dan, 1984, Comments and Reply on "Mount St. Helens 1980 and Mont Pelee 1902—Flow Or Surge?": *Geology,* vol. 12, pp. 692-695.

Holden, Edward S., 1898, *Catalogue of Earthquakes on the Pacific Coast 1769 to 1897: Smithsonian Miscellaneous Publication 1087.*

Holmes, Kenneth L., 1955, Mount Saint Helens' Recent Eruptions: *Oregon Historical Quarterly,* vol. 56, pp. 197-210.

Holmes, Kenneth L., 1980, *Mount St. Helens, Lady with a Past,* Salem Press, Salem, Oregon, 48 p.

Hooper, P.R., Herrick, I.W., Laskowski, E.R., and Knowles, C.R., 1980, Composition of the Mount St. Helens Ashfall in the Moscow-Pullman Area on May 18, 1980: *Science,* vol. 209, no. 4461, pp. 1125-1126.

Hopson, C.A., 1971, Eruptive Sequence at Mt. St. Helens, Washington: *Geological Society of America Abstracts with Programs,* vol. 3, no. 2, p. 138.

Hopson, C.A., and Melson, W.G., 1984, Eruption Cycles and Plug-domes at Mount St. Helens: *Geological Society of America Abstracts with Programs,* vol. 16, no. 6, p. 544.

Hopson, C.A., Waters, A.C., Bender, V.R., and Rubin, M., 1962, The Latest Eruptions from Mount Rainier Volcano: *Journal of Geology,* vol. 70, pp. 635-646.

Hyde, J.H., 1970, *Geologic Setting of Merrill Lake and Evaluation of Volcanic Hazards in the Kalama River Valley near Mount St. Helens, Washington:* U.S Geological Survey Open-File Report, 17 p.

Hyde, J.H., 1973, *Late Quarternary Volcanic Stratigraphy, South Flank of Mount St. Helens, Washington* Ph.D. Thesis, University of Washington, Seattle, 114 p.

Hyde, Jack H., 1975, Upper Pleistocene Pyroclastic-flow Deposits and Lahars South of Mount St. Helens Volcano, Washington: *U.S. Geological Survey Bulletin,* no. 1383B, 20 p.

Hyde, J.H., and Crandell, D.R., 1972, Potential Volcanic Hazards Near Mount St. Helens, southwestern Washington [abstract]: *Northwest Science Programs and Abstracts.*

Jillson, W.R., 1917a, New Evidence of a Recent Volcanic Eruption on Mt. St. Helens, Washington: *American Journal of Science,* vol. 44, no. 259, pp. 59-62.

Jillson, Willard Rouse, 1917b, The Volcanic Activity of Mount St. Helens and Mount Hood in Historical Time: *Geographical Review,* vol. 3, pp. 482-483.

Jillson, W.R., 1921, Physiographic Effects of the Volcanism of Mt. St. Helens: *Geographical Review,* vol. 11, pp. 398-405.

Kane, Paul, 1925, *Wanderings of an Artist,* Radisson Society, Toronto (reprint of the 1859 edition).

Kerr, Richard A., 1980, Mount St. Helens: An Unpredictable Foe, *Science,* vol. 208, no. 4451, pp. 1446-1448.

Korosec, Michael A., Rigby, James G., Stoffel, Keith L., 1980, *The 1980 Eruption of Mount St. Helens, Washington, Part I: March 20-May 19, 1980, Information Circular No. 71,* Washington State Department of Natural Resources, Division of Geology and Earth Resources, 27 p.

Landerholm, Carl, ed., 1956, *Notices and Voyages of the Famed Quebec Mission to the Pacific Northwest,* Oregon Historical Society.

Landes, Henry, 1901, The Volcanoes of Washington, *Northwest Journal of Education,* vol. 12.

Lawrence, D.B., 1938, Trees on the March: *Mazama,* vol. 20, no. 12, pp. 49-54.

Lawrence, D.B., 1939, Continuing Research on the Flora of Mt. St. Helens: *Mazama,* vol. 21, no. 12, pp. 49-59.

Lawrence, D.B., 1941, The "Floating Island" Lava Flow of Mt. St. Helens: *Mazama,* vol. 23, no. 12, pp. 56-60.

Lawrence, D.B., 1954, Diagrammatic History of the Northeast Slope of Mt. St. Helens, Washington: *Mazama,* vol. 36, no. 13, pp. 41-44.

Lawrence, D.B., and Elizabeth G., 1958, Bridge of the Gods Legends, Its Origin, History and Dating: *Mazama,* vol. 40, no. 13, pp. 33-41.

Lawrence, Donald B., and E.G., 1959, Radiocarbon Dating of Some Events on Mount Hood and Mount St. Helens: *Mazama,* vol. 41, no. 13, pp. 10-18.

Lee, Daniel, and Frost, J.H., 1844, *Ten Years in Oregon,* New York, no. 257.

Lipman, Peter W., and Mullineaux, Donal R., eds., 1981, The 1980 Eruptions of Mount St. Helens: *U.S. Geological Survey Professional Paper No. 1250.*

Lombard, R.E., Miles, M.B., Nelson, L.M., Kresch, D.L., and Carpernter, P.J., 1981, *Channel Conditions in the Lower Toutle and Cowlitz Rivers Resulting from the Mudflows of May 18, 1980: Geological Survey Circular 850-C,* 16 p.

Lyman, W.D., 1915, Indian Myths of the Northwest, *Proceedings of the American Antiquarian Society,* vol. XXV, pp. 375-395.

Lyman, W.D., 1963, *The Columbia River, Its History, Its Myths, Its Scenery, Its Commerce,* rev. ed., Portland: Binfords and Mort, 367 p.

Majors, Harry M., 1980, Mount St. Helens Series: *Northwest Discovery,* vol. 1, no. 1, pp. 4-51.

Majors, Harry M., 1980, Mount St. Helens Series: *Northwest Discovery,* vol. 1, no. 2, pp. 68-108.

Malone, S.D., Boyko, C., and Weaver, C.S., 1983, Seismic Precursors to the Mount St. Helens Eruptions in 1981 and 1982: *Science,* vol. 221, pp. 1376-1378.

Malone, S.D., Endo, E.T., Weaver, C.S., and Ramey, J.W, 1981, Seismic Monitoring for Eruption Prediction: In Lipman, P.W, and Mullineaux, D.R., eds, *The 1980 Eruption of Mount St. Helens:* U.S. Geological Survey Professional Paper 1250, pp. 803-813.

McLucas, Glennda, 1980, Petrology of Current Mount St. Helens Tephra: *Washington Geological Newsletter,* vol. 8, no. 3, pp. 7-13.

Moore, James G., and Albee, W.C., 1981, Topographic and Structural Changes, March-July 1980 Photogrammetric Data: In, Lipman, P.W., and Mullineaux, D.R., eds., *The 1980 Eruptions of Mount St. Helens:* U.S Geological Survey Professinal Paper 1250, pp. 123-134.

Moore, James G., and Rice, Carl J., 1984, Chronology and Character of the May 18, 1980, Explosive Eruptions of Mount St. Helens: In *Explosive Volcanism: Inception, Evolution, and Hazards,* National Academy Press, Washington, D.C., pp. 133-142.

Moore, James G., and Sisson, T., 1981, Deposits and Effects of the May 18 Pyroclastic Surge: In Lipman, P.W., and Mullineaux, D.R., eds., *The 1980 Eruptions of Mount St. Helens, Washington:* U.S. Geological Survey Professional Paper 1250, pp. 421-433.

Moxham, R.M., 1970, Thermal Features at Volcanoes in the Cascade Range, as Observed by Aerial Infrared Surveys: *Bulletin Volcanologique,* vol. 34, pp. 77-106.

Mullineaux, D.R., 1977, Volcanic Hazards; Extent and Severity of Potential Tephra Hazard Interpreted from Layer Yn from Mount St. Helens (abs): *Geological Society of America, Abstr. Programs,* vol. 9, no. 4, p. 472.

Mullineaux, D.R., and Crandell, D.R., 1960, Late Recent Age of Mount St. Helens Volcano, Washington: *In U.S. Geological Survey Research 1960: Short Papers in the Geological Sciences,* pp. B307-B308.

Mullineaux, D.R., and Crandell, D.R., 1962, Recent Lahars from Mount St. Helens, Washington: *Geological Society of America Bulletin,* vol. 73, pp. 855-870.

Mullineaux, D.R., and Crandell, D.R., 1981, The Eruptive History of Mount St. Helens: In Lipman, P.W, and Mullineaux, D.R., eds., *The 1980 Eruptions of Mount St. Helens, Washington:* U.S. Geological Survey Professional Paper 1250, pp. 3-15.

Mullineaux, D.R., Hyde, J.H., and Rubin, Meyer, 1972, Preliminary Assessment of Upper Pleistocene and Holocene Pumiceous Tephra from Mount St. Helens, Southern Washington (abstract): *Geological Society of America Abstract with Programs,* vol. 4, no. 3, pp. 204-205.

Mullineaux, D.R., Hyde, J.H., and Rubin, Meyer, 1975, Widespread Late Glacial and Postglacial Tephra Deposits from Mount St. Helens Volcano, Washington, *U.S Geological Survey Journal of Research,* vol. 3, no. 3, pp. 329-335.

Nasmith, H. Mathews, W.H., and Rouse, G.E., 1967, Bridge River Ash and Some Recent Ash Beds in British Columbia: *Canadian Journal of Earth Science,* vol. 4, pp. 163-170.

Nelson, Clifford M., 1980, Mt. St. Helens Before May 18: *Geotimes*, August, 1980, p. 16.

Norgren, J.A., Borchardt, G.A., and Harward, M.E., 1970, Mt. St. Helens Y Ash in Northeastern Oregon and South Central Washington (abstract): *Northwest Science,* vol. 44, p. 66.

North Pacific History Company of Portland, Oregon, 1889, *History of the Pacific Northwest, Oregon and Washington,* vol. 2, pp. 96-97.

Okazaki, Rose, Smith, H.W. Gilkeson, R.A., and Franklin, Jerry, 1972, Correlation of West Blacktail Ash with Pyroclastic Layer T from the 1880 A.D. Eruption of Mount St. Helens: *Northwest Science*, vol. 46, pp. 77-89.

Parker, Samuel, 1846, Journal of an Exploring Tour Beyond the Rocky Mountains (4th edition), Ithaca, pp. 223-229; 293-312; 329-351.

Peterson, D.W., 1986, Mount St. Helens and the Science of Volcanology: *A Five-Year Perspective,* in Keller, S.A.C., ed., *Mount St. Helens: Five Years Later,* Eastern Washington University Press, Cheney, Wash., pp. 3-19.

Peterson, N.V., and Groh, E.A., 1963, Recent Volcanic Landforms in Central Oregon: Oregon Dept. of Geology and Mineral Industries: *The Ore Bin,* vol. 25, no. 3, pp. 33-45.

Parrish, Josiah L., 1906, Letter dated January 13, 1892 to W.G. Steel: *Steel Points,* vol. 1, no. 1, pp. 25-26.

Phillips, K.N., 1941, Fumaroles of Mount St. Helens and Mount Adams: *Mazama,* vol. 23, no. 12, pp. 37-42.

Plummer, F.G., 1893, Western Volcanoes: Chances That Western Washington May See Disastrous Eruptions: *Tacoma Daily Ledger,* February 28, 1893, p. 3, col. 1-4.

Plummer, Frederick G., 1898, Reported Volcanic Eruptions in Alaska, Puget Sound, etc., 1690 to 1896. In Holden, Edwards S., *A Catalogue of Earthquakes on the Pacific Coast 1769 to 1897. Smithsonian Miscellaneous Publication 1087,* Washington, D.C.

Pryde, P.R., 1968, Mount Saint Helens: A Possible National Monument: *National Parks Magazine,* vol. 42, no. 248, pp. 7-10.

Powers, Alfred, 1944, *Legends of the Four High Mountains,* Portland, Oregon Junior Historical Journal.

Ray, V.F., 1932, The Sanpoil and Nespelem: Salishan Peoples of Northeastern Washington: *University of Washington Publications in Anthropology,* vol. 5. Reprinted in 1954 by the Human Relations Area Files, New Haven.

Rosenbaum, J.G., and Waitt, R.B., Jr., 1981, Summary of Eyewitness Accounts of the May 18 Eruption: In Lipman, P.W., and Mullineaux, D.R., eds., *The 1980 Eruptions of Mount St. Helens, Washington:* U.S. Geological Survey Professional Paper 1250, pp. 53-67.

Rosenfeld, Charles, and Cooke, Robert, 1982, *Earthfire,* 2nd printing, Massachusetts Institution of Technology Press, Cambridge, Massachusetts, 155 p.

Sarna-Wojcicki, A.M., Shipley, S., Waitt, R.B., Jr., Dzurisin, D., and Wood, S.M., 1981, Areal Distribution, Thickness, Mass, Volume and Grain Size of Air-Fall Ash from the Six Major Eruptions of 1980: In Lipman, P.W., and Mullineaux, D.R., eds., *The 1980 Eruptions of Mount St. Helens, Washington:* U.S Geological Survey Professional Paper 1250.

Sarna-Wojcicki, A.M., Waitt, R.B., Jr., and Woodward, M.J., 1981, Pyromagmatic Ash Erupted from March 27 through May 14, 1980—Extent, Mass, Volume, and Composition: In Lipman, P.W, and Mullineaux, D.R., eds, *The 1980 Eruptions of Mount St. Helens, Washington:* U.S. Geologcial Survey Professional Paper 1250, pp. 569-576.

Schminky, Bruce, 1954, Records of Eruptions of Mount St. Helens: Geological Society Oregon Country, *Geological News Letter,* vol. 20, no. 9, pp. 81-82.

Shulters, M.V., and Clifton, D.G., 1980, Mount St. Helens Volcanic-Ash Fall in the Bull Run Watershed, Oregon, March-June 1980: *Geological Survey Circular,* no. 850-D, 15 p.

Smith, Henry W., Okazaki, Rose, and Aarsted, John, 1968, Recent Volcanic Ash in Soils of Northeastern Washington and Northern Idaho: *Northwest Science,* vol. 42, pp. 150-160.

Smith, Henry, Okazaki, Rose, and Knowles, Charles R., 1972, Electron Microprobe Analysis of Glass Shards from Tephra Assigned to Set W, Mount St. Helens, Washington: *Quaternary Research,* vol. 7, pp. 207-217.

Steel, William Gladstone, ed., 1899, Mountain Lore: *Oregon Native Son,* vol. 1 p. 276.

Steel, William Gladstone, ed., 1906, Eruption of Mount Saint Helens: *Steel Points,* vol. 1, no. 1, pp. 25-26.

Sterling, E.M., 1968, *Trips and Trails, 2: Family Camps, Short Hikes, and View Roads in the South Cascades and Mt. Rainier.* The Mountaineers, Seattle.

Stevens, Charles, 1936, Letters of Charles Stevens, ed. Nellie Pipes: *Oregon Historical Quarterly,* vol. 37, pp. 250-251.

Stevens, Charles, 1937, Letters of Charles Stevens: *Oregon Historical Quaterly,* vol. 38, p. 73.

St. Lawrence, William and others, 1980, A Comparison of Thermal Observations of Mt. St. Helens Before and During the First Week of the Initial 1980 Eruption: *Science,* vol. 209, no. 4464, pp. 1526-1527.

Stoffel, Dorothy B., and Stoffel, Keith L., 1980, Mt. St. Helens Seen Close up on May 18: *Geotimes,* pp. 16-17.

Stoiber, R.E., Williams, L.L., Malinconico, 1980, Mt. St. Helens, Washington, 1980 Volcanic Eruption: *Science,* vol. 208, no. 4449, p. 1258.

Strong, Emory, 1969, Early Accounts of the Eruption of Mt. St. Helens: *Geological News Letters,* vol. 35, no. 1, pp. 3-5.

Swan, James G., 1857, *The Northwest Coast; or, Three Years Residence in Washington Territory:* Harper & Brothers, New York, p. 395.

Swanson, D.A., 1985, Graben Formation, Thrust Faulting, and Growth of the Dacite Dome, Mount St. Helens, Washington, May-June, 1985: *American Geophysical Union,* in press.

Swanson, D.A., Casadevall, T.J., Dzurisin, D., Malone, S.D., Newhall, C.G., and Weaver, C.S., 1983, Predicting Eruptions at Mount St. Helens, June 1980 through December 1982: *Science,* vol. 221, pp. 1369-1376.

344

Swanson, D.A., Casadevall, T.J., Dzurisin, D., Holcomb, R.T., Newhall, C.G., Malone, S.D., and Weaver, C.S., 1985 Forecasts and Predictions of Eruptive Activity at Mount St. Helens, USA: 1975-1984: *Journal of Geodynamics,* vol. 3, no. 314, pp. 75-89.

Swanson, D.A., Dzurisin, D., Holcomb, R.T., Iwatsubo, E.Y., Chadwick, W.W., Jr., Casadevall, T.J., Ewert, J.W., and Heliker, C.C., 1986, *Growth of the Lava Dome at Mount St. Helens, Washington* (USA): U.S. Geological Survey Professional Paper, in press.

Swanson, D.A., and Holcomb, R.T., 1985, Regulartities in the Growth of Mount St. Helens' Dome: *Geological Society of America Abstracts with Programs,* vol. 17, no. 6, p. 411.

Teit, J.A., 1930, *The Salishan Tribes of the Western Plateau,* Bureau of American Ethnology Annual Report, 1927-1928. In Blinman, et. al., 1977, p. 422.

Thornton, J. Quinn, 1855, *Oregon and California in 1848,* Harper & Brothers, New York, vol. 1, pp. 256-257.

Tilling, Robert I., 1982, *Eruptions of Mount St. Helens: Past, Present, and Future:* U.S. Geological Survey Series of General Interest Publications, 46 p.

Unger, John D., and Mills, Kay F., 1973, Earthquakes Near Mount St. Helens, Washington: *Geological Society of America Bulletin,* vol. 84, no. 3, pp. 1065-1068.

U.S. Geological Survey, 1980, *Preliminary Aerial Photographic Interpretative Map Showing Features Related to the May 18, 1980 Eruption of Mount St. Helens, Washington:* U.S. Geol. Surv., Misc. Feild Study Map no. MF-1254.

U.S Geological Survey, 1980, Volcano Log: Mount St. Helens 1980, *Earthquake Information Bulletin,* vol. 12, no. 4, pp. 142-149.

University of Washington Geophysics Program, 1980, Eruption of Mt. St. Helens: Seismology: *Nature,* vol 285, no. 5766, pp. 529-531.

Verhoogen, Jean, 1937, *Mount St. Helens, A Recent Cascade Volcano: California University Publication Geological Science,* vol. 24, pp. 263-302.

Voight, B.H., 1981, Time Scale for the First Moments of the May 18 Eruption: In Lipman, P.W, and Mullineaux, D.R., eds., *The 1980 Eruptions of Mount St. Helens, Washington:* U.S Geological Survey Professional Paper 1250, pp. 69-86.

Voight, B.H., Glicken, R.J., Janda, and Douglass, P.M., 1981, Rockslide Avalanche of May 18: In Lipman, P.W., Mullineaux, D.R., eds., *The 1980 Eruptions of Mount St. Helens, Washington:* U.S Geological Survey Professional Paper 1250, pp. 347-377.

Walker, G.P.L., and McBroome, L.A., 1983, Mount St. Helens 1980 and Mont Peleé 1902—Flow or Surge?: *Geology,* vol. 11, pp. 571-574.

Westgate, J.A, and Dreimanis A., 1967, Volcanic Ash Layers of Recent Age at Banff National Park, Alberta, Canada: *Candaian Journal of Earth Science,* vol. 4, pp. 155-161.

Wilcox, R.E., 1965, Volcanic-Ash Chronology: In Wright, H.E., Jr., and Frey, D.G., *The Quaternary of the United States,* Princeton University Press, N.J., pp. 807-816.

Wilkes, Charles, 1845, *Narrative of the United States Exploration During the Years 1838, 1839, 1840, 1841, and 1842,* Philadelphia, vol. 4, pp. 439-440.

Williams, C., 1980, *Mount St. Helens: A Changing Landscape,* Graphic Arts Center Publishing Co., Portland, Oregon, 128 p.

Williams, John H., 1912, *The Guardians of the Columbia,* J.H. Williams, Tacoma, Wash., 144 p.

Williams, I.A., 1922, Tree Casts in Recent Lava: *Natural History,* vol. 22, no. 6, pp. 543-548.

Yamaguchi, David K., 1983, New Tree-Ring Dates for Recent Eruptions of Mount St. Helens: *Quaternary Research,* vol. 20, pp. 246-250.

Yamaguchi, David K., 1985, Tree-Ring Evidence for a Two-Year Interval between Recent Prehistoric Explosive Eruptions of Mount St. Helens: *Geology,* vol. 13, no. 8, pp. 354-357.

Yamaguchi, David K., 1986, Tree-Ring Dating of Mount St. Helens Flowage Deposits Emplaced in the Valley of the South Fork Toutle River during the Last 350 Years: In Keller, S.A.C., ed., *Mount St. Helens: Five Years Later,* Eastern Washington University Press, Cheney, Washington.

Youd, T.L., and Wilson, R.C., 1980, Stability of Toutle River Blockage, Mount St. Helens Hazards Investigations: *U.S. Geological Survey Open-File Report 80-898,* 14 p.

XVII. Mt. Rainier

Beckey, Fred, 1973, *Cascade Alpine Guide: Climbing and High Routes Columbia River to Stevens Pass,* The Mountaineers, Seattle.

California Division of Mines and Geology, 1970, Mount Rainier Restless: California Division of Mines and Geology, *Mineral Information Services,* vol. 23, no. 4, pp. 86-87.

Condie, K.C., and Swenson, D.H., 1973, Compositional Variation in Three Cascade Stratovolcanoes: Jefferson, Rainier, and Shasta: *Bulletin Volcanologique,* vol. 37, no. 2, pp. 205-230.

Coombs, Howard A., 1936, *The Geology of Mount Rainier National Park, Washington:* Washington University. (Seattle) Pub. Geol., vol. 3, pp. 131-212.

Coombs, Howard A., 1960, United States of America: in Part 9 of *Catalogue of the Active Volcanoes of the World,* International Volcanological Association, Naples, Italy, pp. vii-xii, 1-58.

Crandell, Dwight R., 1963a, Paradise Debris Flow at Mount Rainier, Washington: In *Short Papers in Geology and Hydrology: U.S. Geological Survey Professional Paper 475-B,* pp. B135-B139.

Crandell, Dwight R., 1963b, Suficial Geology and Geomorphology of the Lake Tapps Quadrangle, Washington: *U.S. Geological Survey Professional Paper 388-A,* pp. A1-A84.

Crandell, Dwight R., 1965, The Glacial History of Western Washington and Oregon: In Wright, H.E., Jr., and Frey, D.G., *The Quaternary of the United States,* Princeton University Press, N.J., pp. 341-354.

Crandell, Dwight R., 1969a, *The Geologic Story of Mount Rainier: U.S. Geological Survey Bulletin 1292,* 43 p.

Crandell, Dwight R., 1969b, *Surficial Geology of Mount Rainier National Park, Washington, U.S. Geological Survey Bulletin 1288,* 41 p.

Crandell, Dwight R., 1971, *Postglacial Lahars from Mount Rainier Volcano, Washington: U.S. Geological Survey Professional paper 677,* 75 p.

Crandell, Dwight R., 1973, *Potential Hazards from Future Eruptions of Mount Rainier, Washington: U.S. Geological Survey Miscellaneous Geologic Investigations,* Map I-836.

Crandell, D.R., and Fahnestock, R.K., 1965, *Rockfalls and Avalanches from Little Tahoma Peak on Mount Rainier, Washington: U.S. Geological Survey Bulletin 1221-A,* 30 p.

Crandell, D.R., and Miller, R.D., 1964, Post-Hypsithermal Glacier Advances at Mount Rainier, Washington: In *Geological Survey Research 1964: U.S. Geological Survey Professional Paper 501-D,* pp. D110-D114.

Crandell, D.R., Mullineaux, D.R., Miller, R.D., Rubin, Meyer, 1962, Pyroclastic Deposits of Recent Age Mount Rainier, Washington: In *Short Papers in the Geologic and Hydrologic Sciences: U.S. Survey Professional Paper 450-D,* pp. D64-D68.

Crandell, D.R., Mullineaux, D.R., 1967, *Volcanic Hazards at Mount Rainier, Washington: U.S Geological Survey Bulletin 1238,* 26 p.

Crandell, D.R., and Waldron, H.H., 1956, A Recent Volcanic Mudflow of Exceptional Dimensions from Mount Rainier, Washington: *American Journal of Science,* vol. 254, p. 349-362.

Crandell, Dwight R., and Waldron, H.H., 1969, Volcanic Hazards in the Cascade Range: In Olson, R.A. and Wallace, M.M., *Geologic Hazards and Public Problems, Conference Proceedings,* Office of Emergency Preparedness, pp. 5-18.

Cullen, J.M., 1978, *Impact of a Major Eruption of Mount Rainier on Public Service Delivery Systems in the Puyallup Valley, Washington,* Unpublished Master's Thesis, Dept. of Urban Planning, University of Washington, Seattle.

Diller, J.S., 1915, The relief of our Pacific coast, *Science* vol. 41 p. 48-57.

Driedger, C.L., and Kennard, P.M., 1986, Ice Volumes on Cascade Volcanoes: Mount Rainier, Mount Hood, Three Sisters, and Mount Shasta: *U.S. Geological Survey Professional Paper 1365.*

Emmons, Samuel Franklin, 1877, Volcanoes of the Pacific Coast of the U.S.: *Journal of the American Geographical Society,* vol. 9 pp. 45-65.

Fiske, R.S., Hopson, C.A. and Waters, A.C., 1963, *Geology of Mount Rainier National Park, Washington: U.S. Geological Survey Professional Paper 444,* 93 p.

Folsom, M.M., 1970, Volcanic Eruptions: The Pioneers' Attitude on the Pacific Coast from 1800 to 1875: Oregon Department of Geology and Mineral Industries: *The Ore Bin,* vol. 32, pp. 61-71.

Frémont, J.C., 1845, *Report of the Exploring Expedition to the Rocky Mountains in the Year 1842, and to Oregon and North California in the Years 1843-1844,* Blair and Rives, Washington, D.C., 693 p.

Friedman, Jules D., 1972, Aerial Thermal Surveillance of Volcanoes of the Cascade Range, Washington, Oregon, and Northern California: *EOS (Amer. Geophys. Union, Trans.)* vol. 53, no. 4, p. 533, 1972.

Grater, R.J., 1948, The Flood that Swallowed a Glacier: *Natural History,* vol. 57, pp. 276-278.

Hague, Arnold and Iddings, Joseph P., 1883, Notes on the Volcanoes of Northern California, Oregon and Washington Territory: *American Journal of Science,* 3rd series, vol. 26, no. 153, pp. 222-235.

Haines, Aubrey L., 1962, *Mountain Fever! Historic Conquests of Rainier:* Oregon Historical Society, Portland, Oregon.

Hazard, Joseph T., 1933, *Snow Sentinels of the Northwest,* Lowman and Hanford, publisher, Seattle.

Hedlund, Gerald C., 1976, Mudflow Disaster, *Northwest Anthropological Research Notes,* vol. 10, pp. 77-89.

Holden, Edward S., 1898, *A Catalogue of Earthquakes on the Pacific Coast, 1769 to 1897: Smithsonian Miscellaneous Collections no. 1087,* Smithsonian Institution, Washington D.C.

347

Hopson, C.A., Waters, A.C., Bender, V.R., Rubin, Meyer, 1962, The Latest Eruptions from Mount Rainier Volcano: *Journal of Geology,* vol. 70 pp. 635-646.

Kirk, Ruth, 1968, *Exploring Mount Rainier,* University of Washington Press, Seattle & London.

Kiver, Eugene P., and Mumma, Marten P., 1971, Summit Firn Caves, Mount Rainier, Washington: *Science,* vol. 173, pp. 320-322.

Kiver, Eugene P., and Steel, William K., 1975, Firn Caves in the Volcanic Craters of Mount Rainier, Washington: *National Speleological Society Bulletin,* vol. 37, no. 1.

Lange, I.M., and Avent, J.C., 1975, Ground-Based Thermal Infrafred Surveys of Mt. Rainier Volcano, Wash. [abst.], *Geological Society of America, Abstracts with Programs,* vol. 7, no. 5, p. 619.

Lange, Ian M., and Avent, Jon C., 1973, Ground-Based Thermal Infrared Surveys as an Aid in Predicting Volcanic Eruptions in the Cascade Range: *Science,* vol. 182, no. 4109, pp. 279-281.

Macdonald, Gordon A., 1972, *Volcanoes,* Prentice-Hall, Englewood Cliffs, N.J., 510 p.

Majors, Harry M., 1981, Mount Rainier—The Tephra Eruption of 1894: *Northwest Discovery,* vol. 2, no. 6, pp. 334-381.

Matthes, F.E., 1914, *Mount Rainier and Its Glaciers,* National Park Service, pp. 1-48.

Miller, M.M., 1961, Wind, Sky and Ice report from Project Crater, 1959-1960: *Havard Mountaineering,* no. 15.

Molenaar, Dee, 1971, *The Challenge of Rainier,* The Mountaineers, Seattle.

Moxham, R.M., 1970, Thermal Features at Volcanoes in the Cascade Range, as Observed Aerial Infrared Surveys: *Bulletin Volcanologique,* vol. 34, no. 1, pp. 77-106.

Moxham, R.M., Boynton, G.R., and Cote, C.E., 1972, Satellite Telemetry of Fumarole Temperatures: *International Union of Geology and Geophysics,* vol. 36, no. 1, pp. 191-199.

Moxham, R.M., Crandell, D.R., and Mariatt, W.E., 1965, Thermal features of Mount Rainier, Washington, as revealed by infrared surveys: *U.S. Geological Survey Professional Paper 525-D,* pp. 93-100.

Mullineaux, D.R., 1974, *Pumice and Other Pyroclastic Deposits in Mount Rainier National Park, Washington: U.S. Geological Survey Bulletin 1326,* 83 p.

Mullineaux, D.R. Sigafoos, R.S., and Hendricks, E.L., 1969, A Historic Eruption of Mount Rainier, Washington: *U.S. Geological Survey Professional Paper 650-B,* pp. 315-318.

Plummer, Frederick G., 1893, Western Volcanoes: Chances That Western Washington May See Disastrous Eruptions: In *Tacoma Daily Ledger,* February 28, 1893, p. 3, col. 1-4.

Plummer, Frederick G., 1898, Reported Volcanic Eruptions in Alaska, Puget Sound, etc., 1690 to 1896: in Holden, E.S., *A Catalogue of Earthquakes on the Pacific Coast, 1769 to 1897:* Smithsonian miscellaneous collection no. 1087, Smithsonian Institution, Washington, D.C.

Russell, Israel C., 1898, Glaciers of Mount Rainier, with a Paper on the Rocks of Mount Rainier, by G.O. Smith: *U.S. Geological Survey 18th Annual Report, Part 2,* pp. 349-423.

Schear, R., 1965, Mount Rainier: Inside Look at a Smoldering Volcano: *Seattle Magazine,* vol. 2, no. 16, pp. 30-3; 46-47.

Sigafoos, R.S. and Hendricks, E.L., 1961, *Botanical Evidence of the modern History of Nisqually Glacier, Washington: U.S. Geological Survey Professional Paper 387-A,* 20 p.

Stagner, Howard R., 1966, *Behind the Scenery of Mount Ranier National Park*, rev. ed., Mt. Rainier National Historic Assoc., 64 p.

Stevens, Hazard, 1876, The Ascent of Takhoma: *Atlantic Monthly*, vol. 38, pp. 513-530.

Unger, John D., and Decker, Robert W., 1970, The Microearthquake Activity of Mount Rainier, Washington: *Bulletin of the Seismological Society of America*, vol. 60, no. 6, pp. 2023-2035.

Veatch, Fred M., 1969, *Analysis of a 24-year Photographic Record of Nisqually Glacier, Mount Rainier National Park, Washington: U.S. Geological Survey Professional Paper 631*, 52 p.

Wickersham, James, 1892, Address before the Tacoma Academy of Science, Feb. 6, 1892 (cited in Plummer, F.G., *Guide Book to Mount Tacoma*, Tacoma, Washington, News Publishing Co., 1893, 12 p.)

Williams, John H., 1911, *The Mountain That Was "God"*, G.P. Putnam & Sons, N.Y.: John H. Williams, Tacoma.

XVIII: GLACIER PEAK

Becky, F., and Rands, G., 1977, *Hikers Climbers Map of the Glacier Peak Area of the North Cascades*, The Mountaineers, Seattle.

Begét, James E., 1979, Late Pleistocene and Holocene Pyroclastic Flows and Lahars at Glacier Peak, Washington: *Geological Society of America Abstracts with Programs*, vol. 11, p. 68.

Begét, James E., 1980a, Tephrochronology of Deglaciation and Latest Pleistocene or Early Holocene Moraines at Glacier Peak, Washington: *Geoloical Society of America Abstracts with Programs*, vol. 12, p. 96.

Begét, James E., 1980b, Tephrochronology of Late-Glacial and Holocene Moraines in the central North Cascade Range, Washington State: *Geological Society of America Abstracts with Programs*, vol. 12, p. 384.

Begét, James E., 1980c, Eruptive History of Glacier Peak Volcano, Washington State: *Geological Society of America Abstracts with Programs*, vol. 12, p. 384.

Begét, James e., 1980d, Minimum Late-Pleistocene Equilibrium Line Atltitudes near Glacier Peak, Washington, after about 12,500 C14 Years Ago: *American Quaternary Association Sixth Biennial Meeting Abstracts and Programs*, pp. 14-15.

Begét, James E., 1981a, Early Holocene Glacier Advance in the North Cascade Range, Washington: *Geology*, vol. 9, pp. 409-413.

Begét, James E., 1981b, Postglacial Eruption History and Volcanic Hazards at Glacier Peak, Washington: unpublished Ph.D. Thesis, University of Washington, Seattle, 195 p.

Begét, James E., 1982a, *Postglacial Volcanic Deposits at Glacier Peak, Washington, and Potential Hazards from Future Eruptions: A Preliminary Report: U.S. Geological Survey Open File Report 82-830*.

Begét, James E., 1982b, Recent Volcanic Activity at Glacier Peak: *Science*, vol. 215, pp. 1389-1390.

Begét, James E., 1984, Tephrochronology of Late Wisconsin Deglaciation and Holocene Glacier Fluctuations near Glacier Peak, North Cascade Range, Washington: *Quaternary Research*, vol. 21, pp. 304-316.

Borchardt, Glenn A., Norgren, Joel A., and Harward, Moyle E., 1973, Correlation of Ash Layers in Peat Bogs at Eastern Oregon: *Geological Society American Abstract Programs*, vol. 84, no. 9, pp. 3101-3108.

Ford, A.B., 1959, *Geology and Petrology of the Glacier Peak Quadrangle, Northern Cascades, Washington*, Ph.D. Thesis, University of Washington, Seattle, 374 p.

Ford, A.B., 1960, Metamorphism and Granitic Instrusion in the Glacier Peak Quadrangle, Northern Cascade Mountains of Washington [abstract]: *Geological Society American Bulletin*, vol. 71, no. 12, part 2, p. 2059.

Fryxell, Ronald, 1965, Mazama and Glacier Peak Volcanic Ash Layers—Relative Ages: *Science,* vol. 147, no. 3663, pp. 1288-1290.

Knowles, Charles R., 1977, Electron Microprobe Data for Tephra Attributed to Glacier Peak, Washington: *Quaternary Research,* no. 7, pp. 197-206.

Lemke, R.W., Mudge, M.R., Wilcox, R.E., and Powers, H.A., 1979, Geologic Setting of the Glacier Peak and Mazama Ash-bed Markers in West-central Montana: *U.S. Geological Survey Bulletin,* 1395-H.

Majors, H.M., 1980, *Northwest Discovery,* vol. 1, p. 109.

Mehringer, P.J., Blinman, E., and Petersen, Kenneth, 1977, Pollen Influx and Volcanic Ash: *Science,* vol. 198, no. 4314, pp. 257-259.

Porter, Stephen C., 1977, Relationship of Glacier Peak Tephra Eruptions to Late-glacial Events in the North Cascade Range, Washington, *Geological Society America Abstracts Programs,* vol. 9, no. 7, p. 1132.

Porter, Stephen C., 1978, Glacier Peak Tephra in the North Cascade Range, Washington: Stratigraphy, Distribution, and Relationship to Late-glacial Events, *Quaternary Research,* vol. 10, no. 1, pp. 30-41.

Powers, H.A., and Wilcox, R.E., 1964, Volcanic Ash from Mount Mazama (Crater Lake) and from Glacier Peak: *Science,* vol. 144, no. 3624, pp. 1334-1336.

Rigg, G.B., and Gould, H.R., 1957, Age of Glacier Peak Eruption and Chronology of Post Glacial Peat Deposits in Washington and Surrounding Areas: *American Journal of Science,* vol. 255, pp. 341-363.

Smith, Henry W., Okazaki, Rose, and Knowles, C.R., 1977, Electron Microprobe Data for Tephra Attributed to Glacier Peak, Washington: *Quaternary Research,* vol. 7, pp. 197-206.

Tabor, R.W., and Crowder, D.F., 1969, On Batholiths and Volcanoes—Intrusion and Eruption of Late Cenozoic Magmas in the Glacier Peak Area, North Cascades, Washington: *U.S. Geological Survey Professional Paper 604.*

Vance, J.A., 1957, *The Geology of the Sauk River Area in the Northern Cascades of Washington:* Seattle, Washington University, Ph.D. Thesis, 312 p.

Westgate, J.A. and Evans, M.E., 1978, Compositional Variability of Glacier Peak Tephra and its Stratigraphic Significance: *Canada Journal of Earth Science,* vol. 15, no. 10, pp. 1554-1567.

Wilcox, R.E., 1965, Volcanic Ash Chronology: In Wright, H.E., Jr. and Frey, D.G., eds., *The Quaternary of the United States,* Princeton University Press, Princeton, N.J., pp. 807-816.

Williams, Hill, ed., 1975, Glacier Peak's Last Debris Spree: *Seattle Times*, September 25, B3.

XIX. MOUNT BAKER

Anonymous, 1859, *Pioneer-Democrat:* Olympia, Washington Territory, vol. 8 no. 1, November 25, p. 2, col. 4.

Anonymous, 1860, *Pioneer-Democrate:* Olympia, Washington Territory, December 28, p. 2, col. 5.

Anonymous, 1863a, *The Daily British Colonist* (Victoria), July 27, p. 3, col. 1.

Anonymous, 1863b, *Washington Sentinel,* August 18.

Anonymous, 1863c, *Washington Standard* (Olympia), August 1, p. 3, col. 1.

Anonymous, 1865, "Mount Baker," *Morning Oregonian/The Daily Oregonian* (Portland), April 18, p. 2, col. 3.

Anonymous, 1880a, "Eruption of Mount Baker," *Morning Oregonian* (Portland), December 14, p. 1, col. 7.

Anonymous, 1880b, "Mount Baker in a State of Eruption," *The Washington Standard* (Olympia), September 17, p. 5, col. 2.

Anonymous, 1880c, *The Washington Standard* (Olympia), April 30, p. 4, col. 3.

Borah, Leo, A., 1933, Washington, The Evergreen State: *National Geographic,* vol. 63, no. 2, p. 144.

Bortleson, G.C., Wilson, R.T., and Foxworthy, B.L., 1977, Water-Quality Effects on Baker Lake of Recent Volcanic Activity at Mt. Baker, Washington: *U.S. Geological Survey Professional Paper, no. 1122-B,* 30 p.

Brewer, William H., 1865, "Volcanic Eruptions in Northern California and Oregon," *The American Journal of Science,* 2nd series, vol. 40, no. 119, p. 264.

Burke, Raymond, 1972, *Neoglaciation of Boulder Valley, Mount Baker, Washington* Master's Thesis, Western Washington State College, Bellingham.

Coleman, Edmund T., 1869, Mountaineering on the Pacific: *Harper's New Monthly Magazine,* November.

Coleman, Edmund T., 1877, Mountains and Mountaineering Far West: *Alpine Journal,* vol. 8, no. 57, pp. 233-242.

Cook, Warren L., 1973, *Flood Tide of Empire: Spain and the Pacific Northwest, 1543-1819,* New Haven & London: Yale University Press.

Coombs, H.A., 1939, Mount Baker, a Cascade Volcano: *Geological Society of America Bulletin,* vol. 50, no. 10, pp. 1493-1510.

Coombs, H.A., 1960, United States of America: *Catalogue of the Active Volcanoes of the World,* Part 9, International Volcanological Association Naples, Italy, p. vii-xii, 1-58.

Crandell, Dwight R., and Waldron, H.H., 1969, Volcanic Hazards in the Cascade Range: In Olson, R.A., and Wallace, M.M., eds., *Geologic Hazards and Public Problems,* Office of Emergency Preparedness, pp. 5-18.

Davidson, George, 1885, Recent Volcanic Activity in the U.S.: Eruptions of Mount Baker: *Science,* vol. 6, no. 138, p. 262.

Diller, Joseph Silas, 1915, The Relief of our Pacific Coast: *Science,* vol. 41, pp. 48-57.

Easterbrook, D.J., 1975, Mount Baker Eruptions: *Geology,* vol. 3, no. 12, pp. 679-682.

Easterbrook, D.J., 1976, Pleistocene and Recent Volcanic Activity of Mt. Baker, Wash. (abs): *Geological Society American Abstracts Programs,* vol. 8, no. 6, p. 849.

Easterbrook, Don J., and Rahm, David A., 1970, *Landforms of Washington,* Department of Geology, Western Washington State College, Bellingham, Washington, pp. 20-23.

Easton, C.F., unpublished, Mt. Baker, Its Trails and Legends: Compilation of photographs, newspaper articles and manuscript commissioned in 1911 by the Mount Baker Club and archived at the Whatcomb County Museum of Natural History, Bellingham, Washington, 262 p.

Edson, Lelah Jackson, 1968, *The Fourth Corner, Highlights from the Early Northwest,* Whatcom Museum of History and Art, Bellingham, pp. 151-156.

Eichelberger, J.C., Heiker, F., Widdicombe, R., et. al., 1976, New Fumarolic Activity on Mt. Baker; Observations During April Through July 1975: *Journal Volcanol. Geothermal Research, vol. 1, no. 1, pp. 35-53.*

Eichelberg, J., Keady, C.J., and Wright, D., 1975, *Mt. Baker: 11 April to July 1975:* 33 p. (Rep. No. EGG1183-5058) available from: Los Alamos Science Lab., Los Alamos, N.M., United States.

Figge, John Thomas, 1980, *Mountaineering on the Nooksack: The History of the Mount Baker Area,* privately printed, 67 p.

Finkboner, C., 1867, "Mount Baker a Burning Volcano," *The Pacific Tribune* (Olympia), March 30.

Frank, David, 1983, *Origin, Distribution, and Rapid Removal of Hydrothermally Formed Clay at Mount Baker, Washington: U.S. Geological Survey Professional Paper 1022-E.*

Frank, David, and Krimmel, R.M., 1978, Mount Baker Thermal Activity Continues—Visual Observations, April 1976 to August, 1977, *EOS* (American Geophys. Union Trans.), vol. 59, no. 4, p. 236.

Frank, David, and Krimmel, R.M., 1980, Progress Report on Chemical Monitorying of the Subglacial Stream Draining Sherman Crater, Mount Baker, Washington: *EOS,* (American Geophys. Union Trans.), vol. 61, no. 6, p. 69.

Frank, D., Meier, M.F., and Swanson, D.A., 1977, Assessment of Increased Thermal Activity at Mt. Baker, Wash., March 1975-March 1976: *U.S Geological Survey Professional Paper, no. 1022-A,* 49 p.

Frank, D., Post, Austin, and Friedman, J.D, 1975, Recurrent Geothermally Induced Debris Avalanches on Boulder Glacier, Mount Baker, Washington: *U.S. Geological Survey Journal of Research,* vol. 3, no. 1, pp. 77-87.

Friedman, Jules D., and Frank, David, 1980, *Infrared Surveys, Radiant Flux, and Total heat Discharge at Mount Baker Volcano, Washington, between 1970 and 1975: U.S. Geological Survey Professional Paper 1022-D.*

Friedman, J.D., Frank, D.G., Preble, Duane, and Painter, J.E., 1973, Thermal Surveillance of Cascade Range Volcanoes Using ERTS-1 Multispectral Scanner, Aircraft Imagining Systems and Ground-Based Data Communication Platforms, in Symposium on Significant Results Obtained from the Earth Resources Technology Satellite-1, New Carrollton, Md., March 5-9, 1973: *NASA SP-327,* p. 1549-1560.

Gibbs, George, 1855, Report on the Geology of the Central Portions of Washington Territory: *U.S. Pacific Railroad Explorations, U.S. 33d. Congress 1st session, H. Ex. Document 129,* vol. 18, part 1, 1:494-512.

Gibbs, George, 1870, Physical Geography of the Northwestern Boundary of the United States: *American Geographic Society Journal,* vol. 3, pp. 134-157.

Gibbs, George, 1874, Physical Geography of the Northwestern Boundary of the United States: *American Geographic Society Journal,* vol. 4, pp. 298-392.

Hodge, David, Sharp, Virginia, and Marts, Marion, 1979, Contemporary Responses to Volcanism: Case Studies from the Cascades and Hawaii, in Sheets, Payson D. and Grayson, Donald K., *Volcanic Activity and Human Ecology,* Academic Press, pp. 221-248.

Holden, Edward, S., 1898, *Catalogue of Earthquakes on the Pacific Coast 1769 to 1897,* Smithsonian Misc. Coll. No. 1087.

Hyde, J.H., and Crandell, D.R., 1975, Origin and Age of Postglacial Deposits and Assessment of Potential Hazards from Future Eruptions of Mount Baker, Washington: *U.S. Geological Survey Open-File Report,* 75-286, 22 p.

Hyde, J.H., and Crandell, D.R., 1976, Potential Hazards from Future Volcanic Eruptions of Mt. Baker [abst.]: *EOS,* vol. 57, no. 2, p. 87.

Hyde, J.H., and Crandell, D.R., 1978, Postglacial Volcanic Deposits on Mt. Baker, Washington, and Potential Hazards from Future Eruptions: *U.S. Geological Survey Professional Paper* , no. 1022-C, 17 p.

Jeffcott, P.B., 1949, *Nooksack Tales and Trails,* Ferndale, Washington, 436 p.

Kiver, E.P., 1978a, Geothermal Ice Caves and Fumaroles, Mount Baker Volcano, 1974-77 [abst.]: *Geo. Soc. Am., Abstr. Program,* vol. 10, no. 3, p. 112.

Kiver, E.P., 1978b, Mount Baker's Changing Fumaroles, *The Ore Bin,* vol. 40, no. 8 , pp. 133-145.

Krimmel, R.M., and Frank, D., 1980, Aerial Observations of Mt. Baker, Washington—1976-1979 Update: *EOS* (Amer. Geophys. Union Trans.), vol. 51, no. 6, p. 69.

Majors, H.M., 1978, *Mount Baker: A Chronicle of its Historic Eruptions and First Ascent:* Seattle: Northwest Press, 221 p.

Malone, Stephen D., 1976, Deformation of Mount Baker Volcano by Hydrothermal Heating [abst.]: *EOS,* vol. 57, no. 12, p. 1016.

Malone, Stephen D., and Frank, David, 1975, Increased Heat Emission from Mount Baker, Washington, *EOS,* (Am. Geophys. Union Trans.), vol. 56, no. 10, pp. 679-685.

Misch, Peter, 1952, Geology of the Northern Cascades of Washington: *The Mountaineer,* vol. 45, no. 13, pp. 4-22.

Nitsan, V., 1976, The Effect of Increased Geothermal Heat Flux on the Flow of Mt. Baker Glaciers, [abst.]: *EOS,* vol. 57, no. 2, p. 89.

Nolf, B., 1976, Tilt-bar stations on Mt. Baker Washington [abst.]: *EOS,* vol. 57, no. 2, p. 88.

Perrey, A., 1832-1856, *Bulletins de L'Academie Royale des Sciences,* First Series, Brussels, Belgium.

Perrey, A., 1832-1856, Notes sur les Tremblements de Terre: in *Bulletins de L'Academie Royale des Sciences,* First Series, Brussels, Belgium.

Phillips, James W., 1971, *Washington State Place Names,* University of Washington Press.

Plummer, F.G., 1893, Western Volcanoes: Chances that Western Washington May See Disastrous Eruptions: *Tacoma Daily Ledger,* February 28, p. 3, col. 1-4

Plummer, F.G., 1898, Reported Volcanic Eruptions in Alaska, Puget Sound, etc., 1690-1896: In Holden, Edward S., 1898, *Catalogue of Earthquakes on the Pacific Coast 1769 to 1897, Smithsonian Institution Misc. Col. 1087,* pp. 24-27.

Rosenfeld, C.L., and Schlicker, H.G., 1976, The Significance of Increased Fumarolic Activity at Mt. Baker, Washington: *The Ore Bin,* vol. 38, no. 2, pp. 23-35.

Rusk, C.W., 1978, *Tales of a Western Mountaineer,* Seattle: The Mountaineers, 309 p.

Smith, G.O., and Calkins, F.C., 1904, A Geological Reconnaissance Across the Cascade Range Near the 49th Parallel: *U.S. Geological Survey Bulletin, no. 235,* pp. 1-99.

Steel, William Gladstone, 1906, Ascent of Mount Baker: *Steel Points,* vol. 1, no. 1, pp. 27-28.

Winthrop, Theodore, 1913, *The Canoe and the Saddle,* ed. by John Williams, Tacoma: Williams, 331 p.

XX. MT. GARIBALDI

Anderson, R.G., 1975, *The Geology of the Volcanics of the Meager Creek Map Area, Southwestern British Columbia* B.S. Thesis, University of British Columbia, 130 p.

Armstrong, J.E., Crandell, D.R., Easterbrook, D.J., and Noble, J.B., 1965, Late Pleistocene Stratigraphy and Chronology in Southwestern British Columbia and Northwestern Washington: *Geological Society of America Bulletin,* vol. 76, pp. 321-330.

Barton, Robert H., Christiansen, E.A., Kupsch, W., Matthews, W.H., Gravenor, C.P. and Bayrock, L.A., 1966, Quaternary, Chapter 14: In *Geological History of Western Canada.* Alberta Society of Petroleum Geologists, Calgary, Alberta, pp. 195-200.

Bevier, M.L., Armstrong, R.L., and Souther, J.G., 1979, Miocene Peralkaline Volcanism in West-central British Columbia—Its Temporal and Plate Tectonic Setting: *Geology,* vol. 7, pp. 389-392.

Crickmay, Colin Hayter, 1930, The Structural Connection between the Coast Range of British Columbia and the Cascade Range of Washington: *Geology Magazine,* vol. 67, pp. 482-491.

Green, N.L., 1981, Geology and Petrology of Quaternary Volcanic Rocks, Garibaldi Lake Area, Southwestern British Columbia: *Geological Society of America Bulletin,* vol. 92, part I, pp. 697-702, and part II, pp. 1359-1470.

Green, N.L., Armstrong, R.L., Harakal, J.E., Souther, J.G., and Read, P.B., 1986, K-Ar Geochronology and Eruptive History of the Garibaldi Volcanic Belt, Southwestern Brisith Columbia, in press.

Lawrence, R.B., 1979, *Garibaldi Group Volcanic Rocks of the Salal Creek Map Area (Northern Half), Southwestern British Columbia,* B.S. Thesis, University of British Columbia, 83 p.

Lowden, J.A., and Blake, W., 1975, Radiocarbon Dates XV, *Geological Survey of Canada, Paper 75-7.*

Lowden, J.A., and Blake, W., 1978, Radiocarbon Dates XVIII, *Geological Survey of Canada, Paper 78-7,* 22 p.

Mathews, W.H., 1951a, Historic and Prehistoric Fluctuations of the Alpine Glaciers in the Mount Garibaldi Map-Area, Southwestern British Columbia. *Journal of Geology,* vol. 59, pp. 357-380.

Mathews, W.H., 1951b, The Table, A Flat-Topped Volcano in Southern British Columbia: *American Journal of Science,* vol. 249, pp. 830-841.

Mathews, W.H., 1952, Mount Garibaldi, A Supraglacial Pleistocene Volcano in Southwestern British Columbia: *American Journal of Science,* vol. 250, pp. 81-103.

Mathews, W.H., 1958, Geology of the Mount Garibaldi Map-Area, Southwestern British Columbia, Canada: *Bulletin of the Geological Society of America,* vol. 69, pp. 161-198.

Mathews, W.H., 1972, *Geology of Vancouver Area of British Columbia: International Geological Congress 24th Montreal 1972, Guidebook 24, part A05-C05,* 47 p.

Mathews, W.H., 1975, *Garibaldi Geology: A Popular Guide to the Geology of the Garibaldi Lake Area,* Geological Association of Canada, Cordilleran Section, Vancouver, British Columbia, 48 p.

McBirney, Alexander, R., 1958, Petrochemistry of the Cascade Volcanoes: In Doles, H.M., ed., *Andesite Conference Guidebook; Oregon Dept. of Geology and Mineral Industries Bulletin 62,* pp. 101-107.

McTaggart, K.C., 1970, Tectonic History of the Northern Cascade Mountains: In *Structure of the Southern Canadian Cordillera, Geological Association of Canada Special Paper no. 6,* pp. 137-148.

Nasmith, H., Mathews, W.H. and Rouse, G.E., 1967, Bridge River Ash and Some Other Recent Ash Beds in British Columbia: *Canadian Journal of Earth Science,* vol. 4, no. 1, pp. 163-170.

Read, P.B., 1979, *Geology of Meager Creek Geothermal Area, British Columbia: Geological Survey of Canada, Open File 603.*

Sivertz, G.W.G., 1976, *Geology, Petrology and Petrogenesis of Opal Cone and the Ring Creek Lava Flow, Southern Garibaldi Park, B.C.* B.S. Thesis, University of British Columbia, 74 p.

Souther, J.G., 1970, Recent Volcanism and its Influence on Early Native Cultures of the Northwestern British Columbia: In *Early Man and Environments:* In Smith, R.A., and Smith, J.W., eds. *Northwest North America,* Calgary University Archaeological Association Annual, Paleoenvironmental Workshop, 2nd Proc., University of Calgary Students Press, Calgary, Alberta, pp. 53-64.

Souther, J.G., 1970a, *Volcano Fly-By, Fieldstrip Guidebooks, GAC, MAC, SEG, CGU, Joint Annual Meetings, Vancouver, B.C., April, 1977, 15 p.*

Souther, J.G., 1970b, Volcanism and Tectonic Environments in the Canadian Cordillera—A Second Look: In Baragar, W.R.A., et al, eds. *Volcanic Regimes in Canada: Geological Association of Canada, Spcial Paper 16,* pp. 3-24.

Sutherland-Brown, A., 1969, Aiyansh Lava Flow, British Columbia: *Canadian Journal of Earth Science,* vol. 6, pp. 1460-1468.

Stevens, R.D., Delabio, R.N., and Lachance, G., 1982, Age Determinations and Geologic Studies, K-Ar Isotope Ages, Report 15: *Geological Survey of Canada Paper 81-2,* 56 p.

Westgate, J.A., and Dreimanis, A., 1967, Volcanic Ash Layers of Recent Age a Banff National Park, Alberta, Canada: *Canadian Journal of Earth Science,* vol. 4, pp. 155-161.

XXI. Volcanic Fire and Glacial Ice

Abelson, Philip H., 1980, Monitoring Volcanism (ed.), *Science,* vol. 289, no. 4454, p. 343.

Anonymous, 1975, Volcanoes and Ice Ages: A Link?: *Science News,* vol. 107, p. 100.

Bacon, Charles R., 1980, Goals Are Set for Research in Cascades: *Geotimes,* vol. 25, no. 8, p. 16.

Beckey, Fred, 1973, *Cascade Alpine Guide: Climbing and High Routes, Columbia River and Stevens Pass,* The Mountaineers, Seattle.

Begét, James, 1982, *Evidence of Pleistocene Explosive Eruptions of Mount Jefferson, Oregon,* in press.

Chesterman, Charles W., 1971, Volcanism in California: *California Geology,* vol. 24, no. 8, pp. 139-147.

Colman, S.M., and Pierce, K.L., 1981, *Weathering Rinds on Andesitic and Basaltic Stones as a Quaternary Age Indicator, Western United States: Geological Survey Professional Paper 1210.,* 56 p.

Crandell, Dwight R., 1965, The Glacial History of Western Washington and Oregon: in Wright, H.E., Jr., and Frey, David G., eds., *The Quaternary of the United States,* Princeton University Press, Princeton, N.J., pp. 341-353.

Crandell, Dwight R., 1969, *The Geologic Story of Mount Rainier,* U.S. Geological Survey Bulletin 1292.

Crandell, Dwight R., 1971, *Postglacial Lahars from Mount Rainier Volcano, Washington:* U.S. Geological Survey Professional Paper 677, 75 p.

Crandell, Dwight R., 1972, *Glaciation Near Lassen Peak, Northern California:* U.S Geological Survey Professional Paper 800-C, pp. C179-C188.

Crandell, Dwight R., 1973, *Potential Hazards from Future Eruptions of Mount Rainier, Washington: U.S. Geological Survey Miscellaneous Geologic Investigations,* Map 1-836.

Crandell, Dwight R., 1980, *Recent Eruptive History of Mount Hood, Oregon, and Potential Hazards from Future Eruptions: U.S. Geological Survey Bulletin 1492,* 81 p.

Crandell, Dwight R., and Mullineaux, Donal R., 1967, *Volcanic Hazards at Mount Rainier:* U.S. Geological Survey Bulletin 1238, 26 p.

Crandell, Dwight R., and Mullineaux, Donal R., 1973, *Pine Creek Volcanic Assemblage at Mt. St. Helens, Washington:* U.S Geological Survey Bulletin 1383-A, 23 p.

Crandell, Dwight R., and Mullineaux, Donal R., 1974, Appraising the Volcanic Hazards of the Cascade Range of the Northwestern U.S.: *Earthquake Inf. Bull.,* vol. 6, no. 5, pp. 3-10.

Crandell, Dwight R., and Mullineaux, Donal R., 1975, Technique and Rationale of Volcanic-Hazards Appraisals in the Cascade Range, Northwestern United States: *Environmental Geology.*

Crandell, Dwight R., Mullineaux, Donal R., 1978, Potential Hazards from Future Eruptions of Mt. St. Helens Volcano: *U.S Geol. Surv. Bull., no. 1383-C,* 26 p.

Crandell, Dwight R., Mullineaux, Donal R., and Miller, C. Dan, 1979, Volcanic Hazards Studies in the Cascade Range of the Western United States, Chapter 7, in Sheets, P.D., and Grayson, D.K., eds. *Volcanic Activity and Human Ecology,* New York, Academic Press, pp. 195-217.

Crandell, Dwight R., Mullineaux, Donal R., and Rubin, Meyer, 1975, Mt. St. Helens Volcano: Recent and Future Behavior: *Science,* vol. 187, no. 4175, pp. 438-441.

Crandell, Dwight R., Mullineaux, Donal R., Sigafoos, R.S., and Rubin, Meyer, 1974, Chaos Crags Eruptions and Rockfall-Avalanches, Lassen Volcanic National Park, California: *U.S. Geological Survey,* vol. 2, no. 1, pp. 49-59.

Crandell, Dwight R., and Waldron, H.H., 1969, *Volcanic Hazards in the Cascade Range:* in Olson, R.A., and Wallace R.M., eds., *Geologic Hazards and Public Problems:* Office of Emergency Preparedness, Region 7 Conf. Proc., Santa Rosa, Calif., May 27-28, pp. 5-18.

Decker, R.W., 1973, State-of-the-Art in Volcano Forecasting: *Bulletin Volcanologique,* vol. 37, no. 3, pp. 372-393.

Dodge, Nicholas A., 1964, Recent Measurements on the Eliot Glacier: *Mazama,* vol. 46, no. 13, pp. 47-49.

Dodge, Nicholas A., 1971, The Eliot Glacier: New Methods and Some Interpretations: *Mazama,* vol. 53, no. 13, pp. 25-29.

Driedger, Carolyn L., 1981, *The Effect of Ash Thickness on Snow Ablation: U.S. Gelogical Survey Preliminary Report.*

Driedger, Carolyn L., 1986, *A Visitor's Guide to Mount Rainier Glaciers*, Longmire, Washington: Pacific Northwest National Parks and Forests Association.

Driedger, Carolyn L., and Kennard, P.M., 1984, *Ice Volumes on Cascade Volcanoes. Mount Rainier, Mount Hood, Three Sisters, and Mount Shasta: U.S. Geological Survey Professional Paper 1365.*

Gorshkov, G.S., 1971, Prediction of Volcanic Eruptions and Seismic Methods of Location of Magma Chambers—A Review: *Bulletin Volcanologique,* vol. 35, pp. 198-211.

Hammond, Paul E., 1973, If Mt. Hood Erupts: *The Ore Bin,* vol. 35, no. 6, pp. 93-102.

Hammond, Paul E., 1974, The Pulse of the Cascade Volcanoes: *Pacific Search,* vol. 8, no. 8, pp. 5-9.

Heiken, G., and Eichelberger, J.C., 1980, Eruptions at Chaos Grags, Lassen Volcanic National Park, Calif., in Gordon A. Macdonald Memorial Volume, McBirney, A.R., ed., *Journal of Volcanology and Geothermal Research,* vol. 7, no. 3-4, pp. 443-481.

Lange, Ian M., and Avent, Jon C., 1973, Ground-Based Thermal Infrared Surveys as an Aid in Predicting Volcanic Eruptions in the Cascade Range: *Science,* vol. 182, no. 4109, pp. 279-281.

Lathrop, T.G., 1968, Return of the Ice Age?: *Mazama,* vol. 50, no. 13, pp. 34-36.

MacLeod, Norman S., Sherrod, D.R., Chitwood, L.A, and McKee, E.H., 1981, Newberry Volcano, Oregon, in Johnston, D.A. and Donnelly-Nolan, Julie, eds., *Guides to Some Volcanic Terranes in Washington, Idaho, Oregon, and Northern California: Geological Survey Circular 838,* pp. 85-103.

McKee, Bates, 1972, Cascadia: *The Geologic Evolution of the Pacific Northwest,* McGraw-Hill, New York.

Miller, C. Dan, 1980, Potential Hazards from Future Eruptions in the Vicinity of Mount Shasta Volcano, Northern California: *Geological Survey Bulletin 1503.*

Molenaar, Dee, 1971, *The Challenge of Rainier,* The Mountaineers, Seattle.

Montague, Malcolm J., 1973, The Little Glacier That Couldn't: *Mazama,* vol. 55, no. 13, pp. 73-75.

Mullineaux, Donal R., 1977, Volcanic Hazards; Extent and Severity of Potential Tephra Hazard Interpreted from Layer Yn from Mount St. Helens [abst.]: *Geol. Soc. Am., Abstr. Programs,* vol. 9, no. 4, p. 472.

Nafziger, Ralph H., 1971, Oregon's Southernmost Glacier: A Three Year Report: *Mazama,* vol. 53, no. 13, pp. 30-33.

Phillips, Kenneth N., 1938, Our Vanishing Glaciers: *Mazama,* vol 20., no. 12, pp. 24-41.

Phillips, Kenneth N., 1939, Farewell to Sholes Glacier: *Mazama,* vol. 21, no. 12, pp. 37-40.

Porter, Stephen C., and Denton, G.H., 1967, Chronology of Neoglaciation in the North America Cordillera, *American Journal of Science,* vol. 265, pp. 171-210.

Post, Austin, 1970, Recent Changes in Glaciers of the North Cascades, Washington: U.S. Geological Survey unpublished preliminary report.

Post, Austin, 1974, Mount Rainier's Glaciers Current Advancing—And Retreating, U.S. Geological Survey, Tacoma, Washington, Press Release, March 26, 1974.

Richards, Carl Price, 1937, Photographic Survey of the Glaciers of Mt. Jefferson and the Three Sisters: *Mazama,* vol. 19, no. 12, pp. 66-75.

Richardson D., 1968, Glacier Outburst Floods in the Pacific Northwest: *U.S. Geological Survey Professional paper 600.*

Russell, Israel C., 1901, *Glaciers of North America:* Ginn and Co., Boston.

Swanson, D.A., 1976, Techniques for Instrumentally Monitoring Active Volcanoes [abst.]: *EOS,* vol. 57, no. 2, p. 87.

Taylor, Edward M., 1981, Roadlog for Central High Cascade Geology: Bend, Sisters, McKenzie Pass, and Santiam Pass, Oregon, in Johnston, D.A., and Donnelly-Nolan, Julie, eds., *Guides to Some Volcanic Terranes in Washington, Idaho, Oregon, and Northern California: Geological Survey Circular 838,* pp. 59-81.

Tilling, R.I., 1978, U.S. Geological Survey's Program in Volcano Studies Present Status and Future Aspirations [abst.]: *Geol. Soc. Amer. Abstr. Programs,* vol. 10, no. 3, pp. 150-151.

Veatch, Fred M., 1969, *Analysis of a 24-Year Photographic Record of Nisqually Glacier, Mt. Rainier National Park, Washington: U.S. Geological Survey Professional Paper 631.*

Warrick, R.A., 1975, *Volcano Hazard in the United States: A Research Assessment,* Boulder: Institute of Behavioral Science, University of Colorado.

Williams, Howel, 1944, Volcanoes of the Three Sisters Region, Oregon Cascades: *University of California Dept. of Geological Sciences Bulletin,* vol. 27, no. 3, pp. 37-84.

Williams, Howel, 1957, *A Geologic Map of the Bend Quadrangle, Oregon, and a Reconnaissance Geologic Map of the Central Portion of the High Cascade Mountains:* Oregon Dept. of Geology and Mineral Industries, in coop. with U.S. Geological Survey.

Wozniak, K.C., and Taylor, E.M., 1981, Late Pleistocene Summit Construction and Holocene Flank Eruptions of South Sister Volcano, Oregon: *EOS,* Trans. Amer. Geophys. Union, vol. 62, no. 6, p. 61.

XXII: When Mt. Shasta Erupts

Christiansen, R.L., 1982, Volcanic Hazard Potential in the Cascades, in Martin, R.J., and Davis, J.F., eds. *Status of Volcanic Prediction and Emergency Resource Capabilities in Volcanic Hazard Zones of California: California Division of Mines and Geology Special Publication 63,* pp. II-43, II-65.

Christiansen, R.L., and Miller, C. Dan, 1976, Volcanic Evolution of Mount Shasta, Californi. *Geological Society of America, Abstracts with Programs, Cordilleran Section Meeting,* vol. 8, no. 3, pp. 360-361.

Christiansen, R.L., Kleinhampl, F.J., Blakely, R.J., Tuchek, E.T. Johnson, F.L., and Conyak, M.D., 1977, Resource Appraisal of the Mt. Shasta Wilderness Study Area, Siskiyou County, California: *U.S. Geological Survey Open-File Report 77-250,* 53 p.

Crandell, D.R., 1973, Hot Pyroclastic-Flow Deposits of Probable Holocene Age West of Mount Shasta Volcano, California: *Geological Society of America, Abstracts with Programs,* 69th Annual Meeting.

Crandell, D.R., Miller, C.D., Glicken, H.X., Christiansen, R.L., and Newhall, C.G., 1984, "Catastrophic Debris Avalanche from Ancestral Mount Shasta Volcano," California: *Geology,* vol. 12, no. 3, pp. 143-146.

Crandell, D.R., Mullineaux, D.R., and Miller C.D., 1979, Volcanic Hazards Studies in the Cascade Range of the Western United States, Chapter 7, in Sheets, P.D., and Grayson, D.K., eds., *Volcanic Activity and Human Ecology,* New York: Academic Press, pp. 195-219.

Diller, J.S., 1895, Mount Shasta: A Typical Volcano: *National Geographic Society Monograph,* vol. 1, no. 8, pp. 237-268.

Eichorn, Arthur F., 1957, *The Mount Shasta Story,* rev. ed., the Mount Shasta Herald, Mount Shasta City, California.

Finch, R.H., 1930, Activity of a California Volcano in 1786: *The Volcano Letter,* no. 308, p. 3.

Harris, Stephen L., 1980, *Fire and Ice: The Cascade Volcanoes,* rev. ed., The Mountaineers and Pacific Search Press, Seattle, Washington.

Miller, C. Dan, 1978, Holocene Pyroclastic-Flow Deposits from Shastina and Black Butte, West of Mount Shasta, California: *U.S Geological Survey, Journal of Research,* vol. 7, no. 5, pp. 611-624.

Miller, C. Dan, 1980, *Potential Hazards from Future Eruptions in the Vicinity of Mount Shasta Volcano, Northern California: Geological Survey Bulletin 1503,* 43 p.

Williams, Howel, 1932, Mount Shasta, A Cascade Volcano: *Journal of Geology,* vol. 40, no. 5, pp. 417-429.

Williams, Howel, 1934, Mount Shasta, California: *Zeitschrift Vulkanologie,* vol. 15, no. 4, pp. 225-253.

Glossary

Aa: Hawaiian word used to describe a lava flow whose surface is broken into angular, jagged fragments.

Agglutinate: a volcanic deposit formed by the accumulation of flattened and welded fragments, typically derived from showers of still-molten rock particles ejected in magma fountains. The liquid fragments may accumulate to form a stream of lava.

Andesite: a lava of intermediate composition, usually light gray or brown in color. Andesite has a silica content ranging from about 54 to 62 percent.

Andesite line: an imaginary line drawn around the boundary of the Pacific Ocean basin, separating continental and oceanic lava rocks according to their chemical composition. On the Pacific side of the line lavas are basaltic. On the continental side, lavas with a higher silica content, such as andesites, commonly occur.

Ar: the element argon.

Ash: fine particles of pulverized rock blown from a volcano. Measuring less than about 0.1 in diameter (under 4 mm), ash may be either solid or molten when first erupted. By far the most common variety is vitric ash, glassy particles formed by gas bubbles bursting through liquid magma. Lithic ash is formed of older rock pulverized during an explosive eruption, while in crystal ash each grain is composed of a single crystal or groups of crystals with only traces of glass adhering to them. Many volcanic ash deposits contain mixtures of all three kinds in various proportions.

Ash fall: a rain of ash from an eruption cloud.

Ash flow: an avalanche of hot volcanic ash and gases that can travel great distances at high speeds from an erupting vent. Large-volume ash flow deposits commonly solidify to form *Ignimbrites.* see *Pyroclastic flow*.

Asthenosphere: a zone of the earth's outer shell beneath the lithosphere. Of undetermined thickness, this is a region of weakness where plastic movements occur.

Basalt: a dark-colored, fine-grained lava rich in iron and magnesium and relatively poor in silica (less than 54 percent). The most common of earth's volcanic rocks, basaltic lavas compose all the ocean floors and many continental formations as well. Typically very fluid because of their low silica content, basaltic lavas can flow great distances from their source, forming broad lava plains such as the Columbia River Plateau. Shield volcanoes are typically formed exclusively of basalt.

Bergschrund: a crevass at the back of a glacier between the glacier and its rock headwall, formed by partial melting and movement of the glacier.

Block: angular fragment of lava rock measuring a minimum of 2.5 inches to several tens of feet in diameter.

Block-and-ash flow: variety of a pyroclastic flow, a turbulent mass of hot dense rock fragments that avalanches downslope as the result of an eruption. Block-and-ash flows are commonly triggered by the disruption or collapse of a dome while still hot.

Blocky lava: a lava flow whose surface is characterized by a jumble of large angular blocks.

Bomb: a lump of plastic or molten lava thrown out during an explosive eruption. Bombs range in size from 2.5 inches to many feet in diameter. Because of their plastic condition when first ejected, bombs are commonly modified in shape during their flight through the air and/or by their impact on the ground. As the outer crust cools and solidifies, continued expansion of the interior by gas pressure sometimes causes cracking, which may form a bomb surface resembling the crust of freshly baked bread (breadcrust bombs).

Breccia: rock composed of many distinct fragments, typically sharp and/or angular, embedded in a matrix of fine material. Breccias are sometimes formed when shattered lava blocks are transported by avalanches or mudflows.

Caldera: the Spanish word for cauldron used to denote a large basin-shaped volcanic depression—by definition at least a mile in diameter. Calderas are usually found at the top of volcanic cones and are caused by the collapse of a former summit. Unlike craters, calderas are always formed by collapse.

Cinders: a general term applied to vesicular rock fragments ejected during explosive eruptions.

Cinder cone: a volcanic cone built entirely of tephra and/or other pyroclastic material, commonly by the mildly explosive ejection of semi-fluid or plastic lava fragments. Most cinder cones are relatively small steep-sided structures built during a single eruptive episode. Cinder cones commonly discharge lava flows from vents at the base of the cone.

Cirque: an amphitheatre-like depression in alpine regions, formed by the plucking action of glacial ice. These large semi-circular basins have been carved into the slopes of most Cascade peaks.

Composite cone: another term for a stratovolcano, a large volcanic cone constructed of both lava flows and pyroclastic material. Most of the world's great continental volcanoes—Vesuvius, Fuji, Shasta, Rainier and St. Helens—are composite cones.

Conduit: the feeding pipe of a volcano, the "throat" through which magma passes on its way to the Earth's surface.

Continental drift: the theory that horizontal movements of the Earth's surface causes the continents slowly to shift their positions relative to each other.

Convection currents: movements of material caused by differences in density, typically the result of heating.

Core (of the Earth): the central region of the earth, thought to be composed largely of molten iron and nickel.

Crater: the bowl or funnel-shaped hollow at or near the top of a volcano, through which volcanic gas, lava, and/or pyroclastic material are ejected. The term derives from the Greek word for "wine-mixing bowl."

Crystalline rock: a hard rock composed of interlocking crystals, typically of igneous origin.

Dacite: a (usually) light-colored lava with a high silica content, about 64 percent or more. Gas-rich dacite magmas are commonly highly explosive, while gas-poor dacites typically form thick, viscous tongues of lava. When particularly stiff and pasty, dacite lavas may form steep-sided domes, such as that presently building in the crater of St. Helens.

Detonation: an explosion made by the combination of gases or by the abrupt release of gases from a volcanic vent.

Dike: a sheetlike body of igneous rock that cuts through, in a generally vertical direction, older rock formations. Dikes form when relatively narrow, thin magma sheets intrude a volcanic cone or other structure, intersecting previously existing strata.

Dip: the angle and direction at which a stratum or any tabular plane is inclined from the horizontal.

Dome: a generally rounded protrusion of lava that, when erupted, was too viscous to flow far laterally and instead piled up over the erupting vent to form a mushroom-shaped cap. When the lava mass is an upheaved, consolidated lava conduit filling, the resultant extrusion is called a plug dome.

Dormant: literally, "sleeping." The term used to describe a volcano that is presently inactive but which may erupt again. Most of the major Cascade volcanoes are believed to be dormant rather than extinct.

D.V.I.: the dust veil index.

Earthquake, volcanic: a shaking of the ground caused by the fracturing of subterranean rock as magma rises toward the surface. Most volcanic earthquakes are relatively small and their effects largely restricted to the immediate vicinity of the volcano.

Effusive eruption: a relatively quiet outpouring of lava flows, with little or no explosive activity.

Engulfment: the inward collapse of a volcano, perhaps as the result of an evacuation of the magma chamber.

Epicenter: the point on the Earth's surface directly above the focus of an earthquake.

Epoch: a division of a geologic time period, such as the Pleistocene Epoch of the Quaternary Period.

Eruption: the geologic process by which solid, liquid, and gaseous materials are ejected onto the Earth's surface by volcanic activity. Eruptions vary in behavior from the quiet overflow of molten rock (effusive type) to the tremendously violent expulsion of pyroclastic material.

Eruption cloud: the column of gases, ash and larger rock fragments rising from a volcanic vent. If it is of sufficient volume and velocity, this gaseous column may reach many miles into the stratosphere, high winds may carry it long distances from its source.

Extensional margin: the edges of tectonic plates that are moving apart.

Extinct: the term used to describe a volcano that is not expected to erupt again; a dead volcano.

Extrusive rock: igneous rock that forms at the Earth's surface, also called volcanic rock.

Fault: a crack or fracture in the Earth's crust along which there has been differential movement. It may represent the juncture between two adjoining crustal blocks or plates into which the Earth's surface is broken. Movement along a fault can cause earthquakes, or, in the process of mountain building (orogeny), can release underlying magma and permit it to rise to the surface, creating a volcanic eruption.

Fissure eruption: a volcanic eruption that occurs along a narrow fissure or line of closely-spaced fractures in the Earth's crust.

Flank eruption: an eruption from the side of a volcano, in contrast to one from the central conduit.

Focus: the point within the Earth at which an earthquake originates.

Fumarole: a vent or opening through which issue steam or other volcanic gases, such as hydrogen sulphide.

Fume: a gaseous volcanic cloud not containing tephra.

Geothermal energy: energy derived from the Earth's internal heat.

Geothermal gradient: rate of temperature increase associated with increasing depth beneath the Earth's surface, normally about 25 degrees C per kilometer.

Geothermal power: power derived from harnessing and exploiting the heat energy of the Earth, as by tapping the heat from a hot spring, geyser, or volcano.

Glacier: a large, dense mass of ice, formed on land by the compaction and recrystallization of snow, which moves downslope because of its weight and gravity. Active glaciers are effective eroding agents and have played an important role in sculpturing the present shape of all the larger Cascade volcanoes.

Granite: a coarse-grained igneous rock composed of quartz and feldspar. In chemical composition, it is equivalent to rhyolite.

Granodiorite: a granular igneous rock intermediate between granite and quartz diorite. Massive intrusions of granodiorite occur adjacent to Mount Rainier and in other parts of the Washington Cascades.

Half-life: the time it takes for a given amount of a radioactive isotope to be reduced by one-half.

Hawaiian eruption: effusive eruptions of basaltic lava typical of the Hawaiian shield volcanoes. Usually non-explosive, Hawaiian eruptive activity commonly produces lava fountaining, formation of a lava lake in the summit caldera, and, during the second stage of the eruption, production of voluminous quantities of fluid lava that issue from lengthy fissures along the volcano's flanks.

Holocene Epoch: the 10,000 to 12,000 year-long span of time that has elapsed since the end of the Pleistocene Epoch (Ice Age). It is the division of geologic time in which we now live.

Horizontal blast: an explosive eruption in which the resultant cloud of hot gas and pyroclastic material moves laterally rather than upward. Lassen Peak's "hot blast" of 1915 was such an eruption.

Hornblende: a (usually) dark-colored mineral (amphibole) commonly found in igneous and metamorphic rocks.

Hot spot: a persistent heat source in the upper mantle unrelated to plate boundaries. Isolated "hot spots" generating magma are believed to underlie the Hawaiian Islands and the Yellowstone region.

Hydrothermal metamorphism: alteration of a rock by hot water passing through it.

Hydrothermal reservoir: an underground region of porous rock containing hot water.

Ice fall: a chaotic jumble of ice blocks and crevasses formed when a glacier flows over a cliff or other steep, uneven declivity.

Igneous rock: from the Latin word for "fire," igneous refers to rocks derived from the solidification of magma. Igneous rocks that congeal beneath Earth's surface are called plutonic, while those formed on the surface are volcanic.

Ignimbrite: a highly silicic volcanic rock formed by the eruption of large-volume ash flows. Dense clouds of incandescent rock fragments erupted at extremely high temperatures settle and congeal so that individual particles are fused together. See *Welded tuff.*

Intrusion: an igneous rock formation created when magma intrudes into a body of surrounding rocks and then solidifies.

Intrusive rock: an igneous rock body that is formed underground when magma is injected into an older body of rock.

Island arc: a curving line of volcanic islands formed along the boundaries of converging plates.

Juan de Fuca Plate: a relatively small segment of the Pacific Ocean plate that is presently being subducted beneath the margin of the North American plate. The Juan de Fuca slab extends from northern California northward to southwest British Columbia.

Juvenile material: fresh magma erupted on the Earth's surface.

Lahar: Indonesian term for a mudflow originating on a volcano.

Lapilli: the Latin term for "little stones," it is applied to round to angular rock fragments measuring 0.1 to 2.5 inches in diameter, which are erupted explosively in either a solid or molten state.

Lava: magma that reaches the Earth's surface through a volcanic eruption. The term is most commonly applied to streams of molten rock flowing from a volcanic vent. It also refers to solidified volcanic rock.

Lava fountain: a gas-charged spray of fluid magma shooting into the air above a fissure or other volcanic vent.

Lava lake: a large pool of molten rock in a crater or other volcanic depression. The term refers to a lake of congealed or solidified lava as well.

Lava tube: a cave or tunnel formed inside a lava flow. The tube is created when the outer crust of the lava stream cools and solidifies but the molten interior continues to flow. After the molten interior has drained away, a long hollow or tube is left behind.

Lithic ash: volcanic ash derived from the explosive pulverizing of previously existing rock.

Lithosphere: the rigid outer shell of the Earth, including the rocky crust and upper mantle.

Lithostatic pressure: the pressure in the solid outer shell of the Earth exerted by the weight of the overlying rocks.

Magma: gas-rich molten rock confined beneath the Earth's surface; when erupted at the surface it is called lava.

Magma chamber: undergound pocket or reservoir of magma, from which the molten material erupted by volcanoes is drawn.

Magmatic differentiation: the subterranean process by which different kinds of igneous rocks are derived from a single parent magma.

Magmatic fluids: volcanic gases, particularly water and carbon dioxide, dissolved in magma.

Magnetic field: a region in which magnetic forces exist.

Magnetic pole: a region where the strength of the Earth's magnetic field is greatest and where the magnetic lines of force apparently enter or leave the Earth.

Magnetic reversal: a change in the polarity of the Earth's magnetic field. The last magnetic reversal occurred about 700,000 years ago.

Magnitude: the amount of energy released during an earthquake or volcanic eruption.

Mantle: that portion of the Earth's interior lying between the molten core and the rigid outer crust, a zone of hot plastic rock extending approximately 1800 miles beneath the surface. This is the region in which magma is generated.

Microearthquake: an earthquake too small to be felt by humans but detectable on a seismometer.

Mid-oceanic ridge: an enormous submarine mountain range that extends around the globe.

Mohorovicic Discontinuity: the boundary separating the Earth's crust from the mantle beneath it, sometimes called the "Moho."

Moraine: a typically linear ridge of rock fragments deposited by a glacier.

Mudflow: a water-saturated mass of rock debris that travels downslope as a liquid under the pull of gravity. Large-volume mudflows are a major potential hazard on many Cascade volcanoes. Generated by heavy rainfall or melting snow and glacial ice, mudflows can fill valleys heading on a volcano to depths of many tens of feet and travel many miles beyond their place of origin.

Nuée Ardente: a "glowing cloud" of hot volcanic ash and gas that typically rises above and extends beyond the margins of a pyroclastic flow.

Obsidian: a dense, black, glassy volcanic rock almost devoid of bubbles or mineral crystals. It is formed from highly silicic magma, ranging from dacite to rhyolite. Young domes and flows of obsidian occur at the Medicine Lake volcano and the Mono Craters in California.

Oceanic crust: the relatively thin slabs of basaltic crust forming the ocean floors.

Olivine: an olive-green mineral commonly found in basalt.

Pahoehoe: Hawaiian word for lava with a smooth, ropy, or billowy crust. Always composed of basalt, pahoehoe flows commonly contain lava tubes.

Paroxysm: a violently explosive eruption of great magnitude.

Pelean eruption: a violently explosive eruption that produces pyroclastic flows and is often associated with dome-building activity.

This type of eruption is named for the 1902 activity of Mont Pelée, a volcano on the Caribbean island of Martinique. That outburst entirely destroyed the city of St. Pierre and killed 30,000 persons. During the recent geologic past, Pelćan eruptions have occurred at many West Coast volcanoes, including Mt. St. Helens, Mt. Hood, Mt. Shasta, and the Mono Craters.

Phreatic eruption: a violent steam explosion that produces little or no new lava. It typically ejects solid fragments of pre-existing rock from the volcanic edifice. Triggered by the conversion of ground water into steam by an underground heat source, small phreatic eruptions characterize the opening stage of many eruptive episodes, as they did at Lassen Peak in 1914 and Mt. St. Helens in 1980.

Pillow lava: a lava flow emplaced underwater that is characterized by interconnected sack-like or "pillow" formations.

Plastic: capable of being molded or bent under stress.

Plate: a large rocky slab of the Earth's crust that slowly moves over the plastic upper mantle beneath it.

Plate tectonics: an important geologic theory stating that the Earth's rigid outer shell is broken up into about a dozen large slabs or plates that are in constant motion. Concentrations of earthquake and volcanic activity occur at the boundaries between plates.

Pleistocene Epoch: the division of geologic time immediately preceding the Holocene Epoch and lasting from about 2,000,000 to 10,000 or 12,000 years ago. It was characterized by repeated development of large continental ice sheets throughout the Northern hemisphere and by the formation of ice caps and valley glaciers in the Cascades, Sierra, Rockies and other mountain ranges. Most of the large composite cones in the Cascades and elsewhere around the Pacific rim were built during this time.

Plinian Eruption: the kind of violently explosive eruption named after Pliny the Younger who wrote of the catastrophic eruption of Mt. Vesuvius, Italy, in A.D. 79. The term is commonly used to denote the phase of an eruption in which violently uprushing gas carries large volumes of fragmental rock high into the stratosphere. It also refers to a paroxysmal outburst like that of Vesuvius, producing large volumes of tephra, pyroclastic flows, and the formation of a caldera, caused by the collapse of the volcano's former summit.

Pliocene Epoch: division of geologic time immediately preceding the Pleistocene and lasting from about 7,000,000 to about 2,000,000 years before the present. During this epoch the Cascade Range underwent a major uplift and a chain of shield volcanoes formed along the present High Cascade axis.

Plug: solidified lava that fills the conduit of a volcano. It is commonly more resistant to erosion than the material composing the surrounding cone and may remain standing as a solitary pinnacle or "neck" when the rest of the original edifice has been eroded away.

Plume: the term designating the column of magma rising from deep within the mantle to produce "hot spot" volcanoes. It can also refer to the column of steam or other gases rising from an erupting volcano.

Pluton: a body of igneous rock intruded into the crust that solidifies at depth.

Plutonic rock: general term for igneous rocks that have crystallized at great depths within the earth.

Pumice: a highly porous volcanic rock fragment formed of glassy lava-foam blown from a vent. Usually light-colored, pumice is typically so full of tiny bubbles or vesicles that it is buoyant.

Pyroclastic: volcanic rock that is erupted in fragments, from the Greek word for "fire-broken."

Pyroclastic flow: an avalanche of incandescent rock fragments and hot gas that travels downslope like a heavy fluid. A pyroclastic flow may be composed of either pumice or dense lithic (non-vesicular) rock debris, or be a combination of both. See *Ash flow.*

Pyroxene: a common mineral group (metasilicate), chiefly of calcium and magnesium, usually in short, thick prismatic crystals or in massive forms, found in igneous rocks.

Quartz: an important rock-forming mineral composed of silicon and oxygen.

Quaternary: the youngest geologic period, encompassing the Pleistocene and Holocene epochs, which began about 2,000,000 years ago, and including the present time.

Radioactive element: an element, such as uranium, capable of changing spontaneously into another element by the emission of charged particles from the nuclei of its atoms.

Radioactive heat: the heat created in rocks by the disintegration of radioactive elements.

Relief: in a landscape, the vertical distance between the highest peaks or ridges and the lowest topographical depressions.

Rhyolite: a generally light-colored lava rock with an extremely high silica content (72 percent or more). Rich in sodium and potassium, it is thick and pasty when erupted. Because gases dissolved in rhyolitic magma cannot escape easily, it can be extremely explosive.

Richter scale: a numerical scale used to measure the size (magnitude) of an earthquake at its source.

Ridge, oceanic: a large submarine mountain range formed by volcanic eruptions along fissures on the ocean floor.

Rift system: linear spreading centers along plate boundaries where crustal plates are separating and moving apart. Magma is erupted along active rifts, creating new crust.

Ring of Fire: term applied to the seismically and volcanically active margin of the Pacific Ocean basin. This circum-Pacific zone contains about 75 percent of the world's active volcanoes.

Scoria: glassy fragments of dark-colored volcanic rock, less porous than pumice, commonly the product of jets of semi-liquid lava shot into the air. Scoria fragments range in size from 0.1 to 2.5 inches.

Sea-floor spreading: term describing the volcanic creation of new sea floor as lava erupts along mid-oceanic ridges, pushing the older ocean floor away from the actively erupting fissures. This process pushes oceanic plates away from the spreading centers toward the edge of continents, where the sea floor is subducted beneath continental margins.

Seismograph: an instrument that detects and records earthquakes, including those too weak to be perceptible to most persons.

Seismometer: an instrument used to detect ground motion, the seismic waves initiated by an earthquake.

Shield volcano: a broad, gently sloping volcanic structure built almost exclusively of thin lava flows. Named for their supposed resemblance to an Icelandic warrior's shield laid down flat with the curved side upward, shield volcanoes typically form by the quiet effusion of fluid basaltic lava.

Silica: the term used for the chemical combination of silicon and oxygen, a primary constituent of volcanic rocks.

Silicic lava: a term describing lava rich in silica (over 62-64 percent) and having a relatively low melting point (about 850 degrees Centigrade). Silicic magma typically emerges as a stiff, viscous mass and does not flow long distances. Silicic lavas may congeal near the eruptive vent to form steep-sided domes, such as Lassen Peak and Chaos Crags.

Solfatara: a term derived from the Solfatara volcano in Italy, which is characterized by the quiet emission of sulphurous gases. A form of fumarolic activity, the solfatara can hydrothermally alter the chemical composition of the rocks surrounding active vents.

Spatter: liquid or plastic fragments of lava blown from a vent.

Stratovolcano: also known as a composite cone, a volcano composed of both lava flows and pyroclastic material. A cross-section through a stratovolcano reveals alternating layers (strata) of lava, ash, breccias, etc.

Strombolian eruption: rhythmic, mildly explosive eruptions of incandescent pyroclastic material that commonly builds cinder cones. Named for the Stromboli volcano, an island off the west coast of Italy, such eruptions may continue with little variation for years or decades. Strombolian activity may also include emission of lava flows, which are usually relatively short and thin.

Subduction: the process by which sea floor is pulled or dragged beneath the margin of a continent or island arc.

Subduction zone: the region of convergence of two tectonic plates, one of which sinks beneath the other, as where part of the Pacific plate is sliding beneath the Pacific Northwest coast of North America.

Tectonics: the study of the major structural features of the Earth and the processes that create them.

Tephra: the term used by Aristotle to describe pyroclastic material that has been thrown into the air above a volcano. Tephra can range in size from fine dust and ash to lava fragments many tens of feet in diameter.

Terminus: the snout or lower edge of a glacier.

Terminal moraine: a ridge-like accumulation of rock debris eroded and deposited at the end of a glacier, marking its farthest advance.

Till: unstratified (unlayered) rock debris carried and deposited by a glacier, typically consisting of unsorted gravel, boulders, clay, and sand.

Tree-mold: a cylindrical hole in a lava flow, created by the combustion or decay of an upright tree when it was surrounded by a lava flow that formed a hollow impression of the tree trunk.

Tree-ring dating: a method of dating events by counting and comparing the annual growth rings in a tree.

Tuff: volcanic rock composed of a fine-grained pyroclastic material, such as the deposit of an ash flow. Welded tuff is a rock formed from fragmental material hot enough to fuse or weld together when emplaced.

Turbidity current: a flowing mass of muddy or sediment-laden water generated by the slumping of loose debris from a steep slope. Heavier than clear water, the muddy stream flows downslope along the bottom of a sea or lake. Turbidity currents at Crater Lake have deposited large quantities of sediment on the caldera floor.

Unconsolidated: term referring to rock particles that are loose, separate, or unattached to each other. Tephra eruptions typically form unconsolidated deposits that are easily eroded.

U-shaped valley: a term referring to the typical U-shaped cross-section of a glacial valley whose walls have been steepened and bottom flattened by glacial erosion.

Vent: an opening, typically cylindrical in form, in the Earth's surface through which volcanic material is ejected.

Vesicle: pores of tiny cavities in a volcanic rock formed by the development of gas bubbles in the liquid magma. Some pumice is vesicular enough to resemble a sponge.

Viscosity: a measure of a liquid's resistance to flow. In magma, viscosity is largely determined by temperature, gas content and the chemical composition of the molten material, particularly its silica content.

370

Vitric: a term describing volcanic material consisting mainly of glassy matter, such as vitric ash, which is at least 75 percent glass.

Volcanic neck: a solidified conduit filling or lava plug that remains standing like a solitary column after the surrounding volcanic cone has been eroded away.

Vulcan: the Roman god of fire and the forge, believed in classical times to have established his workshop beneath an island in the Mediterranean called Vulcano, after which volcanoes are named. His Greek equivalent is Hephaestus.

Vulcanian eruption: the violent ejection of towering cauliflower clouds of dark ash and angular rock fragments. This kind of eruption commonly expells large quantities of old rock and little fresh magma.

Vulcano: the volcanic island in the Mediterranean after which volcanoes are named.

Welded tuff: a fine-grained volcanic rock composed of pyroclastic material that was so hot when emplaced that its fragments fused together.

Index

Check for our books at your local bookstore. Most stores will be happy to order any which they do not stock. We encourage you to patronize your local bookstore. Or order directly from us, either by mail using the enclosed order form or by calling our toll-free number, 1-800-234-5308, and putting your order on your Mastercard or Visa charge card. We will gladly send you a free catalog upon request.

Some other geology titles of interest:

____ROADSIDE GEOLOGY OF ALASKA	14.00
____ROADSIDE GEOLOGY OF ARIZONA	15.00
____ROADSIDE GEOLOGY OF COLORADO	15.00
____ROADSIDE GEOLOGY OF IDAHO	15.00
____ROADSIDE GEOLOGY OF MONTANA	15.00
____ROADSIDE GEOLOGY OF NEW MEXICO	14.00
____ROADSIDE GEOLOGY OF NEW YORK	15.00
____ROADSIDE GEOLOGY OF NORTHERN CALIFORNIA	14.00
____ROADSIDE GEOLOGY OF OREGON	14.00
____ROADSIDE GEOLOGY OF PENNSYLVANIA	15.00
____ROADSIDE GEOLOGY OF TEXAS	16.00
____ROADSIDE GEOLOGY OF UTAH	15.00
____ROADSIDE GEOLOGY OF VERMONT & NEW HAMPSHIRE	10.00
____ROADSIDE GEOLOGY OF VIRGINIA	12.00
____ROADSIDE GEOLOGY OF WASHINGTON	15.00
____ROADSIDE GEOLOGY OF WYOMING	14.00
____ROADSIDE GEOLOGY OF THE YELLOWSTONE COUNTRY	10.00
____AGENTS OF CHAOS	12.95
____COLORADO ROCKHOUNDING	14.00
____FIRE MOUNTAINS OF THE WEST	16.00
____GEOLOGY UNDERFOOT IN SOUTHERN CALIFORNIA	12.00
____ROCKS FROM SPACE: METEORITES AND METEORITE HUNTERS	20.00

Please include $3.00 per order to cover postage and handling.

Please send the books marked above. I enclosed $_____

Name_____

Address_____

City_____State_____Zip_____

☐ Payment enclosed (check or money order in U.S. funds)

Bill my: ☐ VISA ☐ MasterCard Expiration Date:_____

Card No._____

Signature_____

MOUNTAIN PRESS PUBLISHING COMPANY
P.O. Box 2399 • Missoula, MT 59806
Order Toll-Free 1-800-234-5308
Have your MasterCard or Visa ready.